Stem Cell Assays

METHODS IN MOLECULAR BIOLOGY™

John M. Walker, SERIES EDITOR

METHODS IN MOLECULAR BIOLOGY™

Stem Cell Assays

Edited by

Mohan C. Vemuri

Stem Cells, Reprogenetics, West Orange, NJ
Invitrogen Corporation, Grand Island, NY

HUMANA PRESS

Editor:
Mohan C. Vemuri

Series Editor:
John M. Walker

ISBN: 978-1-58829-744-0   e-ISBN: 978-1-59745-536-7

Library of Congress Control Number: 2007930587

Cover illustration: Human embryonic stem cells, stained for Nanog 4 (green) were grown in the presence of mouse feeder cells that were labeled with Qtracker quantum dots (red). Courtesy of Soojung Shin, Stem Cell Group, Invitrogen Corporation.

Cover design: Karen Schultz

10 9 8 7 6 5 4 3 2

Printed on acid free paper

springer.com

Preface

Stem cells and regenerative medicine is a fast emerging field with rapid strides of progress and focus on human health. Successful utilization of stem cells in regenerative medicine depends on two important features. One, stem cells can grow and divide indefinitely, and two, stem cells can differentiate into specialized cell types that make up tissues such as pancreas, heart, liver, blood, and others. Utilizing the property of differentiation, stem cells can be engineered to produce cell types to replace wornout cells and to regenerate damaged tissue.

The scope for improving healthcare using stem cell therapies is thrilling, but considerable technical challenges and methodological constraints need to be addressed. There is a great need for isolation and propagation of "transplant-ready" or "clinical-grade" stem cells and screening methods, well before they are used in therapies. Recent scientific advances have contributed enormously to the development of new methods pertaining to both embryonic and adult stem cell isolation, characterization, expansion in undifferentiated state, customized differentiation, efficient stem cell transduction for gene therapy, quick monitoring of rejection, through polymerase chain reaction approaches following transplantation, and more.

The scientific progress that makes stem cell biology so exciting also threatens to suffocate researchers with an avalanche of information. Keeping with the tradition of Humana Press to bring these developments to the forefront in a timely manner, the book *Stem Cell Assays* aims to present scientific advances in stem cell methods for a wider use by novice and expert scientists, through the *Methods in Molecular Biology* series.

I thank Dr. Alan W. Flake and his laboratory team in the Children's Hospital of Philadelphia for stimulating discussions in stem cell biology. It was in Dr. Flake's laboratory that I came up with the concept for this book. I thank Dr. John Walker, senior editor of the series, the publishing team at Humana Press, and Dr. Abby Kurien for lively and extended discussions on the implications of stem cells, health, and society. Finally, I thank my wife Aruna and children Bindu and Sindu for enduring my continuing obsession with stem cells.

Mohan C. Vemuri

Contents

Contributors

CHRISTOPHER ADAMS • *Epigenetics-R&D, Invitrogen Corporation, Carlsbad, CA*

MICHAL AMIT • *The Bruce Rappaport Faculty of Medicine, Technion - Israel Institute of Technology, Haifa, Israel*

MIGUEL A. ANDRADE-NAVARRO • *Ottawa Health Research Institute, Molecular Medicine Program, Ottawa, Ontario, Canada, and University of Ottawa, Department of Cellular and Molecular Medicine, Ottawa, Ontario, Canada*

SZCZEPAN W. BARAN • *Fred Hutchinson Cancer Research Center, University of Washington, Seattle, WA*

PABLO BOSCH • *Departamento de Biología Molecular, Facultad de Ciencias Exactas Físico-Químicas y Naturales, Universidad Nacional de Río Cuarto, Río Cuarto, Córdoba, Argentina*

SHANNON BUCKLEY • *Stem Cell Institute, University of Minnesota, Minneapolis, MN*

TONG CAO • *Stem Cell Laboratory, Faculty of Dentistry, National University of Singapore, Singapore*

JONATHAN D. CHESNUT • *Stem Cells and Regenerative Medicine, Invitrogen Corporation, Carlsbad, CA*

EMER CLARKE • *StemCell Technologies, Vancouver, British Columbia, Canada*

JACQUES COHEN • *Reprogenetics LLC, West Orange, NJ*

PERE COLLS • *Reprogenetics LLC, West Orange, NJ*

JACKIE DAMEN • *StemCell Technologies, Vancouver, British Columbia, Canada*

MICHA DRUKKER • *Department of Pathology, Stanford University School of Medicine, Stanford, CA*

JENNIFER ELISSEEFF • *Department of Biomedical Engineering, Johns Hopkins University School of Medicine, Baltimore, MD*

ROBERTO ENSENAT-WASER • *Institute for Biomedical Engineering – Cell Biology, University Medical School/Rheinisch-Westfälische Technische Hochschule Aachen, Aachen Germany*

STEPHEN G. HALL • *AlphaGenix, Wright Place, Carlsbad, CA*

BOON CHIN HENG • *Stem Cell Laboratory, Faculty of Dentistry, National University of Singapore, Singapore*

NATHANIEL S. HWANG • *Department of Biomedical Engineering, Johns Hopkins University School of Medicine, Baltimore, MD*

SHIAW-MIN HWANG • *Bioresource Collection and Research Center, Food Industry Research and Development Institute, Hsinchu, Taiwan*

ALISON VENABLE JOHNSON • *Regenerative Bioscience Center, The University of Georgia, Athens, GA*

PAUL M. KRZYZANOWSKI • *Ottawa Health Research Institute, Molecular Medicine Program, Ottawa, Ontario, Canada*

UMA LAKSHMIPATHY • *Stem Cell Institute, University of Minnesota, Minneapolis, MN, and Invitrogen Corporation, Carlsbad, CA*

SANG-HUN LEE • *Department of Biochemistry & Molecular Biology, Institute of Mental Health, Hanyang University, Seoul, Korea*

BRAD LOVE • *Corporate Research Labs, Invitrogen Corporation, 1610 Faraday Ave. Carlsbad, CA 92008*

SANTIAGO MUNNE • *Reprogenetics LLC, West Orange, NJ*

ENRIQUE M. MURO • *Ottawa Health Research Institute, Molecular Medicine Program, Ottawa, Ontario, Canada*

CHRISTINA MUSCAT • *Department of Pathology, Stanford University School of Medicine, Stanford, CA*

STEFANIA A. NOTTOLA • *Department of Anatomy, Laboratory for Electron Microscopy "Pietro M. Motta", La Sapienza University, Rome, Italy*

GARETH A. PALIDWOR • *Ottawa Health Research Institute, Molecular Medicine Program, Ottawa, Ontario, Canada*

REDDANNA PALLU • *Department of Animal Sciences, School of Life Sciences, University of Hyderabad, Hyderabad, India*

MICHAEL PAPAMICHAIL • *Cancer Immunology and Immunotherapy Center, Saint Savas Hospital, Athens, Greece*

CHANG-HWAN PARK • *Department of Microbiology, College of Medicine, Institute of Mental Health, Hanyang University, Seoul, Korea*

CARLA PEREIRA • *StemCell Technologies, Vancouver, British Columbia, Canada*

SONIA A. PEREZ • *Cancer Immunology and Immunotherapy Center, Saint Savas Hospital, Athens, Greece*

CAROLINA PEREZ-IRATXETA • *Ottawa Health Research Institute, Molecular Medicine Program, Ottawa, Ontario Canada*

CHRISTOPHER J. PORTER • *Ottawa Health Research Institute, Molecular Medicine Program, Ottawa, Ontario, Canada*

VINAGOLU K. RAJASEKHAR • *Memorial Sloan-Kettering Cancer Center, New York, NY*

RAJ R. RAO • *Department of Chemical and Life Science Engineering, Department of Human Genetics, Virginia Commonwealth University Richmond, VA*

NISHANTH P. REDDY • *Department of Animal Sciences, School of Life Sciences, University of Hyderabad, Hyderabad, India*

JUAN ANTONIO REIG • *Instituto of Bioengineering, University Miguel Hernandez, Alicante, Spain*

ENRIQUE ROCHE • *Instituto of Bioengineering, University Miguel Hernandez, Alicante Spain*

REATHA SANDIE • *Ottawa Health Research Institute, Molecular Medicine Program, Ottawa, Ontario, Canada*

ALFREDO SANTANA • *Genetic and Cytogenetic Unit, Childhood Hospital of Canary Islands, Las Palmas Spain*

A. HENRY SATHANANTHAN • *Monash Immunology & Stem Cell Laboratories, Monash University, Melbourne, Australia*

TIM SCHIMMEL • *Tyho Galileo Research Laboratories, West Orange, NJ*

IGOR I. SLUKVIN • *Wisconsin National Primate Research Center, University of Wisconsin Graduate School, Madison, WI, WiCell Research Institute, Madison, WI, and Department of Pathology and Laboratory Medicine, University of Wisconsin, Madison, WI*

PANAGIOTA A. SOTIROPOULOU • *Cancer Immunology and Immunotherapy Center, Saint Savas Hospital, Athens, Greece*

STEVEN L. STICE • *Regenerative Bioscience Center, The University of Georgia, Athens, GA*

PATRICK STORDEUR • *Département d'Immunologie-Hématologie-Transfusion, Hôpital Erasme, Université Libre de Bruxelles, Brussels, Belgium*

MEREDITH A. THOMPSON • *Herman B Wells Center for Pediatric Research, Department of Pediatrics, Indiana University School of Medicine, Indianapolis, IN*

BHASKAR THYAGARAJAN • *Stem Cells and Regenerative Medicine, Invitrogen Corporation, Carlsbad, CA*

WEI SEONG TOH • *Stem Cell Laboratory, Faculty of Dentistry, National University of Singapore, Singapore*

SHYNI VARGHESE • *Department of Biomedical Engineering, Johns Hopkins University School of Medicine, Baltimore, MD*

CATHERINE VERFAILLIE • *Stem Cell Institute, University of Minnesota, USA, and Stem Cell Institute, Katholieke University, Leuven, Belgium*

NESTOR VICENTE-SALAR • *Instituto of Bioengineering, University Miguel Hernandez, Alicante, Spain*

MOHAN C. VEMURI • *Stem Cells, Reprogenitics, West Orange, NJ, and Invitrogen, Grand Island, NY*

MAXIM A. VODYANIK • *Wisconsin National Primate Research Center, University of Wisconsin Graduate School, Madison, WI*

CAROL B. WARE • *Department of Comparative Medicine, University of Washington, Seattle, WA*

IRVING L. WEISSMAN • *Department of Pathology, Stanford University School of Medicine, Stanford, CA*

ZHENG YANG • *Stem Cell Laboratory, Faculty of Dentistry, National University of Singapore, Singapore, and Department of Orthopaedic Surgery, Yong Loo Lin School of Medicine, NUS Tissue Engineering Program, National University of Singapore, Singapore*

CHAO-LING YAO • *Bioresource Collection and Research Center, Food Industry Research and Development Institute, Hsinchu, Taiwan*

MERVIN C. YODER • *Herman B Wells Center for Pediatric Research, Department of Pediatrics, Indiana University School of Medicine, Indianapolis, IN*

FRANK ZEIGLER • *Orion Biosolutions, Vista, CA*

MARTIN ZENKE • *Institute for Biomedical Engineering – Cell Biology, University Medical School/Rheinisch-Westfälische Technische Hochschule Aachen, Aachen, Germany*

GANG-MING ZOU • *Herman B Wells Center for Pediatric Research, Department of Pediatrics, Indiana University School of Medicine, Indianapolis, IN*

1

Derivation of Human Embryonic Stem Cells in Xeno-Free Conditions

Mohan C. Vemuri, Tim Schimmel, Pere Colls, Santiago Munne, and Jacques Cohen

Summary

Human embryonic stem cells (hESC) have the potential to treat a wide range of diseases. Currently, the use of existing hESC lines in human clinical applications is limited, as they are derived from blastocysts subjected to immunosurgery with animal derived antibodies, and are maintained on mouse embryonic feeder (MEF) cells, in the presence of either fetal calf serum (FCS) or on Matrigel or with conditioned media from MEFs. Successful derivation of hESCs in xeno-free conditions is crucial in advancing stem cell therapy applications. Two hESC lines, one from chromosomally abnormal embryos and another cell line from normal embryos from the inner cell mass of human blastocysts are derived using a culture media that had 20% serum replacement (SR) and human FGF2 on human foreskin fibroblasts as feeder cells. Derivation and characterization of such xenofree hESCs suitable for clinical studies is described in this chapter.

Key Words: Human embryonic stem cells; Serum-free cultures; Derivation; N2B27CDM.

1. Introduction

Human embryonic stem cell (hESC) biology is rapidly progressing toward clinical transition and potential therapies. Despite the increasing interest and demand for hESCs, the available number of hESC lines is limited and their use in clinical cell therapy is restricted, as existing lines express the non-human antigen Neu5Gc *(1)*. In addition, prolonged culturing of hESCs induces genomic alterations *(2)*, and the presence of high levels of major histocompatibility complex-1 (MHC-1) antigens makes them susceptible to immunological

From: *Methods in Molecular Biology, vol. 407: Stem Cell Assays*
Edited by: M. C. Vemuri © Humana Press, Totowa, NJ

rejection following transplantation *(3)*. Hence, derivation of hESCs in xeno-free conditions is crucial to develop successful stem cell therapy approaches. Recent advances in the long-term maintenance of hESCs in feeder-free and serum-free conditions *(4)*, as well as medium optimization that enables hESC growth in an undifferentiated state, have contributed toward a new generation of hESCs grown in animal component-free systems. The molecular events taking place during hESC derivation must be better understood, and development of feeder-free and serum-free media will facilitate their scrutiny *(5)*. Existing hESCs have been derived in the presence of fetal bovine serum, maintained in conditioned medium from mouse embryonic feeder cells, or using Matrigel substratum. Use of these complex media systems with ingredients of animal origin not only makes it difficult to understand the molecular mechanisms that control hESC pluripotency but also renders them unfit for clinical use. KnockOut Serum Replacement, a serum-free proprietary formulation of GIBCO-BRL, Grand Island (Patent WO 98/30679, 1998, Invitrogen, Carlsbad, CA, USA), has presented a major advance in hESC culture medium development. However, serum replacement still contains animal-derived proteins, and its composition remains proprietary in nature.

The Inzunza et al. *(6)* team derived two hESC lines, HS293 and HS306, using human foreskin fibroblasts (HFFs) feeder cells and a medium containing serum replacement. Here, a reliable method is described for the derivation and culture of hESCs from human blastocysts, using HFF feeder cells. Following passages, hESCs are propagated in feeder-free conditions through clonal selection in a chemically defined medium *(7)*. This method is simple, as it requires minimal blastocyst manipulation and obviates the need for inner cell mass isolation. Because these hESCs are derived and maintained in xeno-free conditions, they will be suitable for use in clinical disease models of tissue repair and tissue engineering.

2. Materials

1. Human embryos: Derivations of hESC were made possible through donation of surplus human embryos following in vitro fertilization procedures. These donations complied with local regulations, received institutional review board clearance and specific donor consent.
2. Global medium: A medium containing reduced glucose and amino acids (cat. no. LGGG-050, LifeGlobal) was supplemented with 5% human serum albumin (IISA) (cat. no. ART3003, Cooper Surgical, TrumBull, CT, USA).
3. Pronase: 10 U/mL (cat. no. P6911, Sigma, St. Louis, MI, USA).
4. Human tubal fluid (HTF) in 4-(2-hydroxyethyl)-1-piperayineethanesulfonic acid (HEPES) (cat. no. GMHH-100, LifeGlobal) supplemented with 5%.

5. HFFs (cat. no. CRL-1634, ATCC, Manassas, VA, USA).
6. Feeder cell culture medium: D-MEM/F-12 (cat. no. 11330-032, Invitrogen) containing 20% KnockOut Serum Replacement (GIBCO-BRL; cat. no. 10828-028, Invitrogen), 100 μM non-essential amino acids (cat. no. 11140-050, Invitrogen), 2 mM L-glutamine (cat. no. 25030-081, Invitrogen), 50 U/mL penicillin, 50 μg/mL streptomycin (cat. no. 15070-063, Invitrogen) and 100 μM 2-mercaptoethanol (cat. no. 21985-023, Invitrogen).
7. Mitomycin C (cat. no. M-0503, Sigma): Prepared in feeder cell medium at 10 μg/mL.
8. N2B27 Complete Defined Medium (N2B27 CDM): The composition of N2B27CDM was adopted with slight modification of the medium developed by Yao et al. *(8)*. N2B27CDM contained D-MEM/F-12 (cat. no. 11330-032, Invitrogen) supplemented with $1 \times$ N-2 Supplement (cat. no. 17502-048, Invitrogen), $1 \times$ B-27 Supplement (cat. no. 17504-044, Invitrogen), 2 mM L-glutamine (cat. no. 25030-081, Invitrogen), 110 μM 2-mercaptoethanol (cat. no. 21985-023, Invitrogen), 100 μM non-essential amino acids (cat. no. 11140-050, Invitrogen), 20 ng/mL basic fibroblast growth factor (bFGF), (cat. no. 13256-029, Invitrogen) and 50 mg/mL HSA (cat. no. ART3003, TrumBull, CT, USA).
9. Stem cell cutting tool (Swemed, Sweden, http://www.swemed.com).
10. Stereomicroscope.
11. Stem cell marker kit: Primary antibodies specific for hESC markers TRA-1-60, TRA-1-81, SSEA-4, SSEA-3 and Oct-4 (cat. no. SCR002, Chemicon, Temecula, CA, USA) and isotype controls. Alkaline phosphatase activity kit (cat. no. SCR004, Chemicon).
12. TRI Reagent for total RNA extraction (cat. no. TR118, Molecular Research Center Inc., Cincinnati, OH, USA).
13. Primers for Oct-4, reverse transcription and polymerase chain reaction (PCR) amplication. Taq polymerase (0.5 μL) (Promega Corp., Madison, WI, USA). The primers used were specific for the *Oct-4* gene (5′-CTTGCTGCAGAAGTGGGTG GAGGAA-3′, 5′-CTGCAGTGTGGGTTTCGGGCA-3′), with human β-actin as control.
14. Four-well Nunclon tissue culture dishes (cat. no. 144444, Nalge Nunc International, Rochester, NY, USA).

3. Methods

3.1. Human Embryo Culture

Embryos were cultured to the blastocyst stage using Global medium. Blastocyst morphology was graded for symmetry, cytoplasmic appearance, extent of fragmentation and blastomere nuclear status (*see* **Note 1**). Preimplantation genetic diagnosis (PGD) samples that scored as relatively high quality were selected for use.

3.2. Preparation of Human Material

Previous studies have isolated the inner cell mass from the trophoectoderm, using complement and laser surgery or micromanipulation. In the current method, manipulations were minimized to retain the inner cell mass in its native configuration (1–3). The zona pellucida was removed by treatment with pronase at 37 °C for 2 min, and the embryo was then washed in global medium supplemented with 5% HSA (*see* **Note 2**). The use of pronase to digest the zona was closely monitored so as to retain the inner cell mass, while still allowing the removal of zona and trophoectoderm cells (*see* **Fig. 1A and B**). The cellular mass retained following digestion was carefully transferred to pre-warmed wash medium consisting of HEPES-buffered HTF supplemented with 5% HSA. Following two washes, the cell mass was pipetted through a sterile glass capillary tube and transferred to a HFF feeder cell layer (*see* **Fig. 1C**).

3.3. Human Embryo Culture

Embryos were cultured to the blastocyst stage in a single complex formulation termed Global medium. PGD samples with relatively high quality blastocysts were used following morphological grading based upon features of blastomere symmetry, cytoplasmic appearance, extent of fragmentation and blastomere nuclear status (*see* **Note 1**).

3.4. Preparation of Human Material

In order to efficiently derive hESCs, previous studies have isolated the inner cell mass from the trophoectoderm using complement and laser surgery or micromanipulation. In the current method, minimal manipulations were done to retain inner cell mass in its native configuration. The zona pellucida was removed by treatment with pronase at 37 °C for 2 min, and the embryo was then washed in global medium supplemented with 5% HSA (*see* **Note 2**). The process of pronase digestion was closely monitored so as to retain the inner cell mass but still allow the removal of zona and trophoectoderm cells (*see* **Fig. 1A and B**). The cellular mass retained after pronase digestion was carefully transferred into pre-warmed wash medium, consisting HTF in HEPES and supplemented with 5% HSA. Following two washes, the cell mass was pipetted through a glass capillary and placed on to human feeder cell layer (*see* **Fig. 1C**).

3.5. Feeder Cell Preparation and Adaptation to Xeno-Free Conditions

HFF need to be pre-adapted to growth in feeder cell culture medium. When the cells were about 70–80% confluent, HFFs were mitotically inactivated

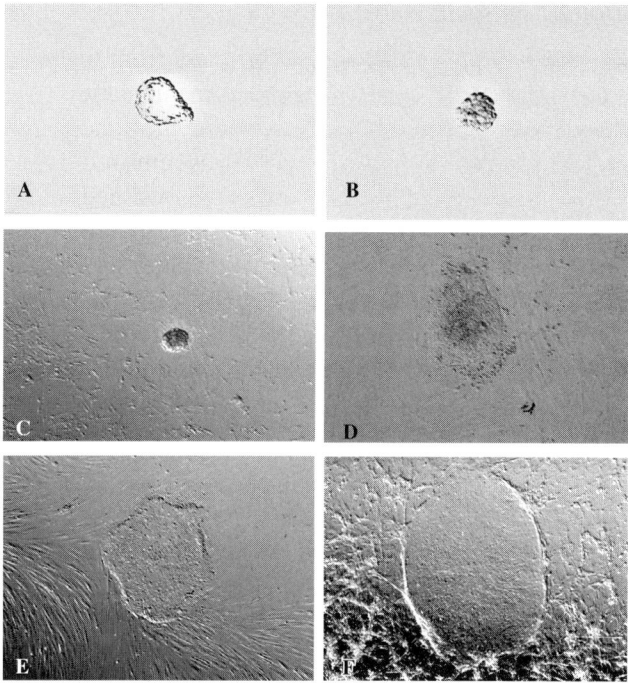

Fig. 1. De novo process of human embryonic stem cell derivation and cell morphology in xeno-free culture system. (**A**) Image of a day 5 human blastocyst at early hatching stage with inner mass of cells and outer zona pellucida. (**B**) Mass of cells isolated following pronase treatment of blastocyst to remove zona and other cells. (**C**) Pronase-treated cell mass transferred onto inactivated human foreskin fibroblast feeders. (**D**) Appearance of cells with stem cell-like morphology by day 7. Note the appearance of apoptosis of several cells in the culture during initial attachment, settlement of cells and proliferation of stem cell like colony. (**E**) Embryonic stem-like colony acquires characteristic embryonic stem cell-like colony morphology between days 12 and 15. (**F**) Subsequent passaging of this colony results in the formation of a typical embryonic stem cell colony that expresses stem cell markers.

by exposure to feeder cell medium containing 10 µg/mL mitomycin C for 3 hours at 37 °C (*see* **Note 3**). Following inactivation, the medium was replaced with fresh feeder cell culture medium, and cells were incubated overnight in order to recover from treatment. Cells were trypsinized and transferred to 4-well Nunclon tissue culture dishes at a density of 70,000 cells per well, in N2B27CDM. These mitotically inactive HFF dishes were used only after 24 h of plating (*see* **Note 4**).

3.6. Culture Initiation to Derive hESCs

The initial hESC culture was established by transferring a zona-free blastocyst to inactivated HFF feeder cells, using N2B27CDM. Culture medium was changed each day, and the cell/colony morphology was recorded daily. The first stem-cell like colony (*see* **Fig. 1D**) was observed approximately 7–12 days after initial plating. (*see* **Note 5**).

3.7. Early Steps of hESC Clonal Expansion

After an initial growth period of 12 days, the cell aggregates were removed mechanically from the original plate and transferred to fresh inactivated human feeder cell layers. Mechanical passages were performed by cutting the colony into six to eight pieces using a Swemed stem cell cutting tool, while observing cultures under the stereomicroscope. Mechanical passaging was carried out every week. Undifferentiated cells, as judged by morphology (*see* **Note 6**), were further passaged by clonal expansion (*see* **Note 7**). Culture kinetics, population doubling times and growth variation from passage to passage were evaluated by assessing colony and cell numbers.

3.8. Analysis of Stem Cell Marker Expression and Immunocytochemistry

Stem cell phenotype and differentiation status of hESCs were evaluated using an immunocytochemical stem cell marker kit. Antigen expression specific for the hESC markers TRA-1-60, TRA-1-81, SSEA-4, SSEA-3 and Oct-4 was analyzed. The immunocytochemical profiles were captured using confocal microscopy (*see* **Fig. 2**). Alkaline phosphatase activity was assayed in hESC cultures using an alkaline phosphatase activity kit, following manufacturer's instructions.

3.9. Expression of Oct-4 as a Pluripotency Marker

Cells from different hESC passages were assessed for pluripotency by examining Oct-4 gene expression. Oct-4 is expressed in hESCs and thought to be involved in maintaining their pluripotent state.

Total RNA was extracted with TRI reagent. Four micrograms of total RNA was reverse-transcribed in a 30 μL reaction mixture containing 6 μL of Reverse Transcription (RT) buffer, 1 μL 10 mmol/L dNTP, 1 μL 0.5mg/mL Oligo (dT)15, 0.5 μL 50 IU/mL ribonuclease inhibitor and 1 μL 200 IU/mL M-MLV transcriptase (Promega Corp.). After reverse transcription, 2 μL of cDNA was amplified in 50 μL of PCR mixture containing 5 μL PCR buffer,

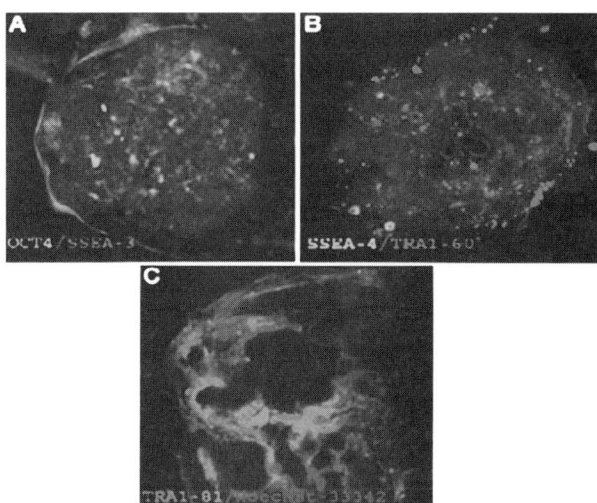

Fig. 2. Immunofluorescence staining to characterize undifferentiated state of human embryonic stem cell line RGK230 with a panel of antibodies specific for human stem cell markers. (**A**) positive staining for Oct-4 (green) and weak staining with SSEA-3 (red). (**B**) Positive staining for SSEA-4 (green) and TRA-1-60 (red). (**C**) Nuclear counter staining with Hoechst 33342 (blue) and positive staining with TRA-1-81(red) antibody. View the figures in color in the accompanying CD.

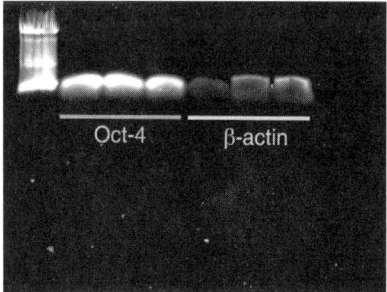

Fig. 3. Analysis of pluripotent marker gene, Oct-4 expression by polymerase chain reaction in human embryonic stem cell line RGK230. Lane 1, (from right) DNA marker; lanes 2–4, expression of Oct-4; lanes 5–7, β-actin.

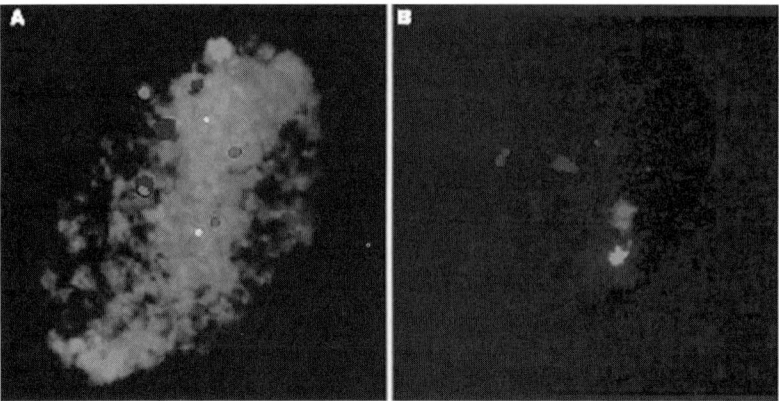

Fig. 4. Fluorescence in situ hybridization analysis of human embryonic stem cell line RGK230 (**A**) for the chromosomes 13 (red), 16 (aqua), 18 (blue), 21 (green) and 22 (yellow) and (**B**) chromosomes 15 (red), 17 (blue), X (green) and Y (pink). View the figures in color in the accompanying CD.

$3\,\mu L$ 25 mmol/L $MgCl_2$, $1\,\mu L$ 10 mmol/L dNTP, 5 IU/mL Taq polymerase $0.5\,\mu L$ and $1\,\mu L$ 10 pmol/mL of each primer pair. The primers used were specific for the *Oct-4* gene (5′-CTTGCTGCAGAAGTGGGTGGAGGAA-3′ and 5′-CTGCAGTGTGGGTTTCGGGCA-3′), with human β-actin as control (*see* **Fig. 3**).

3.10. Fluorescence In Situ Hybridization

Chromosomal analysis of hESC cells was carried out using fluorescence in situ hybridization (FISH) with probes specific for chromosomes X, Y, 13, 15, 16, 17, 18, 21 and 22 in two consecutive hybridizations, following previously published protocols *(7)*. The first hybridization was performed using a cocktail of probes for chromosomes 13, 16, 18, 21 and 22. The second hybridization used a homemade cocktail of probes for chromosomes X, Y, 15 and 17 prepared as previously described *(9)*. Slide preparations were viewed using a fluorescent microscope. Images were photographically captured and adjusted for signal intensity through image analysis software (*see* **Fig. 4**). Additional information on hESC derivation from trisomic embryos is described *(10)*.

4. Notes

1. In the first phase of the study, a total of 13 blastocysts with variable inner cell mass quality were used for embryonic stem cell derivation.

2. On day 5 or 6 after fertilization, blastocysts selected for hESC isolation were processed for zona pellucida removal by short-term exposure (2–5 min) to pronase in embryo culture medium. We intentionally avoided laser biopsy in view of possible failure of laser-biopsied cells to establish cell lines. We also avoided the procedure of immunosurgery, as this involves the use of complement, a product of animal origin. The inner cell mass was isolated by mechanical micromanipulation, and the cells were collected as aggregates rather than as single cells. The aggregates were plated onto mitotically inactivated HFF cells.

3. Mitomycin C undergoes rapid degradation in acidic solutions (when pH < 6). It is stable in solutions and buffers at a pH of 6–9. It is better to prepare a working solution and pre-warm it to 37 °C, just prior to exposing HFF cells to mitomycin C treatment.

4. It is essential to have inactive feeder cell cultures prepared at least one day prior to starting hESC derivation or for routine hESC passaging.

5. Following the initial plating of blastocyst cells, half-medium changes were carefully conducted daily in a sterile environment, with minimal disturbance to the cultures for the first 3 days. After day 3, full-medium changes were carried out daily. Cell and colony morphology were recorded daily. Appearance of the first stem cell-like colonies occurred approximately 7 days after plating (*see* **Fig. 1**).

6. Differentiated cells were first noticed around the periphery of the colony, while undifferentiated cells were primarily located toward the center of the colony. This has been observed consistently by hESC laboratories. Using the Swemed tool or custom fire-polished sterile Pasteur pipettes, these differentiated cells can be removed at the time of passaging.

7. Early hESCs are heterogeneous in nature, as they are derived from the inner cell mass of the blastocyst. This heterogeneity can be addressed by subcloning and by evaluating ESC-specific surface marker expression (*see* **Fig. 1E and F**)

Acknowledgment

The authors thank Mary Lynn Tilkins for helpful comments and suggestions.

References

1. Martin MJ, Muotri A, Gage F et al. Human embryonic stem cells express an immunogenic nonhuman sialic acid. Nat Med 2005: 11: 228–32.

2. Inzunza J, Sahlen S, Holmberg K et al. Comparative genomic hybridization and karyotyping of human embryonic stem cells reveals the occurrence of an isodicentric X chromosome after long-term cultivation. Mol Hum Reprod 2004: 10: 461–6.

3. Drukker M, Katchman H, Katz G et al. Human embryonic stem cells and their differentiated derivatives are less susceptible for immune rejection than adult cells. Stem Cells 2006: 24: 221–9.

4. Li Y, Powell S, Brunette E et al. Expansion of human embryonic stem cells in defined serum-free medium devoid of animal-derived products. Biotechnol Bioeng 2005: 91: 688–98.

5. Rajasekhar, V.K. and Vemuri. M.C. Molecular insights into the function, fate, and prospects of stem cells. Stem Cells 2005: 23: 1212–20.

6. Inzunza J, Gertow K, Stromberg MA et al. Derivation of human embryonic stem cell lines in serum replacement medium using postnatal human fibroblasts as feeder cells. Stem Cells 2005: 23: 544–9.

7. Munne S, Marquez C, Magli C et al. Scoring criteria for preimplantation genetic diagnosis of numerical abnormalities for chromosomes X, Y, 13, 16, 18 and 21. Mol Hum Reprod 1998: 4: 863–70.

8. Yao S, Chen S, Clark J et al. Long-term self-renewal and directed differentiation of human embryonic stem cells in chemically defined conditions. Proc Natl Acad Sci USA 2006: 103: 6907–12.

9. Munne S, Magli C, Bahce M et al. Preimplantation diagnosis of the aneuploidies most commonly found in spontaneous abortions and live births: XY, 13, 14, 15, 16, 18, 21, 22. Prenat Diagn 1998: 18: 1459–66.

10. Munne S. Velilla E, Colls P et al. Self-correction of chromosomally abnormal embroys in culture and implications for stem cell production. Fertil Steril. 2005: 84: 1328–34.

2

Feeder-Layer Free Culture System for Human Embryonic Stem Cells

Michal Amit

Summary

Human embryonic stem cells (hESCs) are pluripotent stem cells derived from the inner cell mass of the blastocyst. Due to their unique properties, hESCs might be used for research fields such as self-renewal, specific lineage differentiation, human developmental biology, and teratology. hESCs also have outstanding potential to serve for clinical purposes as a source for cell-based therapies. Traditionally, these cells are cultured and derived with mouse embryonic fibroblast as supportive layer, using a medium supplemented with fetal bovine serum. Future industrial and clinical implementation of hESCs will require the use of a defined medium and an animal-free culture method that will prevent their possible exposure to animal pathogens. This chapter discusses the advancements in the development of methods for the defined culture of hESCs and describes a simple method for animals serum-free and feeder layer-free culture of hESCs.

Key Words: Embryonic stem cells; Transforming growth factor β1; Basic fibroblast growth factor; Fibronectin.

1. Introduction

Human embryonic stem cells (hESCs) are pluripotent stem cells derived from the inner cell mass of embryos at the blastocyst stage, first isolated at 1998 *(1)*. As is evident from the extensive research with mouse ESCs during the last 25 years, hESCs could serve for the study of self-renewal processes, identification of early human developmental events, and for the induction of differentiation into specific cell types. During the first few years of study, hESCs were cultured using the methods developed in the 1970s for culture

From: *Methods in Molecular Biology, vol. 407: Stem Cell Assays*
Edited by: M. C. Vemuri © Humana Press, Totowa, NJ

of embryonal carcinoma cell lines, that is, co-culture with mitotically inactivated mouse embryonic fibroblasts (MEFs) using a medium supplemented with 20% fetal bovine serum (FBS). Although these traditional culture conditions allowed the propagation of hESCs to reasonable amounts *(2)*, they presented two major disadvantages: (i) exposure to animal pathogens through the MEFs or FBS and (ii) variations between different batches of MEFs and serum in their ability to support culture of undifferentiated hESCs. Thus, although the traditional culture methods facilitated 25 years of fruitful research of mouse ESCs, the possible future clinical and industrial applications of hESCs require their culture in defined conditions, preferably animal serum-free products, without the utilization of a feeder layer, and in a reproducible manner of the culture system.

The development of defined culture systems for the successful propagation of hESCs requires either the substitution of the MEF feeder layer with feeder layers of human origin or the detection of conditions for the culture of ESC without the need for a supportive cell line. Both options require animal serum substitutes such as the existing commercial serum replacements. Indeed, in recent years, extensive research to accomplish the goal of developing defined culture systems for hESCs has yielded four major achievements: (i) propagation of hESCs in serum-free culture conditions *(3)*; (ii) maintenance of hESCs as undifferentiated cells using a feeder layer-free culture method based on Matrigel matrix combined with 100% MEF-conditioned medium *(4)*; (iii) substitution of MEFs with human supportive layers such as embryonic fibroblasts, adult Fallopian tube epithelium *(5)*, or foreskin fibroblasts *(6–8)*; and (iv) culture of hESCs without any feeder layer while using a medium supplemented with serum replacement, selected growth factors, and substitute matrix *(9–12)*.

The easiest resolution to avoid the presence of animal products in hESCs culture is using human supportive layer and human serum *(5)*. This culture method had been demonstrated to promote both the prolonged culture of hESCs and the isolation of new hESC lines *(5,8)*. Although these methods were demonstrated to efficiently promote hESC culture while maintaining low background differentiation rates, stable karyotype, and cell pluripotency, the culture remains exposed to variations between serum and supportive cell batches and to the complexity of co-culture.

In view of the possible need for large-scale culture and defined culture conditions required for industrial or clinical uses, the ideal culture method for hESCs would be one using human serum and feeder layer-free conditions. The first step toward this goal was achieved by Xu et al. *(4)*, who introduced for the first time a culture system for hESCs in which no direct contact exists between the ESCs and MEFs, using Matrigel matrix and 100% MEF-conditioned medium supplemented with serum replacement and basic fibroblast growth factor (bFGF). Although several misgivings exist in this culture system, which include the need for parallel culture of both hESC and MEFs, the exposure of the cultured ESCs to variations between MEF batches, and the risk of exposure to animal pathogens, the method is nevertheless widely used by researchers due to its simplicity and efficiency. This chapter focuses on the culture method offered by Amit and colleagues *(9)* that is a feeder-free culture system where no conditioned medium is used. The method is based on the implementation of human fibronectin matrix together with medium supplemented with serum replacement, bFGF, and transforming growth factor β1 (TGFβ1). When cultured under these conditions, hESCs were able to retain their stem cell characteristics for over a year while maintaining a stable karyotype and demonstrating both the expression of genes specific to their undifferentiated state and Embryoid body (EB) and teratoma formation. A hESC colony propagated under these culture conditions is illustrated in **Fig. 1**. In addition to the described method, other techniques for feeder layer-free cultures were developed based on a non-conditioned medium supplemented with high concentrations of bFGF (40–100 ng/ml) and the utilization of either Matrigel, Laminin, or fibronectin as matrix *(10,11)*. Although these methods were able to support prolonged undifferentiated culture of hESCs, these culture systems are based on the use of mediums supplemented with serum replacement, containing "Albumax," which is a non-defined material derived from lipid-enriched bovine serum albumin.

The majority of the reported hESC lines were isolated while using MEFs as a supportive layer *(1,13–16)*, whereas some were derived using human cell lines as feeder layers *(5,8)*. The first account on the derivation of new hESC lines without the use of any supportive layer whatsoever, reported by Klimanskaya and colleagues *(17)*, described the isolation of the hESCs while using MEF-manufactured matrix together with a medium supplemented with a high dose of bFGF (16 ng/ml), Leukemia inhibitory factor (LIF), serum replacement, and plasmanate. This new hESC line preserve hESC characteristics including retainment of stable normal karyotypes for over 30 passages of continuous culture *(17)*. Thus for the first time, the principle of supportive layer-free derivation of hESC lines was demonstrated. A current publication by Ludwig and colleagues overcomes the main obstacle of the Klimanskaya study. While Klimanskaya and colleagues used both non-defined and animal-derived materials such as MEF matrix and konockout serum replacement, Ludwig and colleagues *(12)* demonstrated utilization of defined serum and an animal-free

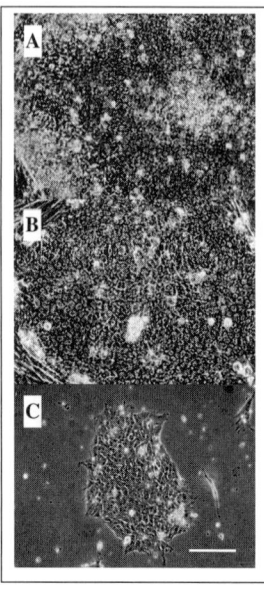

Fig. 1. Human embryonic stem cell (hESC) colonies of I4 cell line cultured (*A*) with mouse embryonic fibroblasts (MEFs), (*B*) with human foreskin fibroblasts and medium supplemented with growth factors, (*C*) and with fibronectin and serum-free medium supplemented with growth factors. Bar = 50 μm.

medium suitable both for the prolonged culture of hESCs and for the isolation of new lines under culture conditions that are feeder layer free. Interestingly, the medium combination offered by Ludwig et al. requires the addition of both bFGF and TGFβ1. However, one disturbing difficulty regarding the karyotypic stability of the cultured hESCs still needs to be answered. The authors described two new hESC lines isolated while using these defined conditions: one was reported to harbor 47 chromosomes with an XXY abnormality and the second carried a trisomy of chromosome 12 after 4 and 7 months of continuous culture respectively. Whether these karyotype abnormalitics arc exceptional occasions known to occur during prolonged culture or if the offered method does not sustain karyotype stability is yet to be clarified.

Although the technique to culture and isolate hESCs under defined and supportive layer-free conditions is available, the mechanisms controlling hESC self-maintenance are still unrevealed. This chapter discusses a culture system based on medium supplemented with TGFβ1 and bFGF. Recently, the involvement of members of the TGFβ superfamily in maintaining undifferentiated culture of hESCs was offered, and increasing evidences indicate that TGFβ1, activin, and nodal might be involved in mechanisms supporting hESC self-renwal through the action of the transcription factor SMAD2/3 *(18–20)*. James and colleagues *(18)* demonstrate that SMAD2/3 signaling is increased in undifferentiated cells, decreased at early stages of differentiation, and that the expression of some markers specific to the undifferentiated stage depends on its activation. Additional studies suggested that TGFβ1/activin/nodal and Wnt signal transduction pathways have a positive effect on Nanog or Oct 4 activities in the nucleus through SMAD2/3 and therefore might participate in hESC self-renewal mechanisms *(19)*. These growth factors, TGFβ1, activin, or Bio (an activator of the Wnt signaling, a GSK3β inhibitor), were indeed demonstrated to be essential supplements to the medium for maintaining the hESCs as undifferentiated cells in feeder layer-free culture systems *(9,12,21–23)*. All existing feeder layer-free culture methods require bFGF as an essential factor at the high concentration of 100ng/ml solely to support the undifferentiated state of hESC culture *(10)*. However, none of the offered factors was proven yet to be directly involved in the hESC mechanism of self-renewal. Thus, extensive research is needed to reveal the mechanism(s) controlling hESCs self-maintenance.

A major component of supportive layer-free culture systems for hESCs is the matrix. One of the commonly used matrix is Matrigel, containing collagen, laminin, fibronectin, and growth factors. In the described culture system fibronectin is used as a substitute matrix *(9)*. Fibronectin, a basal lamina component, is known to increase cell adhesion to the culture dishes and to mediate cell adhesion to extracellular matrix proteins, through integrin receptor. Integrin receptors participate in pathways which activate a variety of intracellular signal transduction pathways which might be involved in cell proliferation, apoptosis, shape formation, polarity, motility, gene expression profiles, and differentiation *(24)*. The fibronectin-specific integrin receptor, $\alpha_5\beta_1$, was demonstrated to be expressed on undifferentiated hESCs *(9)*. Further complementary research is required to clarify the possible role of extracellular proteins such as fibronectin in hESCs maintenance. In this chapter, a simple method for feeder-free and serum-free propagation of hESCs is described.

2. Materials

2.1. Culture Medium

1. Feeder layer-free culture medium—85% knockout Dulbecco's Modified Eagle's Medium (DMEM) (cat. no. 10829018, Invitrogen Corporation, Carlsbad, CA, USA), 15% knockout serum replacement (Invitrogen Corporation, cat. no. 10828028), 1%

non-essential amino acids (cat. no. 11140035, Invitrogen Corporation; *see* **Note 1**), 1 mM L-glutamine (cat. no. 25030024, Invitrogen Corporation; *see* **Note 1**), 0.1 mM β-mercaptoethanol (cat. no. 31350010, Invitrogen Corporation; *see* **Note 1**), 4 ng/ml bFGF (cat. no. 13256029, Invitrogen Corporation; *see* **Note 2**), and 0.12 ng/ml TGFβ1 (cat. no. 240-B, R&D Systems, Minneapolis, MN, USA; *see* **Note 2**).

2. Freezing medium: 60% DMEM (cat. no. 41965039, Invitrogen Corporation), 20% dimethyl sulfoxide (DMSO) (cat. no. D-2650, Sigma-Aldrich Inc., St Louis, MO, USA), and 20% defined FBS (cat. no. SH30070.03, HyClone, Logan UT, USA).

3. Serum-free freezing medium: 50% DMEM (cat. no. 41965039, Invitrogen corporation; *see* **Note 1**), 20% DMSO (cat. no. D-2650, Sigma-Aldrich Inc.; *see* **Note 3**), 30% serum replacement (Invitrogen Corporation, cat. no. 10828028).

4. Splitting medium: DMEM (cat. no. 41965039, Invitrogen Corporation; *see* **Note 1**), supplemented with 1.5 mg/ml type IV collagenase (cat. no. 17104019, Invitrogen Corporation; *see* **Note 1**).

5. Freezing box (cat. no. 5100-0001, Nalgne, Rochester, NY, USA).

6. Fibronectin (human): $5 \mu g/cm^2$ human plasma fibronectin (cat. no. FC010-10, Chemicon International Temecula, CA, USA; *see* **Note 4**).

3. Methods

3.1. Medium Preparation

1. Culture medium, 500 ml: Add 416.5 ml of knock out-(ko)-DMEM, 75 ml of serum replacement, 5 ml of non-essential amino acids, 2.5 ml of L-glutamine, 1 ml of β-mercaptoethanol, 2000 ng of bFGF, and 60 ng of TGFβ1 into 0.22-μm filter unit and filter (*see* **Note 5**).

2. Freezing medium, 10 ml: Add 6 ml of DMEM, 2 ml of DMSO, and 2 ml of defined FBS into a tube and filter through a 0.22-μm filter (*see* **Note 6**).

3. Serum-free freezing medium, 10 ml: Add 5 ml of DMEM, 2 ml of DMSO, and 3 ml of serum replacement into a tube and filter through 0.22-μm filter (*see* **Note 6**).

4. Splitting medium, 100 ml: Add 100 ml of DMEM into filter unit, add 150 mg of type IV collagenase, let set for 2 min for the powder to be dissolved and filter (*see* **Note 5**).

5. Preparation of fibronectin-coated plates: Dilute 1 mg of fibronectin in 10 ml of sterile water to create a fibronectin stock solution. Coat plates according to the amounts described in **Table 1** to achieve desirable concentration of $5 \mu g/cm^2$. The plates should be placed at room temperature for at least 1 h before hESC plating (*see* **Notes 7** and **8**).

3.2. Splitting hESCs

The following protocol is suitable for hESCs plated on fibronectin-covered plates. Cells should be split in a ratio of 1:2 or 2:3 every 4 to 5 days. The medium

Table 1
Recommended Amount of Fibronectin Stock
Solution per Well

Plate/dish	Volume of fibronectin stock solution per well (ml)
4 wells ($2.5\,cm^2$)	0.3
6 wells ($10\,cm^2$)	0.5
35 mm	0.5

should be changed on a daily basis. It is recommended to scrape differentiating colonies every 5–7 passages.

1. Remove medium from well. Add 0.5 ml of splitting medium (for one well in a six-well plate) and incubate for 25 min or until most colonies float (*see* **Note 9**).
2. Add 1 ml of culture medium and gently collect cells with a 5-ml pipette (differentiated cells will remain attached to the plate). Transfer collected cells into a conical tube.
3. Centrifuge for 3 min at $90\,g$ at $4\,°C$.
4. Re-suspend cells in fresh culture medium and plate directly on a ready-to-use fibronectin-covered culture plate.

3.3. Freezing hESCs

1. The recommended freezing ratio is all cells seeded on one 10-cm^2 well per vial (one well in six-well plates) or 1–2 million cells/vial.
2. Remove medium from well. Add 0.5 ml of (for one well of a six-well plate) splitting medium and incubate for 25 min or until most colonies float (*see* **Note 9**).
3. Add 1 ml of culture medium, gently scrape the cells using a 5-ml pipette and transfer cells into a conical tube.
4. Centrifuge cells for 3 min at $90\,g$ at $4\,°C$.
5. Resuspend cells in culture medium (*see* **Note 10**).
6. Drop by drop, add an equivalent volume of freezing medium and mix gently (*see* **Note 11**).
7. Pour 0.5 ml into a 1-ml cryogenic vial (*see* **Note 12**).
8. Freeze overnight at $-80\,°C$ in a freezing box (*see* **Note 13**).
9. Transfer to liquid nitrogen on the following day (*see* **Note 14**).

3.4. Thawing hESCs

1. Remove a vial from the liquid nitrogen.
2. Gently swirl the vial in a water bath warmed to $37\,°C$.
3. When only a small clump of unthawed ice remains, wash the vial in 70% ethanol.

4. Pipette the content of the vial up and down once to mix.
5. Place the content of the vial into a conical tube and add, drop by drop, 2 ml of culture medium (*see* **Note 11**).
6. Centrifuge cells for 3 min at 90 g at 4 °C.
7. Remove the supernatant and resuspend the cells in 2 ml of fresh medium.
8. Place the cell suspension in one well of a six-well plate (or on a four-well plate) pre-covered with fibronectin (*see* **step 5** in **Subheading 3.1.**).

4. Notes

1. The material can be replaced with other manufacturer's products.
2. The growth factors are crucial supplements of the medium; it is not recommended to change the source of the factors as different manufacturers might produce factors with reduced efficiency. In case of change, the concentration of the factors should be adjusted according to the materials $ED_{50} \leq 0.5 \text{ng/ml}$.
3. It is not recommended to replace the DMSO source, not all available DMSOs are suitable for freezing.
4. Human foreskin fibroblast cellular fibronectin (cat. no. F6277, Sigma-Aldrich Inc.) and human plasma fibronectin (cat. no. F2006, Sigma-Aldrich Inc.) were also found to support hESC feeder layer-free culture.
5. Should be kept at 4–8 °C for no more than 5 days.
6. It is better to prepare fresh freezing medium. If kept, it should be stored at 4–8 °C for no more than 3 days.
7. Fibronectin-coated plates can be prepared in advance and stored in a clean place at room temperature or in a 37 °C incubator.
8. It is not necessary to remove fibronectin residues before hESC plating.
9. Incubating the cells in the splitting medium for more than 1 h might harm the cells.
10. Do not break the cells into small clumps.
11. Adding the medium in this stage drop by drop is highly important. If the medium is added all at once the survival rates decrease dramatically.
12. A volume of 250 µl per tube can also be successfully used.
13. In our experience, the use of Nalgene special freezing boxes increases cell survivability.
14. It is not recommended to leave the vials at −80 °C for less than 24 h, or more than 2 days.

Acknowledgments

The author thank Dr. Ilana Goldberg-Cohen for critically reading the manuscript and Mrs. Hadas O'Neill for editing. The described research was partly the Technion Research Fund (TDRF).

References

1. Thomson, J.A., Itskovitz-Eldor, J., Shapiro, S.S., Waknitz, M.A., Swiergiel, J.J., Marshall, V.S., Jones, J.M. (1998). Embryonic stem cell lines derived from human blastocysts. *Science* **282**, 1145–1147 (erratum in *Science* 1998;282:1827).
2. Reubinoff, B.E., Pera, M.F., Vajta, G., Trounson, A.O. (2001). Effective cryopreservation of human embryonic stem cells by the open pulled straw vitrification method. *Hum Reprod* **16**, 2187–2194.
3. Amit, M., Carpenter, M.K., Inokuma, M.S., Chiu, C.P., Harris, C.P., Waknitz, M.A., Itskovitz-Eldor, J., Thomson, J.A. (2000). Clonally derived human embryonic stem cell lines maintain pluripotency and proliferative potential for prolonged periods of culture. *Dev Biol* **227**, 271–278.
4. Xu, C., Inokuma, M.S., Denham, J., Golds, K., Kundu, P., Gold, J.D., Carpenter, M.K. (2001). Feeder-Layer free growth of undifferentiated human embryonic stem cells. *Nat Biotechnol* **19**, 971–974.
5. Richards, M., Fong, C.Y., Chan, W.K., Wong, P.C., Bongso, A. (2002). Human feeders support prolonged undifferentiated growth of human inner cell masses and embryonic stem cells. *Nat Biotechnol* **20**, 933–936.
6. Amit, M., Margulets, V., Segev, H., Shariki, C., Laevsky, I., Coleman, R., Itskovitz-Eldor, J. (2003). Human feeder layers for human embryonic stem cells. *Biol Reprod* **68**, 2150–2156.
7. Hovatta, O., Mikkola, M., Gertow, K., Stromberg, A.M., Inzunza, J., Hreinsson, J., Rozell, B., Blennow, E., Andang, M., Ahrlund-Richter, L. (2003). A culture system using human foreskin fibroblasts as feeder cells allows production of human embryonic stem cells. *Hum Reprod* **18**, 1404–1409.
8. Inzunza, J., Gertow, K., Stromberg, M.A., Matilainen, E., Blennow, E., Skottman, H., Wolbank, S., Ahrlund-Richter, L., Hovatta, O. (2005). Derivation of human embryonic stem cell lines in serum replacement medium using postnatal human fibroblasts as feeder cells. *Stem Cells* **23**, 544–549.
9. Amit, M., Shariki, C., Margulets, V., Itskovitz Eldor, J. (2004). Feeder and serum free culture system for human embryonic stem cells. *Biol Reprod* **70**, 837–845.
10. Xu, R.H., Peck, R.M., Li, D.S., Feng, X., Ludwig, T., Thomson, J.A. (2005). Basic FGF and suppression of BMP signaling sustain undifferentiated proliferation of human ES cells. *Nat Methods* **2**,185–190.
11. Xu, C., Rosler, E., Jiang, J., Lebkowski, J.S., Gold, J.D., O'Sullivan, C., Delavan-Boorsma, K., Mok, M., Bronstein, A., Carpenter, M.K. (2005). Basic fibroblast growth factor supports undifferentiated human embryonic stem cell growth without conditioned medium. *Stem Cells* **23**, 315–323.
12. Ludwig, T.E., Levenstein, M.E., Jones, J.M., Berggren, W.T., Mitchen, E.R., Frane, J.L., Crandall, L.J., Daigh, C.A., Conard, K.R., Piekarczyk, M.S., Llanas, R.A., Thomson, J.A. (2006). Derivation of human embryonic stem cells in defined conditions. *Nat Biotechnol* **24**, 185–187.

13. Reubinoff, B.E., Pera, M.F., Fong, C., Trounson, A., Bongso, A. (2000). Embryonic stem cell lines from human blastocysts: somatic differentiation in vitro. *Nat Biotechnol* **18**, 399–404.

14. Amit, M., Itskovitz-Eldor, J. (2002). Derivation and spontaneous differentiation of human embryonic stem cells. *J Anat* **200**, 225–232.

15. Cowan, C.A., Klimanskaya, I., McMahon, J., Atienza, J., Witmyer, J., Zucker, J.P., Wang, S., Morton, C.C., McMahon, A.P., Powers, D., Melton, D.A. (2004). Derivation of embryonic stem-cell lines from human blastocysts. *N Engl J Med* **350**, 1353–1356.

16. Verlinsky, Y., Strelchenko, N., Kukharenko, V., Rechitsky, S., Verlinsky, O., Galat, V., Kuliev, A. (2005). Human embryonic stem cell lines with genetic disorders. *Reprod Biomed Online* **10**, 105–110.

17. Klimanskaya, I., Chung, Y., Meisner, L., Johnson, J., West, M.D., Lanza, R. (2005). Human embryonic stem cells derived without feeder cells. *Lancet* **365**, 1636–1641.

18. James, D., Levine, A.J., Besser, D., Hemmati-Brivanlou, A. (2005). TGFbeta/activin/nodal signaling is necessary for the maintenance of pluripotency in human embryonic stem cells. *Development* **132**, 1273–1282.

19. Valdimarsdottir, G., Mummery, C. (2005). Functions of the TGFbeta superfamily in human embryonic stem cells. *APMIS* **113**, 773–789.

20. Besser, D. (2004). Expression of nodal, lefty-a, and lefty-B in undifferentiated human embryonic stem cells requires activation of Smad2/3. *J Biol Chem* **279**, 45076–45084.

21. Beattie, G.M., Lopez, A.D., Bucay, N., Hinton, A., Firpo, M.T., King, C.C., Hayek, A. (2005). Activin A maintains pluripotency of human embryonic stem cells in the absence of feeder layers. *Stem Cells* **23**, 489–495.

22. Vallier, L., Alexander, M., Pedersen, R.A. (2005). Activin/nodal and FGF pathways cooperate to maintain pluripotency of human embryonic stem cells. *J Cell Sci* **118**, 4495–4509.

23. Sato, N., Meijer, L., Skaltsounis, L., Greengard, P., Brivanlou, A.H. (2004). Maintenance of pluripotency in human and mouse embryonic stem cells through activation of Wnt signaling by a pharmacological GSK-3-specific inhibitor. *Nat Med* **10**, 55–63.

24. Hynes, R.O. (2002). Integrins: bidirectional, allosteric signaling machines. *Cell* **110**, 673 687.

3

Digital Imaging of Stem Cells by Electron Microscopy

A. Henry Sathananthan and Stefania A. Nottola

Summary

This chapter deals with basic techniques of scanning and transmission electron microscopy applicable to stem cell imaging. It is sometimes desirable to characterize the fine structure of embryonic and adult stem cells to supplement the images obtained by phase-contrast and confocal immunofluorescent microscopy to compare with the microstructure of cells and tissues reported in the literature. This would help confirm their true identity whilst defining their surface and internal morphology. The intention is to put a face on stem cells during their differentiation.

Key Words: SEM; TEM; Stem cells; Embryonic; Adult, Microstructure; Imaging.

1. Introduction

The electron microscope (EM) has been an invaluable tool in the study of the fine structure of cells and tissues in the past 75 years of biomedical research. This includes both scanning electron microscopy (SEM) and transmission electron microscopy (TEM) in exploring the surface structure and the internal structure of cells and tissues, respectively, at high magnifications *(1–4)*, well beyond the resolution of the light microscope (LM) and fluorescent microscope (FM). It is well known that stem cell research is a continuously and rapidly expanding field of investigation that could gain considerable momentum from ultra structural studies. Most stem cell researchers publish their images using phase-contrast and confocal microscopy applying immunofluorescence at the LM level *(5)*. They use specific surface markers or stains to image cells by FM. Others use routine histological techniques, as well, to demonstrate the

From: *Methods in Molecular Biology, vol. 407: Stem Cell Assays*
Edited by: M. C. Vemuri © Humana Press, Totowa, NJ

microstructure of stem cells, both embryonic and adult. It is desirable, however, to examine the fine structure of these cells to compare with cells and tissues widely documented by SEM and TEM in the literature. This can be done using simple, basic techniques, if there are established facilities in the institution. Although time-consuming, these techniques can be used to document the fine structure of stem cells to confirm the results obtained by LM and FM *(6–9)*. Since its pioneering application to the study of biological material, SEM has been regarded as an useful tool in both basic and clinical research, in that it reveals the microstructure of whole cells. In fact, due to a combination of the three-dimensional (3D) perspective and surface morphology, magnification, and depth of field, SEM can provide vivid, attractive, and readily interpretable images *(2,10)*. 3D images, combined with LM and TEM images, and possible visualization of biomolecules, allow a better insight into important micro-topographical features of tissues and cells, even in minute cell microdomains. In fact, the combined use of high-resolution SEM and specific techniques in sample preparation are capable of revealing biological surfaces, including the intracellular organization of a cell or tissue *(11)*.

2. Materials

1. Dulbecco's phosphate-buffered saline (PBS).
2. Sodium cacodylate buffer (pH 7.2–7.4) (*see* **Note 1**).
3. Glutaraldehyde: Stock solution (EM grade) in buffer solution (v/v) = Fixative solution (*see* **Note 2**).
4. Primary fixative for SEM: 1.5–2.5% glutaraldehyde in 0.1 M PBS.
5. Primary fixative for TEM (Fixative A): 3% glutaraldehyde in 0.1 M cacodylate buffer (pH 7.3). Stock solutions: glutaraldehyde 25% solution EM grade (store at 4 °C in dark); 0.2 M cacodylate buffer.

   (a) Prepare fixative A: 110 ml 0.2 M cacodylate buffer, 110 ml distilled water (DW), 30 ml of 25% glutaraldehyde (Keeps for 3–6 months at 4 °C).

6. Post-fixative for SEM and TEM (Fixative B): Osmium tetroxide. Crystals are contained in glass ampoules. It is usually used in aqueous solution (*see* **Note 3**).

   (a) Prepare 1% Osmium tetroxide in DW: 1 g Osmium tetroxide crystals sealed in a glass vial, 95 ml DW.

7. Tannic acid: Usually in the form of a yellowish-white or pale brown powder, soluble in water (*see* **Note 4**).
8. Ethyl alcohol or ethanol: anhydrous, purity > 99.9%.
9. Acetone: Colourless liquid used as a solvent (*see* **Note 5**).

3. Methods

3.1. Preparation of Stem Cells for SEM

Cells have to be prepared carefully for EM studies. Hence, the principles of the routine preparation of isolated cells, cell aggregates and observation of specimens, and different approaches in the preparation of cultured stem cells are reviewed and summarized.

3.1.1. Principles and Routine Preparation of Specimens for SEM

Conventional SEM utilizes a focused beam of high-energy electrons that systematically scans across the sample surface, influenced and converged by electromagnetic lenses. The interaction of the beam with the sample produces a large number of signals at or near the surface of the sample. These signals include secondary electrons, which become concentrated and are drawn to a positively biased detector system. The electron signal is converted to an electronic signal that is portrayed on a cathode ray tube *(12)*.

Biological samples have to be adequately prepared for SEM. The following steps require care and special attention: specimen selection, surface preparation, fixation, post-fixation, conductive staining, dehydration, mounting, coating, and, finally, observation.

1. Specimen selection: A small specimen is generally desirable, although specimen size mostly depends upon the size of the specimen holder and the size of the specimen chamber of the microscope. Of course, this problem is less critical when preparing isolated cells or monolayered cultured cells.
2. Surface preparation: Loosely adhering natural materials (such as mucoid substances or serum) or contaminants from the surrounding environment (including salts and other chemicals from fixatives and solutions) may obscure surface details. Thus, gentle and careful rinsing eventually followed by enzymatic or mechanical cleaning techniques (performed without injuring the specimen surface itself) are needed in order to expose a clean specimen surface to the electron beam. A routine buffer solution or DW can be used.
3. Fixation: The aim of fixation is to preserve the cellular integrity during the subsequent treatments. Fixation can be mechanical (by rapid freezing) or chemical (usually with glutaraldehyde). Chemical fixation can be performed by vapour, immersion, and in vivo, by the perfusion method.
4. Post-fixation: We usually use osmium tetroxide, which enhances specimen conductivity. For high-resolution SEM observations, however, a conductive staining method should be applied, especially when observing the samples with a reduced working distance or "in lens."
5. Conductive staining: There are several methods available. For isolated cells and cultured cell monolayers, the method of choice is the so-called tannin-osmium

method *(13)*. It involves the sequential immersion of the samples in osmium tetroxide, then in tannic acid, and then in osmium again, in order to obtain the best specimen conductivity. If this method is applied properly and low accelerating voltage SEM is used during observation (for example when using field emission SEM), the metal coating may be very thin (around 4 nm with platinum) or in some instances even avoided, when immuno-SEM is applied.

6. Dehydration: Water dispersion from wet specimens in the evacuated column of the microscope can result in both contamination of the column and dramatic alteration of the specimen surface; thus, it is necessary to remove water from biological samples during preparation. Soft specimens cannot withstand air drying and need to be subjected to freeze drying or critical point drying. By freeze drying, specimens are quickly frozen and then subjected to ice sublimation at low temperature and high vacuum. Chemical substitution of water with increasing concentrations of organic dehydrating (intermediate) solvents, such as acetone and ethanol, is commonly used as a prerequisite to critical point drying. The critical point drying method is based upon the property of liquids to change from a liquid phase to a gaseous phase without a latent heat of vaporization or density change. This occurs at a given temperature and pressure for each liquid ("critical point"). Applying this method, the specimen becomes dry without being exposed to surface tension forces, which may damage the tissue. The final (transitional) solvent commonly used for this procedure is the carbon dioxide (CO_2), chosen for its acceptable critical point. Parameters (temperature and pressure not harmful for the specimen) *(12)*.

7. Mounting: After dehydration, specimens are mounted on stubs made of conductive materials (aluminium for secondary electron imaging, carbon when using x-ray microanalysis or backscattered electron probes). The sample is glued to the stub using an adhesive (possibly electrically conductive) material, such as silver paint, double stick tape, or carbon-based adhesives.

8. Coating: The primary aim of the coating procedure is to conduct electrical charge and heat away from the specimen to the stub. Coating also leads to an improvement in the strength of secondary electron signal from the specimen surface. Specimens are generally coated with a heavy metal such as gold, platinum, or gold–palladium or with carbon when using x-ray microanalysis or backscattered electron probes. The conductive film must be continuous, uniform, and stable, and it should be thin enough to avoid obscuring the surface details.

9. Observation: Evaluation of SEM micrographs is of course related to the power of the instruments. Today, a huge variety of microscopes are available, and the selection is mostly depending upon the financial availability. High resolution is needed for studying cell surfaces. However, it is also important to visualize the 3D micro-topographic relationships among cells or even cell-surface nanostructures. Therefore, a compromise between best resolution and greatest depth of field should be obtained (*see* **Note 6**) (for further information *see* refs *(12,14)*).

3.1.2. Preparation of Isolated Cells and Cell Aggregates for SEM

1. Fix cells in 1.5% gultraldehyde in 0.1M PBS for at least 48 h.
2. Rinse in 0.1 M PBS.
3. Post-fix in osmium tetroxide (1%) in 0.1 M PBS for 20 min.
4. Rinse in 0.1 M PBS.
5. Treat with tannic acid (1%) in DW for 30 min.
6. Rinse in 0.1 M PBS.
7. Treat again with osmium tetroxide (1%) in 0.1 M PBS for 20 min.
8. Rinse in 0.1 M PBS.
9. Place in polyethylene microporous specimen capsules (meshes = 30 μm).
10. Dehydrate in ascending series of graded ethyl alcohols.
11. Critical point dry in CO_2 atmosphere.
12. Mount on aluminum stubs and coat with platinum (4 nm in thickness) in a sputter coater.

SEM observations were performed in a field emission scanning electron microscope operating at low accelerating voltage (5–10 kV).

The spongy structure of the zona pellucida (ZP) that surrounds oocytes and early cleaving embryos as well as the general surface organization of the cells attached to the outer ZP (shape, dimensions, presence of microvilli and blebs, localization of intercellular contacts, and relationship with the ZP) were demonstrated by applying this technique (*see* **Figs 1** and **2**) (*see* **Notes 7** and **8**) (for further information *see* refs *15–18*).

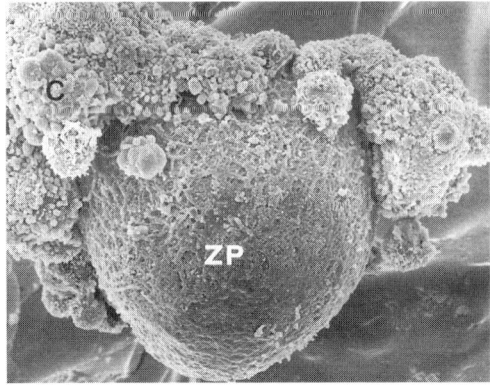

Fig. 1. Human mature oocyte. Round/oval cumulus-corona cells (C) provided with numerous blebs and microvilli are seen in close contact with the zona pellucida (ZP). SEM × 1750 (Reproduced from **ref. 17**).

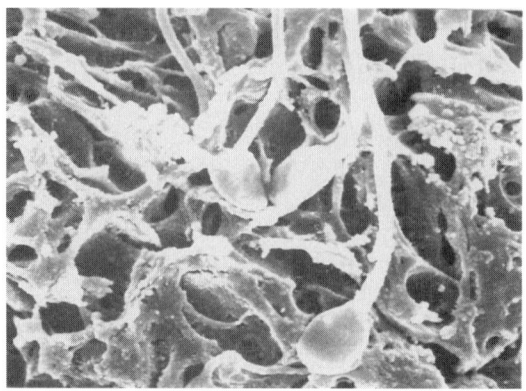

Fig. 2. Human mature oocyte, inseminated *in vitro* with spermatozoa. Sperm heads are seen attached to the spongy surface of the zona pellucida. SEM × 5000 (Reproduced from **ref.** *15*).

3.1.3. Preparation of Cultured Cells for SEM

According to our recent experience (unpublished data), the following technique for SEM could be applied to various types of cultured cells, including stem cells.

1. Transfer the cells with a sterile pipette from the stock bottle to a plastic Petri dish (60 mm in diameter) and allow a period of adjustment of cells (it depends on the cell type) until confluence.
2. Gently remove the medium.
3. Rinse in 0.1 M PBS.
4. Fix with 2.5% glutaraldehyde in 0.1 M PBS for 1–4 h at 0–4 °C.
5. Rinse in 0.1 M PBS.
6. Post-fix in osmium tetroxide (1%) in 0.1 M PBS for 1 h at 0–4 °C.
7. Rinse in 0.1 M PBS.
8. Dehydrate in ascending series of graded ethyl alcohols.
9. Cut the Plastic Petri dish into pieces small enough to be successively mounted on the stubs.
10. Critical point dry.
11. Mount on stud.
12. Coat by SEM.
13. Observe wat with metal as detailed above.

When observed by SEM, cultured cell colonies generally appear as a monolayer of cells often showing different shapes and exhibiting microvilli, ruffles, and blebs on their surface, as well as long prolongations sometimes occurring between cells. Dividing elements can be also found in the colonies.

Peculiarities in behavior and related surface appearance of cells are actually dependent upon the cell type and the culture technique *(19)* (*see* **Notes 9–13**).

3.2. Preparation of Stem Cells for TEM

We present a rapid and proven method of preparing stem cells for TEM, which was used extensively to image gametes and embryos in assisted reproductive technology *(20–22)*.

This can be used for human embryonic stem cells (ESCs) growing in colonies or embryoid bodies (EBs) or neurospheres (NSs) and also human adult stem cells cultured in the laboratory. Combined with advanced digital microscopy, images showing fine details of microstructure can easily be documented *(6,7,9)*.

If a TEM is unavailable, a great deal of information could still be obtained by examining epoxy-resin sections (1 μm thick) by advanced digital LM. They show more structural details than routine paraffin or frozen sections (*see* **Figs 3–6**). These sections could be cut with glass knives or histological diamond knives, if there is an ultra-microtome. Microtomes are now available for cutting resin sections with disposable blades, which might be an useful alternative.

3.2.1. Principles and Methods of Specimen Preparation for TEM

The reader is referred to textbooks in electron microscopy for principles and details of procedure *(14,23)*.

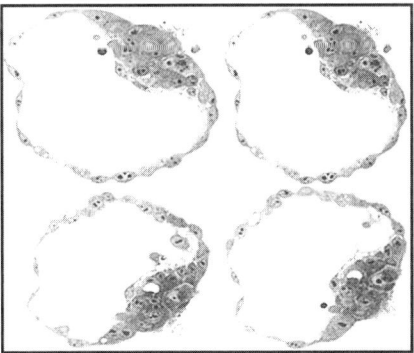

Fig. 3. Hatched human blastocyst (serial sections). Note outer trophoblast and inner cell mass at one pole. LM × 400 (Reproduced from **ref**. *7*).

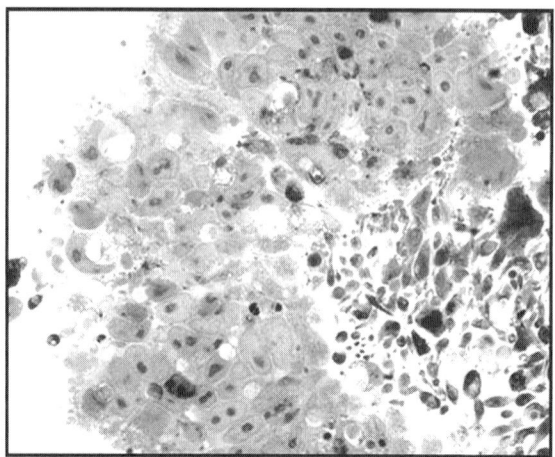

Fig. 4. Human ES cell colony in culture after 35 passages. The cells on the right are differentiating. LM × 1000 (Reproduced from **ref. *9*, *40***).

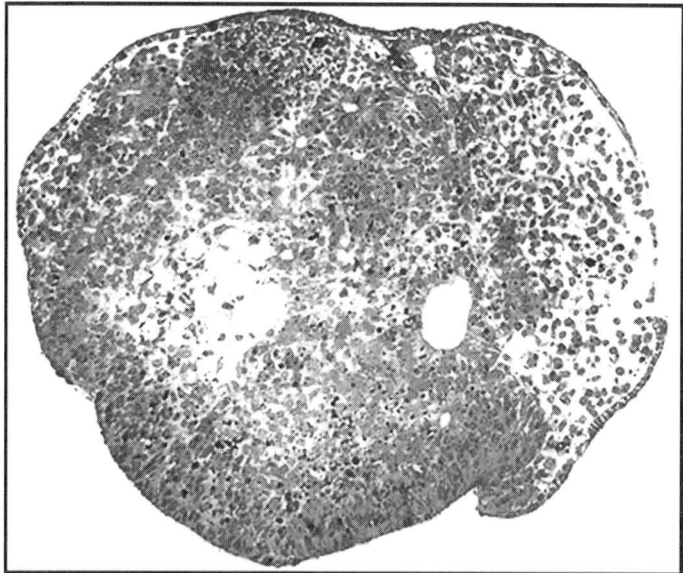

Fig. 5. Human EB cultured in vitro. Note surface epithelium and stem cells within. Some cells have differentiated into mesenchyme (right), neural-like tube (center), and lipid cells (left), Note neural rosette (top). LM × 100 (Reproduced from **ref. *40***).

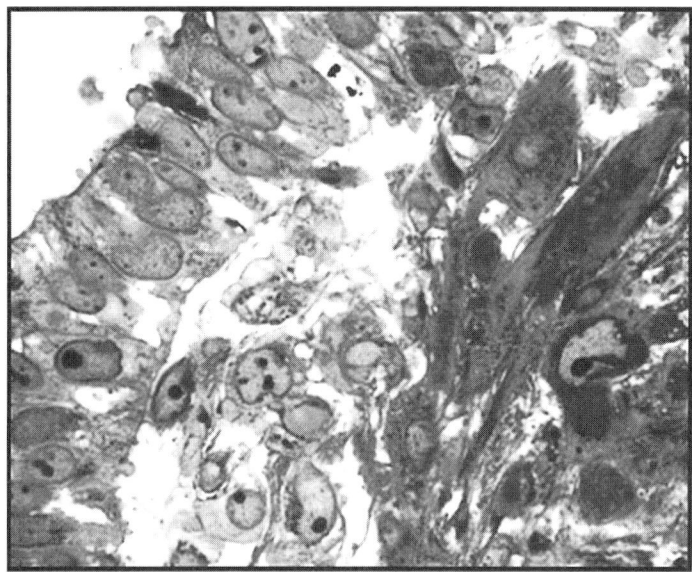

Fig. 6. Human ES cell colony showing differentiated cells–epithelium (left), fibroblasts (center), and cardiac muscle (right). LM × 1000 (Reproduced from ref. *40*).

The steps involved in specimen preparation, examination and imaging are the following:

1. Fixation
2. Processing and embedding
3. Sectioning
4. Staining
5. Microscopy

1. Fixation: The main objective of fixation is to preserve the structure of cells in a more or less life-like state. Simple chemical fixation in buffered glutaraldehyde and osmium tetroxide is usually used. The cells must be alive and fresh and in the medium they were cultured. The ESC, EB, or NS can be fixed whole at the end of the desired culture sequence or passage, whereas cells growing in monolayers can be disloged, centrifuged at 160 or 350 g, pelleted in chemical-proof Eppendorf tubes, and fixed, as we do for sperm samples *(20,24)*. If the pellet breaks it can be centrifuged again at 600 g after glutaraldehyde fixation. This applies to most stem cells growing in flat Falcon tubes or plastic dishes. In all cases, the specimen must be fixed with minimal medium but never dried. Fixation is the most critical step in specimen preparation that will ensure preservation of the microstructure of

the cells. No water should be used for washing before fixation, but a physiological saline may be used for washing, if necessary.

2. Processing and embedding: The specimen has to be dehydrated and embedded in a supporting medium such as a hard epoxy resin. Araldite or Epon resins give good results. The resin imparts a hard texture to the soft specimen required for thin sectioning. Dehydration involves the step-wise, progressive removal of water from the specimen to make it penetrable with the resin. Ethyl alcohol and acetone are used in our laboratory, and the specimen is infiltrated with the resin, embedded in resin, and polymerized to produce a hard block for sectioning.

3. Sectioning: The block is mounted on a ultramicrotome holder, trimmed under a binocular with a sharp razor blade, and sectioned with an ultramicrotome in a histology or TEM laboratory. Glass knives and diamond knives are used for thick sectioning and thin sectioning, respectively. Glass knives have to be made with a knife maker, and diamond knives are available commercially and are very expensive. Beginners should use glass knives for thin sectioning, as well. Thick, survey sections (1 µm) are examined by LM, whereas thin sections (~70 nm) are examined by TEM. Thick sections are very useful to identify specific cells or tissues for examination by TEM. One of the limitations of TEM is that only a small region of the specimen could be examined. Hence, we use serial sectioning, alternating a series of thick sections with thin sections, till we find the desired cell or tissue.

4. Staining: Thick sections are stained with Toluidine or Methylene blue on a hot plate. Sections are simply mounted on clean, glass slides, dried, and stained. Electron stains like uranyl acetate and lead citrate are used routinely for thin sections. These stains contain heavy metals that enhance the electron density of the cells and make them more visible under the EM. The image is formed by scattering of electrons, which produces translucent and dense areas. The cell membranes, granules, and inclusions scatter electrons and appear dense, whereas the ground cytoplasm appears translucent producing a black and white image. Unfortunately, color imaging is not possible with TEMs.

5. Microscopy: Thick sections are examined with an LM, whereas advanced imaging is done with a research microscope with a digital camera, hooked to a computer for image processing and editing. We use Leica QWin or Olympus digital microscopes. Alternatively the sections can be photographed on film and printed or mounted on 35-mm slides. The resolution of these images are superior to the digital images and could be scanned onto a computer for editing. Thin sections are examined by TEM. We have used Jeol or Philips microscopes but others like Zeiss and Hitachi are equally good. The TEM has to be maintained by an experienced technician. Lower magnifications are more useful to image whole cells, whereas high magnifications help identify small cell structures like ribosomes, centrosomes, filaments, and membranes in cells (*see* **Figs. 7–10**).   Higher contrast could be obtained with

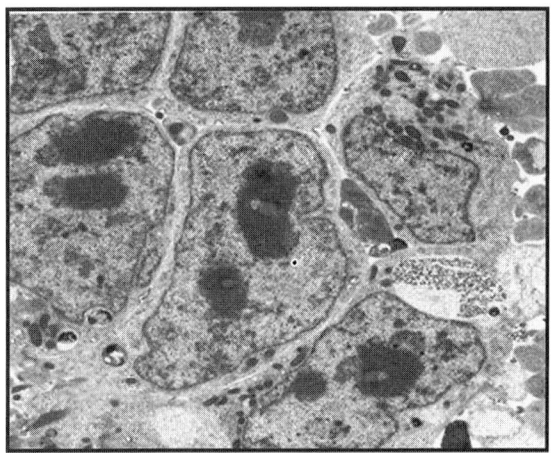

Fig. 7. Undifferentiated ES cells after 35 passages. Note large nuclei and scanty cytoplasm. TEM × 35, 000 (Reproduced from **ref. *40***).

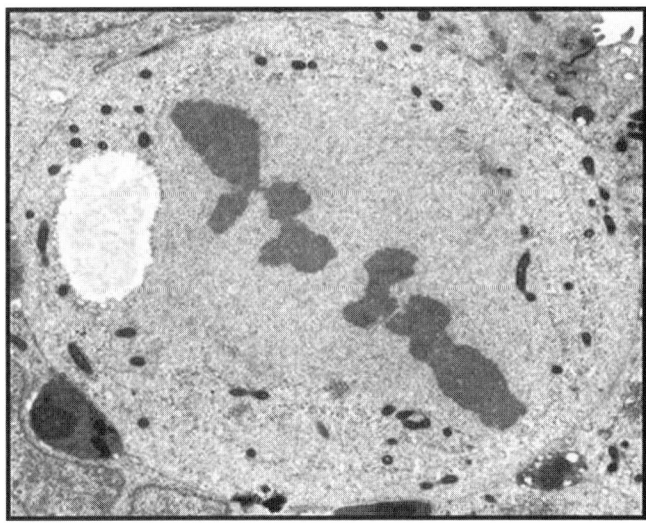

Fig. 8. Mitotic ES cell showing metaphase chromosomes and a minute centriole at right spindle pole. Note dense mitochondria and secretory vacuole outside spindle zone. TEM × 8750 (Reproduced from **ref. *6***).

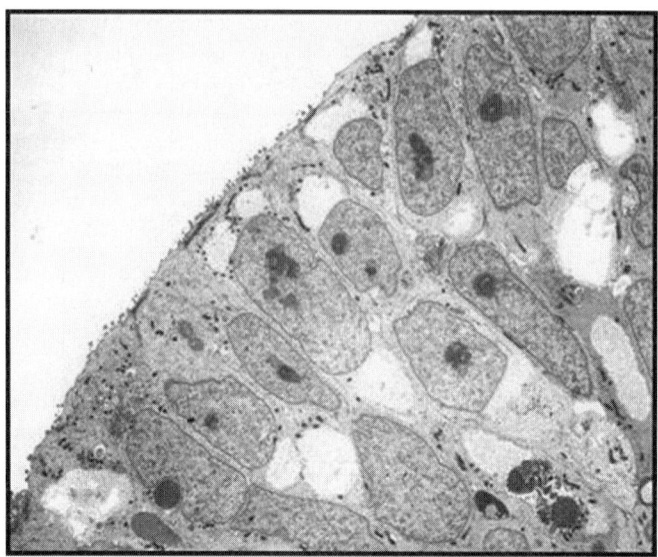

Fig. 9. Colony of differentiated ES cells showing goblet-like cells with secretory vacuoles, probably of endodermal origin. TEM × 3500 (Reproduced from **ref. 6**).

voltages of 60 or 80 kV and by using smaller apertures in the microscope. The electron beam needs to be always aligned before imaging. Do not focus on the same spot for long periods of time. Images are photographed on plate or on 35-mm film. Developing and printing is done in any dark room using chemicals and paper. Printing has been replaced by negative scanners that can produce digital images for editing. The images are then edited, cropped, or colored using the latest versions

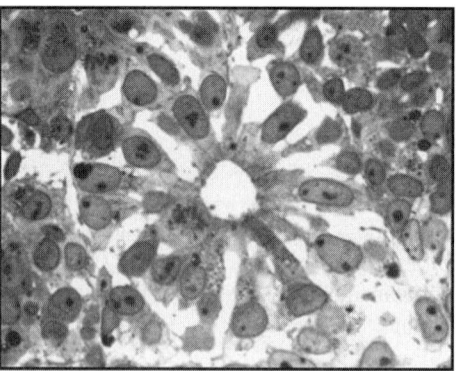

Fig. 10. Neurogenic rosette in an embryoid body with radiating neural stem cells. One cell is at metaphase near its lumen. LM × 1000 (Reproduced from **ref. 9, 40**).

of Adobe Photoshop or Paint Shop Pro and presented on Microsoft Power Point. The latter is very useful for labeling and annotating for presentations in class or conferences. Most of our images are saved in Tiff format and converted to JPEG or GIF for publications online, transmission by email or the web.

3.2.2. Specimen Preparation for TEM (see **Notes 14–37**)

3.2.2.1. FIXATION (*see* **Notes 14–22**)

1. Detach ESCs, EBs, or NS from Falcon dish with a fine needle or forceps.
2. Pick up with forceps or wide-bore Pasteur pipette with minimal medium.
3. Drop into glass vial with 5 ml Fixative A.
4. Fix for 1 h and store at 4 °C (keeps several weeks to months).
5. Rinse briefly in DW (5–10 min).
6. Post-fix in 1 ml of Fixative B for 1 h in dark (process immediately and rapidly).
7. Remove Fixative B with a glass pipette and add 5 ml of 70% ethyl alcohol.

Stem cell pellets (2–5mm), EBs, and NS can be fixed and processed in Eppendorf tubes, chemical-proof plastic tubes, or glass vials, as well. For the composition of Fixatives A and B *see* **Subheading 2**.

3.2.2.2. DEHYDRATION (*see* **Note 23**)

Use ethyl alcohol and dry acetone (AR grade).

1. 70% alcohol (10 min).
2. 90% alcohol (10 min).
3. Absolute alcohol—2 changes 15 min each (dehydrant).
4. Acetone—2 changes 15 min each (dehydrant and resin solvent).
5. Acetone/Araldite mixture (1:1), 30 min.

3.2.2.3. EMBEDDING (*see* **Note 24**)

1. Araldite (Durcupan) Fluka from Sigma, Switzerland. Preparation of Araldite mixture (Epoxy resin): Use a 10-ml plastic measuring cylinder to measure 5 ml of Durcupan ACM, 5 ml of Durcupan B, 0.3 ml of Durcupan C, and 0.3 ml of Durcupan D. Mix in a plastic vial by vigorous agitation (10–20 min). Use immediately or store in a freezer (hardens on keeping).
2. Epon resin or Epon/Araldite may be used instead of Araldite for embedding.
3. Gently transfer colonies into solid watch glasses (embryological dishes).
4. Process under binocular microscope, if specimen is small.
5. Remove Acetone/Araldite mixture with a pipette.
6. Add Araldite mixture (1–2 ml).
7. Infiltrate 3–6 h, preferably overnight (use shaker or rotator if available).
8. Embed in fresh Araldite in rubber or plastic moulds with wells (use stainless steel forceps or fine mounted needles).

9. Orientate each colony at the tip of each well with a fine needle.
10. Polymerize in 60 °C oven (48–72 h).

Glass vials may be used for processing if the specimen is large and visible. Plastic vial caps can be used for flat embedding. Beam capsules may be used but orientation is difficult. Monolayers of cells can be grown on coverslips, scaffolds, or collagen membranes, fixed, and processed in situ *(23)*.

3.2.2.4. SECTIONING—THICK AND THIN SECTIONS (*see* **Notes 25–32**)

THICK SECTIONING

1. Remove hardened block from the mould.
2. Trim the cone on all sides with a sharp razor blade under binocular.
3. Shape a rectangle or trapezium with parallel upper and lower edges.
4. Fix the block onto microtome holder (we use Reichert or LKB ultramicrotomes).
5. Align and orientate block with upper and lower edges parallel to knife edge.
6. Make fresh glass knives with a knife maker and attach a boat or use a histological diamond knife (we used Diatome diamond knives).
7. Cut sections 1 μm thick (manual setting).
8. Mount 5–10 sections onto clean glass slides with a drop of water.
9. Place slide on a hot plate at 60 °C.
10. Dry to attach sections onto slide (1–2 min).
11. Add 2 or 3 drops of 1%Toluidine blue stain in 1% Borax in DW.
12. Leave on hot plate 1–2 min till stain steams (do not dry out).
13. Cool and wash by squirting DW from a wash bottle in a sink.
14. Wipe excess water and allow to dry in air.
15. Examine sections with an LM (we do not mount).
16. When the desired cells are obtained cut thin sections.

THIN SECTIONING

The following procedure has to be done by an experienced technician.

1. Replace glass knife with a diamond knife.
2. Fill boat with DW and focus knife-edge.
3. Re-align block and avoid cutting thick sections.
4. Use automatic cutting mode and cut silver–gold sections (∼ 70 nm thick).
5. Cut a ribbon of 10–15 sections.
6. Rinse copper grids (200 hexagonal mesh) in 1% HCl acid to etch.
7. With a fine forceps place grid under sections in boat and lift the ribbon upwards. If ribbon breaks group sections with a hair mounted on a match.
8. Blot grids of excess water and place on a filter paper placed in a Petri dish.
9. Dry in 60 °C oven for immediate staining (15 min) or leave to dry in air.
10. Wash diamond knife immediately with a jet of DW.

3.2.2.5. STAINING (*see* **Notes 33–35**)

1. Alcoholic uranyl acetate (saturated solution): Make the solution fresh just before staining. Dissolve approximately 600 mg of stain in 5 cc of 70% ethyl alcohol, shake vigorously in a vial for about 20 min, and store in the dark (Stain A).
2. Reynold's lead citrate: 1.33 g of lead nitrate, 1.76 g of Sodium citrate, and 30 ml of CO_2-free DW (Stain B)–follow steps 3–8 below.
3. Boil 100 ml DW for 30 mins to expel CO_2 and air and cool.
4. Shake lead salts vigorously in 30 ml of DW in a volumetric flask (30 min).
5. Add 8 ml of 1 N NaOH to milky suspension while shaking—clears up.
6. Dilute to 50 ml with DW.
7. Store in the dark at 4 °C, wrapped in foil (keeps up to 3 months).
8. Fill 10-ml syringe with stain and filter before use. Use a 0.2-µm Arcodisc filter from Gelman Sciences Michigan, USA or alternative.
   A: Alcoholic uranyl acetate
9. Using a syringe, filter 2 ml of uranyl stain into a plastic cap. Use a 0.2-µm Arcodisc filter.
10. Immerse grids in stain using a fine forceps, sections facing downwards.
11. Stain for 10–15 min in the dark.
12. Rinse grids in 50% alcohol.
13. Rinse in DW.
14. Dry grids on filter paper, sections on top.
    B: Reynold's lead citrate
15. Filter 2 ml of lead stain into a plastic cap.
16. Place cap in a Petri dish with LiOH or few pellets of NaOH (removes CO_2).
17. Immerse grids, sections facing downwards.
18. Cover dish and stain for 10–15 min. Do not breathe onto dish.
19. Rinse (2 changes of DW).
20. Dry grids on filter paper, sections on top.

Grids may be stained on stain droplets but uranyl acetate must be made in DW. Store grids in grid boxes for permanent storage and examination by TEM.

3.2.2.6. TEM EXAMINATION

Refer details in TEM manual. The following procedure has to be done by an experienced operator.

1. Check vacuum and switch on TEM at 60 or 80 kV.
2. Align electron beam each time before use.
3. Insert grid into the column and pump down to restore vacuum.
4. Switch the filament on when high voltage comes on.
5. Remove apertures and examine sections at low magnifications.
6. Select areas of sections to be examined (*see* Step 3 in **Subheading 3.2.1.** sectioning).

7. Insert medium or small apertures for better contrast.
8. Switch to higher magnifications and photograph desired cells.
9. Use magnifications of ×2000 to ×10, 000 for routine work.
10. Remove grid, pump down, off filament and high voltage and leave microscope on or shut down.

4. Notes

1. Cacodylate buffer may be fatal if swallowed. Known carcinogen in humans. Harmful if inhaled, may be harmful by skin contact. Long-term exposure may lead to kidney and liver damage. Eye and skin irritant.
2. Glutaraldehyde fixative solution can be prepared just before using it but it is possible to store the solution in the refrigerator (4 °C, in firmly tight bottles) up to several months. Please avoid the contact with skin and mucous membranes. The solution should be transparent; if, after a long storage, it looks yellow, we suggest you prepare a fresh solution.
3. Osmium telroxide: in order to prepare the aqueous solution, clean the ampoule and remove the label, file the neck with a diamond pen but not completely, place in a suitable flask (amber-colored, thick-sided glass reagent bottle) and break the ampoule in the flask. Add DW (sometimes PBS is used) and seal the lid of the bottle with parafilm. Dissolve for 1–2 days by occasional swirling; final solution will be pale yellow; store at 4 °C in the dark. If, after a long storage, the solution becomes gray or black, do not use it but prepare a fresh one. Osmium is extremely expensive and poisonous. Avoid any contact, the exposure to osmium tetroxide is very dangerous and known to produce ocular effects and respiratory irritation. Also avoid metal contact and light. All chemicals must be handled in a fume cupboard, preferably with gloves.
4. Tannic acid: in order to obtain the aqueous solution, weigh the needed amount of powder, mix it in water (appropriate dilution) and stir until the solution becomes clear, pale yellow in color.
5. If acetone is used for dehydration during TEM procedures, propylene oxide is not needed as transition solvent.
6. In our experience a good compromise at a reasonable cost may be obtained using a field emission SEM, with a short working distance and a low accelerating voltage ($< 7–8$ kV). In fact, the field emission tip allows the preservation of a good depth of field even at a short working distance; in addition, it allows to reach very high resolution (around 4–5 nm) even at a low accelerating voltage, if a short working distance is associated to a very good specimen conductive staining. Finally, the low accelerating voltage permits to scan the most superficial electrons avoiding the loss of details caused when applying high accelerating voltage (> 10kV).
7. The use of the osmium-dimethylsulphoxide-osmium (ODO) method *(25)*, causing the extraction of the soluble cytoplasmic matrices from the freeze-cracked surface of the cells, allowed the visualization of the intracellular organization of both

germ and somatic cells. Applying the ODO method, the 3D micro-topography of organelles and cytoskeletal elements could be traced *(4,26,27)*.

8. Furthermore, the combined use of detergents, osmium-thyocarbohydrazide-osmium (OTO) treatment, and ruthenium red staining made it possible to reveal the finest 3D structure of the extracellular matrices, including the peculiar micro-filamentous texture of the ZP *(28)*.

9. In another study specifically addressed to define the ultrastructural morphology of mouse ESCs by TEM and SEM, these cells were cultured on mouse embryonic fibroblasts, cultivated on glass coverslips in 48-well dishes, and processed for SEM observations after 60 h of culture. In detail, these cells were subjected to primary fixation in 2% glutaraldehyde in 0.1 M sodium cacodylate buffer (pH 7.4), then washed with the same buffer, post-fixed in 1% osmium tetroxide in the same buffer, washed again, and dehydrated in ethanol series. Afterwards, the coverslips were critical point dried, mounted on aluminium stubs, and finally coated with gold using standard techniques *(8)*.

10. According to other protocols, after fixation, mouse mast cells plated onto 12-mm round coverslips coated with Cell Tak have been subjected to OTO treatment before undergoing dehydration *(29)*. For immuno-SEM, properly immuno-stained mouse ESCs were fixed with 2.5% glutaraldehyde/50 mM cacodylate-HCl (pH 7.2), washed with the same buffer, post-fixed in 1% osmium tetroxide, dehydrated and immersed in isoamyl acetate, CO_2 critical point dried, mounted and coated to 3 nm thickness with an osmium plasma coater, and observed with a SEM equipped with a backscatter electron detector. Immuno-SEM has been used to identify cell adhesion-related molecules on the stem cell surface *(30)*.

11. Even embryonic body development and morphology can be studied by SEM *(31)*.

12. SEM has been also employed in investigating stem cell interactions with other cell types in specific culture systems *(32)*. Furthermore, the ability of different supports to sustain stem cell growth and proliferation in culture has been evaluated using conventional SEM *(33–36)*, environmental SEM *(36,37)*, and cryo-SEM *(38)*.

13. SEM has been also proven to be a useful tool in investigating the performance of 3D self-assembling peptide scaffolds that have been suggested to improve culture conditions *(39)*.

14. Keep the specimen as small as possible (2–5 mm)

15. Do not damage the specimen, particularly ESC, EB, or NS.

16. Never dry the specimen at any stage of processing.

17. Do not wash in water before fixation.

18. Fix at room temperature to preserve microtubules.

19. Never freeze specimen in storage. Store at 4 °C.

20. Process in a fume cupboard and wear gloves.

21. Most chemicals are hazardous, some carcinogenic.

22. Avoid breathing chemicals and vapors.

23. Swirl often to promote penetration of chemicals at every stage.
24. Mix resin mixtures thoroughly. Store in freezer.
25. Trim specimen block as small as possible, remove excess resin.
26. Make sure the upper and lower edges are parallel to knife edge.
27. Tighten all screws properly on specimen holder and microtome.
28. Glass knives should be clean and dust free, keep closed.
29. Do not cut thick sections with the diamond knife.
30. Use fine stainless steel needles and forceps for handling specimens.
31. Forceps need to be sharpened with emery paper under a binocular.
32. Hold grids by the edge and bend gently with a forceps to collect thin sections.
33. Hold grids vertically when immersing in electron stains or rinses.
34. Rinse forceps with DW between stains and dry with tissue.
35. Do not overstain in electron stains.
36. All glassware and vessels used must be spotlessly clean.
37. Collect waste chemicals in a bottle for proper disposal.

References

1. Fawcett, D.W. (ed.) (1981) *The Cell*. W.B. Saunders, Philadelphia.
2. Motta, P.M. (ed.) (1989) *Cells and Tissues. A Three-dimensional Approach by Modern Techniques in Microscopy*. Prog. Clin. Biol. Res., Vol. 295. Alan R. Liss, Inc., New York.
3. Sathananthan, A.H. (ed.) (1996) *Atlas of Human Cell Ultrastructure*. CSIRO, Australia.
4. Makabe, S., Van Blerkom, J., Nottola, S.A., and Naguro, T. (eds.) (2006) *Atlas of Human Female Reproductive Function. Ovarian Development to Early Embryogenesis after In Vitro Fertilization*. Taylor & Francis, London.
5. Bongso, A. and Lee, E.H. (eds.) (2005) *Stem Cells From Bench to Bedside*. World Scientific, Singapore.
6. Sathananthan, A.H., Pera, M., and Trounson, A.O. (2002) The fine structure of human embryonic stem cells. *Reprod. BioMed. Online* **4**, 56–61.
7. Sathananthan, A.H., Gunasheela, S., and Menezes, J. (2003) Critical evaluation of human blastocysts for assisted reproduction and embryo stem cell biotechnology. *Reprod. Biomed. Online* **7**, 219–227.
8. Baharvand, H. and Matthaei, K.I. (2003) The ultrastructure of mouse embryonic stem cells. *Reprod. Biomed. Online* **7**, 330–335.
9. Sathananthan, A.H. and Trounson, A. (2005) Human embryonic stem cells and their spontaneous differentiation. Ital. J. Anat. Embryol. **110 (2 Suppl. 1)**, 151–158.
10. Motta, P.M., Andrews, P.M., and Porter, K.R. (eds.) (1977) *Microanatomy of Cell and Tissue Surfaces. An Atlas of Scanning Electron Microscopy*. Lea & Febiger, Philadelphia.

11. Motta, P.M., Nottola, S.A., Familiari, G., Makabe, S., Stallone, T., and Macchiarelli, G. (2003) Morphodynamics of the follicular-luteal complex during early ovarian development and reproductive life. *Int. Rev. Cytol.* **223**, 177–288.

12. Postek, M.T., Howard, K.S., Johnson, A.H., and McMichael, K.L. (eds.) (1980) *Scanning Electron Microscopy. A Student's Handbook*. Ladd Research Industries, Burlington.

13. Murakami, T. (1974) A revised tannin-osmium method for non-coated scanning electron microscope specimens. *Arch. Histol. Jpn.* **36**, 189–193.

14. Hayat, M.A. (ed.) (1989) *Principles and Techniques of Electron Microscopy*. 3rd Edition. CRC Press Inc., Boca Raton, Florida.

15. Familiari, G., Nottola, S.A., Micara, G., Aragona, C., and Motta, P.M. (1988) Is sperm-binding capability of the zona pellucida linked to its surface structure? A scanning electron microscopic study of human in vitro fertilization. *J. In Vitro Fert. Embryo Transf.* **5**, 134–143.

16. Motta, P.M., Nottola, S.A., Micara, G., and Familiari, G. (1988) Ultrastructure of human unfertilized oocytes and polyspermic embryos in an in vitro fertilization program. *Ann. N.Y. Acad. Sci.* **541**, 367–383.

17. Nottola, S.A., Familiari, G., Micara, G., Aragona, C., and Motta, P.M. (1991) The ultrastructure of human cumulus-corona cells at the time of fertilization and early embryogenesis. A scanning and transmission electron microscopic study in an *in vitro* fertilization program. *Arch. Histol. Cytol.* **54**, 145–161.

18. Nottola, S.A., Makabe, S., Stallone, T., Familiari, G., Correr, S., and Macchiarelli, G. (2005). Surface morphology of the zona pellucida surrounding human blastocysts obtained after *in vitro* fertilization. *Arch. Histol. Cytol.* **68**, 133–141.

19. Wheeler, M.B. (1994) Development and validation of swine embryonic stem cells: a review. *Reprod. Fertil. Dev.* **6**, 563–568.

20. Sathananthan, A.H. (ed.) (1996) *Microscopic Atlas of Sperm Function for ART*. National University Hospital and Serono, Singapore.

21. Sathananthan, A.H., Ng, S.C., Bongso, A., Trounson, A., and Ratnam, S.S. (1993) *Visual Atlas of Early Human Development for Assisted Reproductive Technology*. National University Hospital and Serono, Singapore.

22. Website of AHS (2002–2007) "Human Embryo genesis" http://www.sathembryoart. com

23. Glauert, A.M. (ed.) (1974) *Practical Methods in Electron Microscopy*, Vol. 3. North-Holland, Amsterdam.

24. Sathananthan, A.H. (2000) Ultrastructure of human gametes, fertlization, and embryo development, in *Handbook of In Vitro Fertilization* (Trounson, A.O. and Gardner, D.K., eds.), 2nd Edition. CRC, Boca Raton, London, pp. 431–464.

25. Tanaka, K. and Naguro, T. (1981) High resolution scanning electron microscopy of cell organelles by a new specimen preparation method. *Biomed. Res.* **2**, 63–70.

26. Makabe, S., Naguro, T., and Motta, P.M. (1992) A new approach to the study of ovarian follicles by scanning electron microscopy and ODO maceration. *Arch. Histol. Cytol.* **55 (Suppl)**, 183–190.

27. Makabe, S., Naguro, T., Nottola, S.A., and Motta, P.M. (2001) Ultrastructural dynamic features of *in vitro* fertilization in humans. *Ital. J. Anat. Embryol.* **106 (2 Suppl. 2)**, 11–20.

28. Familiari, G., Nottola, S.A., Macchiarelli, G., Micara, G., Aragona, C., and Motta, P.M. (1992) Human zona pellucida during *in vitro* fertilization: an ultrastructural study using saponin, ruthenium red, and osmium-thiocarbohydrazide. *Mol. Reprod. Dev.* **32**, 51–61.

29. Vial, D., Oliver, C., Jamur, M.C., Davila Pastor, M.V., da Silva Trindade, E., Berenstein, E., Zhang, J., and Siraganian, R.P. (2003) Alterations in granule matrix and cell surface of focal adhesion kinase-deficient mast cells. *J. Immunol.* **171**, 6178–6186.

30. Cui, L., Johkura, K., Yue, F., Ogiwara, N., Okouchi, Y., Asanuma, K., and Sasaki, K. (2004) Spatial distribution and initial changes of SSEA-1 and other cell adhesion-related molecules on mouse embryonic stem cells before and during differentiation. *J. Histochem. Cytochem.* **52**, 1447–1457.

31. Weinhold, B., Schratt, G., Arsenian, S., Berger, J., Kamino, K., Schwarz, H., Ruther, U., and Nordheim, A. (2000) Srf -/- ES cells display non-cell-autonomous impairment in mesodermal differentiation. *EMBO J.* **19**, 5835–5844.

32. Akino, K., Mineda, T., and Akita, S. (2005) Early cellular changes of human mesenchymal stem cells and their interaction with other cells. *Wound Repair Regen.* **13**, 434–440.

33. Toquet, J., Rohanizadeh, R., Guicheux, J., Couillaud, S., Passuti, N., Draculsi, G., and Heymann, D. (1999) Osteogenic potential *in vitro* of human bone marrow cells cultured on macroporous biphasic calcium phosphate ceramic. *J. Biomed. Mater. Res.* **44**, 98–108.

34. Schratt, G., Weinhold, B., Lundberg, A.S., Schuck, S., Berger, J., Schwarz, H., Weinberg, R.A., Ruther, U., and Nordheim, A. (2001) Serum response factor is required for immediate-early gene activation yet is dispensable for proliferation of embryonic stem cells. *Mol. Cell. Biol.* **21**, 2933–2943.

35. Srouji, S. and Livne, E. (2005) Bone marrow stem cells and biological scaffold for bone repair in aging and disease. *Mech. Ageing Dev.* **126**, 281–287.

36. Stojkovic, P., Lako, M., Przyborski, S., Stewart, R., Armstrong, L., Evans, J., Zhang, X., and Stojkovic, M. (2005) Human-serum matrix supports undifferentiated growth of human embryonic stem cells. *Stem Cells* **23**, 895–902.

37. Horak, D., Kroupova, J., Slouf, M., and Dvorak, P. (2004) Poly(2-hydroxyethyl methacrylate)-based slabs as a mouse embryonic stem cell support. *Biomaterials* **25**, 5249–5260.

38. Modin, C., Stranne, A.L., Foss, M., Duch, M., Justesen, J., Chevallier, J., Andersen, L.K., Hemmersam, A.G., Pedersen, F.S., and Besenbacher, F. (2006)

QCM-D studies of attachment and differential spreading of pre-osteoblastic cells on Ta and Cr surfaces. *Biomaterials* **27**, 1346–1354.

39. Zhang, S., Gelain, F., and Zhao, X. (2005) Designer self-assembling petide nanofiber scaffold for 3-D tissue cell cultures. *Semin. Cancer. Biol.* **15**, 413–420.

40. Sathananthan, A.H. (2003) Origins of human embryonic stem cells and their spontaneous differentiation. *First National Stem Cell Centre Scientific Conference*, Melbourne, Poster 225.

4

A Controlled-Cooling Protocol for Cryopreservation of Human and Non-Human Primate Embryonic Stem Cells

Carol B. Ware and Szczepan W. Baran

Summary

Freeze storage of human embryonic stem (hES) cells has not proven effective using the methods employed for mouse ES (mES) cells, while rhesus ES (rhES) cells are only modestly effectively frozen using common mES cell methods. Because human and rhES cells are passaged and frozen in clusters that approximate the size of embryos, we employed a mammalian embryo freezing method to cryopreserve primate ES cells. This protocol involves freezing in a dimethyl-sulfoxide cryoprotectant using straws. An ice crystal seed is induced at −10 °C followed by controlled cooling at −1 °C per minute down to −33 °C with a plunge from there directly into liquid nitrogen (LN_2) at −196 °C. Thaw is effected rapidly by moving the frozen cells directly from LN_2 into a water bath and placing directly into culture medium without step-wise cryoprotectant removal. Using this protocol, we have increased the survival of human ES cells from ≤ 1 to ∼ 80% and rhES cells from ∼ 30 to ≥ 90%. Thus, this protocol describes a technically simple but effective means of long-term storage of primate ES cells.

Key Words: Human; Primate; Embryonic stem cells; Cryopreservation; Freezing.

1. Introduction

Successful long-term storage is crucial for any lab growing mammalian cells. When cell quality is linked to passage number, as it is for embryonic stem (ES) cells, effective cryopreservation of low passage cells takes on even more importance. The usual method of tissue culture cryopreservation involves placing the cells in a vial and cooling in a container approximating a temperature drop of −1 °C/min down to −80 °C prior to storage in liquid nitrogen (LN_2)

From: *Methods in Molecular Biology, vol. 407: Stem Cell Assays*
Edited by: M. C. Vemuri © Humana Press, Totowa, NJ

or LN$_2$ vapor phase. This method, though simple and reasonably effective, is much less successful for human ES (hES) cells, where survival on thaw is around 1% relative to the same cells that were passaged without freezing, whereas rhesus ES (rhES) cell survival approaches roughly 30% *(3,4)*. The inefficiency of primate ES cell survival following freezing is concerning for three major reasons: (i) the time it takes to get the cultures growing robustly following freeze, (ii) the chance of inadvertent selection of a subpopulation of cells through freezing and (iii) loss of valuable cells effectively necessitating more cell doublings to complete experiments.

An open-pulled straw method of hES cell vitrification has been described *(1)*. The pulled-straw increases the surface area allowing ultra-rapid cooling. Vitrification is attractive because it is inexpensive and quick. Because of the high level of cryoprotectant used which can cause toxicity, the primary drawback with vitrification is the need for tight adherence to technique including strict time constraints upon both freeze and thaw, which can make the resulting survival variable within a lab and means that transfer of cells to other laboratories can meet with unpredictable success. This is exacerbated by the large surface to volume ratio of the pulled straws causing unintentional warming when the straws are being transferred between LN$_2$ tanks.

Slow controlled-rate cooling is often employed for mammalian embryo cryopreservation. The advantages of controlled-rate freeze are that it is much less technically demanding than vitrification, freezing containers with a smaller surface to volume ratio can be used allowing less stringently rapid transfer between LN$_2$ reservoirs and thaw is simple allowing reproducible transfer of cells between laboratories. The primary drawback is that controlled-rate freezing most commonly requires the use of an expensive programmable freezer. However, an easy and inexpensive apparatus *(2)* can be assembled that is just as effective as a programmable freezer *(3)*, but which requires some technical attention during the freezing process (an estimated 10–15 min above the time needed to use a programmable freezer) and a ready source of approximately 3 l of LN$_2$ must be available for each freeze run.

There are two primary points where the controlled-rate embryo freezing technique differs from more conventional mammalian cell culture freezing. The first is the inclusion of a hold of the freezing process at −7 to −10 °C to allow the introduction of an internal ice crystal seed through supercooling a small external portion of the freezing container. The introduced seed travels rapidly, freezing the cryopreservation medium. In the presence of a cell permeable cryoprotectant, like dimethylsulfoxide (DMSO), the cell interior is left in a liquid state. As the extracellular medium is frozen, water exits the cells in an effort to equilibrate the internal and external salt concentrations. In the absence of an ice crystal seed, the cells remain hydrated and both the internal and external liquid spontaneously freeze at–20 °C, effectively slicing the cells. When the freezing process continues after the seed, the intracellular milieu continues to dehydrate so that intracellaluar ice crystals do not form at –20 °C. The second difference is the controlled rate of subsequent temperature drop, which assures it will neither be too fast to exceed the transit of water across the cell membrane nor too slow, which can lead to high salt and/or cryoprotectant toxicity (for reviews of embryo cryobiology *see* **refs 5–7**).

2. Materials

2.1. Equipment

A BioCool III programmable freezer (FTS Kinetics, Stone Ridge, NY, USA) is routinely used to freeze primate ES cells in our laboratory (*see* **Note 1**).

2.2. Reagents

2.2.1. ES Cell Culture Reagents

1. Dulbecco's Modified Eagle's Medium/F12 (DMEM/F12, cat. no.11320-33, Invitrogen, Carlsbad, CA, USA).
2. GlutaMAX, 2 mM (cat. no.35050-061, Invitrogen).
3. Sodium pyruvate, 1 mM (cat. no. 11360-070, Invitrogen).
4. Nonessential amino acid, 0.1 mM (cat. no. 11140-050, Invitrogen).
5. Penicillin, 50 U/ml; streptomycin, 50 µg/mL (cat. no. 15070-063, Invitrogen); 20% Knockout serum replacer (cat. no. 10828-018, Invitrogen).
6. Basic fibroblast growth factor-2 ng/ml (FGF-2); cat. no. 100-18B, Peprotech, Rocky Hill, NJ, USA).
7. β-mercaptoethanol, 0.1 mM (cat. no. M7522, Sigma, St. Louis, MO, USA).
8. Tissue culture plates treated with 0.1% gelatin (cat. no. G1890, Sigma).
9. Tissue culture grade water (cat. no. 15230-162, Invitrogen).
10. Phosphate-buffered saline (PBS) containing 10% ES-Qualified fetal bovine serum (FBS) (cat. no. 14190-144, Invitrogen).
11. Dispase, 1.2 U/ml (cat. no.17105-041, Invitrogen).

All culture reagents are stored following manufacturer's recommendations and when thawed are stored at 4 °C.

2.2.2. Freezing Reagents

1. DMEM/F12 supplemented with 10% DMSO (DMSO hybridoma grade, cat. no.D2650, Sigma) and 30% FBS (*see* **Note 2**).

2. Cassou straws (0.25 ml) were used as the freezing container (cat. no.04170, Veterinary Concepts Spring Valley, WI, USA; *see* **note 3**). Straws were loaded by fixing the straw to the end of a microliter pipette loader (*see* **Note 4**).

3. Methods

3.1. Cryopreservation Protocol for Primate ES Cells

1. Remove cells for cryopreservation from the plate in clusters (*see* **Note 5**).
2. Pellet cells by centrifugation and aspirate supernatant.
3. Resuspend in culture medium without FGF (DMEM/F12 was supplemented with 2 mM GlutaMAX, 1mM sodium pyruvate, 0.1 mM nonessential amino acid, 50 U/ml penicillin, 50 μg/mL streptomycin, 20% Knockout serum replacer and 0.1 mM β-mercaptoethanol) and pellet.
4. Aspirate supernatant and add freezing medium (DMEM/F12+30% FBS+10% DMSO) and note time.
5. Load straws by drawing in a small column of freezing medium without cells, a small air bubble, followed by a long column of cells in freezing medium. Draw the cell-free column into the plug end of the straw creating an air bubble at the open end of the straw. Seal both ends by passing through a flame and squeezing between gloved thumb and finger (*see* **Fig. 1**).
6. Allow cells to equilibrate in freezing medium for 15 min from first exposure (*see* **Note 6**).
7. Place in –10 °C ethanol bath with the plug end up for 1–5 min.
8. Seed ice crystal (*see* **Note 7**).
9. Hold at –10 °C for a minute or two to allow the extracellular medium to freeze.
10. Begin controlled rate cooling at –1 °C per minute down to –33 °C (*see* **Note 8**)
11. Plunge into LN$_2$ for storage (*see* **Note 9**).

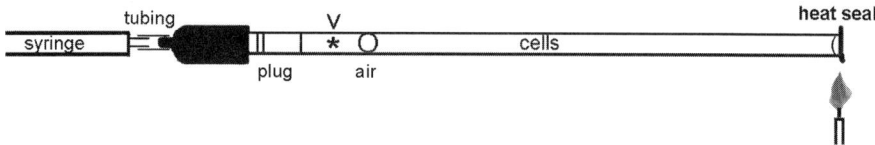

Fig. 1. Loading the freezing straw. The straw is attached to a syringe by the plug end. A small column of freezing medium without cells is drawn up followed by a small air bubble. A long column of cells in suspension is drawn in until the entire contents of the straw are close to the plug end. The cell-free column is pulled into the plug leaving a small air bubble at the end of the straw (right). A flame is used to heat seal both ends of the straw. The heated ends are squeezed shut with fingers or forceps. Once the straws are cooled to –10 °C a seed is initiated at the point indicated by "V" while the seed "*" becomes apparent within the straw as an opaque area adjacent to the straw wall where the cooled forceps touched.

3.2. Thaw Method

1. Thaw in a water bath warmed between 22 and 37 °C (*see* **Note 10**).
2. Empty straw into a waiting tube of culture medium, centrifuge, resuspend pellet in culture medium and plate for standard culture.

3.3. Primate Culture Methods

1. Tissue culture plates treated with 0.1% gelatin are seeded with primary mouse embryonic feeder layers inactivated using 3000 Rad γ-irradiation (*see* **Note 11**).
2. hES cells are passaged using PBS containing 10% ES-Qualified FBS and 1.2 U/ml dispase (*see* **Note 5**).

4. Notes

1. Many options are available for controlled-rate freezing equipment. Considerations for the appropriate machine or device include the ability to hold the temperature during freeze somewhere between –7 and –12 °C and the ability to cool at –1 °C per minute down to a temperature ranging between –30 and –35 °C. Straws and vials should be accessible to introduce the ice crystal seed.
2. The DMEM/F12 medium serves as a buffered physiological salt solution when freezing. Other culture media can substitute for DMEM/F12 with equal effect *(8)*.
3. Either 0.25-cc or 0.5-cc gas-sterilized Cassou straws can be used. When 0.25-cc straws are used, more can be held in the same volume LN_2 reservoir relative to 0.5-cc straws or vials. On the other hand, 0.5-cc straws have the advantage of a smaller surface to volume ratio than 0.25-cc straws. This serves as a buffer against transfer through room temperature air between LN_2 tanks. The much smaller surface to volume ratio of vials allows a relatively leisurely transfer between storage containers, but vials take up a tremendous amount of LN_2 space relative to straws and have a tendency to leak LN_2, which can cause transmission of pathogens resident in the LN_2 reservoir, most notably mycoplasma. Poorly sealed straws are not likely to remain intact when LN_2 has leaked in, self-selecting the straws where the contents have been protected from LN_2 contamination. Thus, this protocol recommends the use of 0.25-cc straws, while freezing container is, in fact, a matter of personal choice.
4. The tubing attached to the microliter pipette holder was cut to about 1.5 cm and fixed to a 1-ml syringe. Thus, the straw could be aspirated by manipulating the plunger of the syringe.
5. Cells can be removed from the plate by enzymes (dispase or collagenase) or by manual cutting. We use dispase and expose the cells until neighboring feeder cells begin to separate as the cells lift from the plate. The cells are removed along with the feeder by gentle washing with the enzyme solution to pull the monolayer away in small strips. They are pelleted and washed one time before being exposed to freezing medium. When suspending in the wash medium they are broken into

clusters by aspirating up and down vigorously using a 5-ml pipette. This technique relies on practice to break up the cell clusters. It is worthwhile assessing the resulting clusters in the microscope to be sure they are disrupted appropriately. We have only tested the freezing ability of cells in clusters using this protocol. Cluster size averages around 100 μM, with a range of single cell (~ 20 μM) up to clusters exceeding 300 μM.

6. The actual time for equilibration was not explored and there is likely much leeway in this parameter as DMSO controls exposed for more than 30 min did not lose viability relative to unfrozen cells and cells frozen after 5 min of equilibration did not suffer noticeably upon thaw.

7. Dip forceps in LN_2 until the tips are supercooled. The straw is pulled out of the –10 °C ethanol bath just to the point of the air bubble near the plug end. The cooled forceps are used to touch the outside of the straw until an opaque crystal is seen inside the straw. If this does not take cool the forceps again until the seed is apparent, keeping the straw cool throughout. The ice crystal seed travels rapidly through the straw and causes almost instantaneous crystallization of the extracellular freezing medium in vials. This step is crucial for success.

8. The final temperature following controlled cooling should reach at least –30 °C. However, the warmest possible temperature was not explored. Controlled cooling can be stopped and held at any point from –30 °C down to –196 °C prior to LN_2 plunge.

9. It is important to devise a rapid method of transferring straws from the freezing machine to the LN_2 tank. An open thermos holding the labeled cane in LN_2 works well. The sleeve containing the freshly transferred straws then holds a reservoir of LN_2 to buffer the transfer from the thermos into the tank.

10. Thaw temperature is another important factor in survival and varies depending on the freezing protocol employed. hES cells frozen by this protocol in 0.25-cc straws thaw well in a 22–37 °C water bath, whereas non-human primate ES cells survive a room temperature water bath thaw most effectively. The final temperature at thaw is less relevant than the rate of thaw. Thus, surface to volume ratio will determine the rate. Use of vials requires that the temperature of the thaw bath be 37 °C.

11. Cells grown on Matrigel without feeders can be frozen successfully following this protocol, but they are not broken as vigorously as clusters passaged on mouse embryonic fibroblasts.

References

1. Reubinoff, B.F., Pera, M.F., Vajta, G., and Trounson, A.O. (2001) Effective cryopreservation of human embryonic stem cells by the open pulled straw vitrification method. *Hum Reprod* **16**, 2187–2194.

2. Pye, J., Holmes, M.C., and Crawford, M. (1980) A simple piece of equipment for controlled cooling and freezing of mouse embryos, in *Frozen Storage of*

Laboratory Animals: Proceedings of a Workshop at Harwell, UK (Zeilmaker, G.H., ed.), Gustav Fischer Verlag, Stuttgart, NY, pp. 61–66.

3. Ware, C.B., Nelson, A.M., and Blau, C.A. (2005) Controlled-rate freezing of human ES cells. *Biotechniques* **38**, 879–883.

4. Baran, S.W. and Ware, C.B. (in press) Cryopreservation of rhesus macaque: Technical note, *Stem Cells Develop* **16**.

5. Mazur, P. (1977) Slow freezing injury in mammalian cells, in *The Freezing of Mammalian Embryos, Ciba Found. Symp. 52* (Elliott, K. and Whelan, J., eds.), Elsevier Excerpta Medica, Amsterdam, pp. 19–42.

6. Leibo, S.P. and Mazur, P. (1978) Methods of the preservation of mammalian embryos by freezing, in *Methods of Mammalian Reproduction* (Daniel, J.C. Jr., ed.), Academic Press, New York, pp. 179–201.

7. Whittingham, D.G. (1981) Sensitivity of mouse embryos to the rate of thawing, in *Frozen Storage of Laboratory Animals* (Zeilmaker, G.H., ed.), Gustav Fischer Verlag, Stuttgart, pp. 221–232.

8. Whittingham, D.G. (1974) Embryo Banks in the future of developmental genetics. *Genetics* **78**, 395–402.

5

Cell Surface Markers in Human Embryonic Stem Cells

Raj R. Rao, Alison Venable Johnson, and Steven L. Stice

Summary

The pluripotent nature of human embryonic stem cells (hESCs) is based on their potential to form every cell type in the body. Prior to use in directed differentiation strategies, these cells need to be thoroughly characterized. The large number of glycoproteins and carbohydrates that exist on the cell surface provide an excellent opportunity for characterizing hESCs and a means to delineate pluripotent and differentiated cell types. A panel of 14 lectins, based on their specificity for a variety of carbohydrates and carbohydrate linkages, along with stage-specific embryonic antigen-4 (SSEA-4), have been chosen to examine hESCs for other potential pluripotent markers. These studies have been achieved by binding quantitation by flow cytometry and binding localization in adherent colonies by immunocytochemistry. We have shown that certain lectins may be used as markers that are associated with the pluripotent state of hESCs because binding percentages and binding localization of these lectins are similar to those of SSEA-4. This presents options for systematic classification of pluripotent hESCs and for distinguishing differentiated hESC types based on glycan presentation that accompanies differentiation.

Key Words: Lectin; Human embryonic stem cells; Immunocytochemistry; Flow cytometry; Pluripotency.

1. Introduction

Human embryonic stem cells (hESCs) are purported to serve as valuable models for studying basic human development in addition to their proposed potential for cell-based therapies. The large number of glycoproteins and carbohydrates that reside on the hESC surface (glycolsignatures) can be used to better characterize hESCs. The most common hESC surface markers are the stage-specific embryonic antigen-3 and 4 (SSEA-3 and SSEA-4). SSEA-3 and SSEA-4 are globoseries cell

From: *Methods in Molecular Biology, vol. 407: Stem Cell Assays*
Edited by: M. C. Vemuri © Humana Press, Totowa, NJ

surface glycoproteins that were first used to delineate embryological changes in the developing mouse embryo *(1,2)*. The glycol signature of hESCs is altered during development and in vitro differentiation *(3,4)*.

Lectins are carbohydrate-binding proteins that recognize diverse sugar structures and have been extensively used to identify and characterize cell surface glycosylation patterns. Lectin studies have led to the delineation of embryologic developmental stages in some species. For example, lectins have been used to investigate and identify cell types based on presentation of specific cell surface carbohydrates *(5–10)*. Additionally, many developmentally regulated glycans identified as lectin receptors on (mouse ESCs) are displayed on cell surfaces at the preimplantation and implantation stages of development. Examples include Concanavalin A (Con A), Peanut agglutinin (PNA), Wheat Germ Agglutinin, Dolichos biflorus agglutinin (DBA), and Ricinus communis agglutinin (RCA) (*see* **Table 1**). These results indicate that glycans may contribute to specific developmental function and also indicate that they can be used as markers to define stages of mouse embryogenesis.

Using enriched high SSEA-4-expressing hESCs, the binding percentages of selected lectins was monitored by flow cytometry and immunocytochemistry *(10)*. Our findings indicate that there are many surface carbohydrate antigens that could be exploited to further characterize hESCs, and these lectins could also provide a source of unique markers for characterizing subpopulations that exist in colonies of adherent hESCs.

2. Materials

2.1. Cell Culture and Expansion

1. NIH approved human embryonic stem cell lines BG01 and BG02 (http://.stemcells.nih.gov/research/registry) *(11)*
2. hESC medium: Dulbecco's Modified Eagle's Medium (DMEM/F12) supplemented with 15% fetal bovine serum (FBS, HyClone, Ogden, UT, USA), 5% knockout serum replacer (KSR), $1\times$ non-essential amino acids, 20 mM L- glutamine, 0.5 U/ml penicillin, 0.5 U/ml streptomycin, 0.1 mM β-mercaptoethanol (Sigma, St. Louis, MO, USA), 4 ng/ml fibroblast growth factor-2 (Sigma), (all from Gibco/Invitrogen, Carlsbad, CA, USA, unless otherwise labeled). Store formulated medium at 4 °C and use within 1 week.
3. Mouse embryonic fibroblast (MEF) medium: DME Medium High Glucose (DMEM-HiGlu) supplemented with 10% FBS (HyClone), 2 mM L-glutamine, 0.5 U/ml penicillin, 0.5 U/ml streptomycin, (all from Gibco/Invitrogen, unless otherwise labeled). Store formulated medium at 4 °C and use within 2 weeks.
4. Collagenase (1 mg/ml) preparation: Weigh out 10 mg Collagenase Type IV (Gibco/Invitrogen) and dissolve in 10 ml of DMEM/F12 medium supplemented

with 15% FBS, 5% KSR, at 37 °C. Filter sterilize and store at 4 °C and use within 1 week.

5. Trypsin solution (0.05%) and ethylenediamine tetraacetic acid (EDTA) (1 mM) from Gibco/Invitrogen.

2.2. Antibodies for Immunocytochemical Analysis

1. SSEA-4 [Developmental Studies Hybridoma Bank (DSHA); Iowa City, IA, USA; 1:100 dilution].

2. Biotinylated lectins: Con A; Phaseolus vulgaris erythro-agglutinin (PHA-E); Phaseolus vulgaris leuco-agglutinin (PHA-L); Sambucus nigra agglutinin; Arachis hypogea peanut (PNA); Vicia villosa agglutinin; Maackia amurensis; RCA; Wisteria floribunda agglutinin; Ulex europaeus agglutinin; Lotus tetragonolobus lectin; DBA; Hippeastrum hybrid lectin; Lycopersicon esculentum tomato. All lectins used obtained from Vector Laboratories, Burlingame, CA, USA; 10 μg/ml; 1:2 dilution for staining). Details of lectins are shown in **Table 1**.

3. Streptavidin conjugated Alexafluor 594 (Molecular Probes, Eugene, OR, USA; 1:250 dilution) and antigoat Mouse IgG conjugated Alexa 488 (Molecular Probes; 1:2000 dilution).

2.3. Magnetic Bead Sorting

1. Staining buffer (SB): 50 ml of SB consists of 0.5 ml of 0.5 U/ml penicillin, 0.5 U/ml streptomycin, 0.5 ml of 100 mM EDTA, 46.5 ml phosphate buffered saline (PBS) and 2.5 ml FBS.

2.4. Immunocytochemical and Flow Cytometry Analysis

1. Paraformaldehyde (PFA) (Fisher, Pittsburgh, PA, USA): Prepare a 4% (w/v) solution in PBS, fresh for each experiment. Work in a fume hood and wear gloves, as PFA is toxic. Weigh out 4 g PFA and add to glass beaker. Weigh out 4 g sucrose and add to PFA in glass beaker. Add 75 ml distilled water and place on heated stirrer to dissolve. The solution needs to be carefully heated (use a stirring hot-plate at temperature ~56 °C), in a fume hood, to dissolve. Add 2 drops of 1 M sodium hydroxide, and once all has gone into solution, add 10 ml of 10X PBS^{++} (with Ca and Mg; Hyclone Labs). Make sure that pH is between 7.2 and 7.4. Make up volume to 100 ml with distilled water. Store at 4 °C and use within 1 week.

2. Blocking solution: 3% (w/v) goat serum (Hyclone) in PBS^{++}. Store at 4 °C and use within 48 h.

2.5. Fluorescence Microscopy

1. Microscope coverslips (22 × 40 × 0.15 mm) from Fisher and Lab-Tek four-well glass chamber slides from Nalge Nunc, Naperville, IL, USA.

2. Nuclear stain: 300 nM 4,6-diamidino-2-phenylindole (DAPI) in water.

3. Mounting medium: Antifade (Molecular Probes).

Table 1
Comparison of the Specificity for Monosaccharides and Oligosaccharides of a Panel of 14 Biotinylated Lectins Used in Immunocytochemistry and Flow Cytometry

Lectin origin	Monosaccharide specificity	inhibitor
Concanavalin A	Man or Glc	200 mM α-methylmannoside or α-methyl glucoside
Phaseolus vulgaris erythro-agglutinin	Gal	Galactose
Phaseolus vulgaris leuco-agglutinin	Gal	Galactose
Sambucus nigra agglutinin	Sialic Acid	500 mM lactose in acetic acid
Arachis hypogea peanut	Gal	200 mM galactose
Vicia villosa agglutinin	GalNAc	200 mM N-acetylgalactosamine
Maackia amurensis	Gal	200 mM lactose
Ricinus communis agglutinin	Gal	200 mM galactose or lactose
Wisteria floribunda agglutinin	GalNAc	200 mM N-acetylgalactosamine
Ulex europaeus agglutinin	Fuc	50–100 mM L-fucose
Lotus tetragonolobus lectin	Fuc	50–100 mM L-fucose
Dolichos biflorus agglutinin	GalNAc	200 mM N-acetylgalactosamine
Hippeastrum hybrid lectin	Man	100 mM mannose
Lycopersicon esculetum tomato	GlcNAc	Chitin Hydrolysate

Source: Reproduced from **ref.** *11* (BioMed Central Ltd.).

3. Methods
3.1. Passage of Human Embryonic Stem Cells

1. Using at least 3-day-old MEF plates (1.2×10^5 cells/cm^2), aspirate off medium and replace with 2 ml of hESC medium (*see* **Note 1**). Place dish at 37 °C until ready to plate out cells.
2. Add 1ml of collagenase solution per 35-mm dish containing hESC colonies. Place on 37 °C stage for 2–3 min. Colonies can be observed rounding under dissection scope.
3. Aspirate off collagenase solution and add 1 ml of 0.05% trypsin solution.
4. Allow trypsin to contact cells for no more than 40 s, then aspirate off the trypsin solution.

5. Add 1 ml of 10% FBS in DMEM/F12 to 35-mm dish and begin to gently pipette up and down to dislodge or knock off and break up cell clumps, while continuously observing cells under microscope.
6. Place harvested cells in 15-ml tube containing 8 ml hESC medium.
7. Add fresh medium to the dish and wash off and collect any remaining trypsinized cells, add to 15-ml tube. (*see* **Note 2**).
8. Spin harvest cells for 4 min at 200 g at room temperature.
9. Resuspend trypsinized pellet in 2 ml of hESC medium per 35-mm dish used.
10. Count cells using hemocytometer by taking 10 μl cell suspension and mixing with 10 μl Trypan blue.
11. Aspirate medium from pre-equilibrated dishes and plate cells at 150,000 cells per 35-mm dish in 2 ml medium. Place at 37 °C in a 5% CO_2 incubator. Evenly distribute cells on the plate through a uni-directional quick movement of the plate on shelf the incubator (do not swirl cells in the plate) (*see* **Note 3**).
12. Feed every day with a 50% medium change until ready to passage again in 3–4 days. The cells at this stage should exhibit distinct colony morphology (high nuclear to cytoplasmic ratio) characteristic of hESCs.

3.2. Enrichment for SSEA-4-Positive Cells (*see* **Note 4***)*

1. Cultures of hESCs are grown in 100-mm MEF dishes and trypsin passaged into single cell suspensions as described above (*see* **Subheading 3.1.**).
2. Cells are then incubated on ice for 15 min in 1:10 dilution of SSEA-4 (MC 813-70, DSHA) in 1 ml of SB in a 15-ml screw-capped tube. Flick the tube every 5 min to resuspend the cells.
3. After incubation, 10 ml SB is added and cells resuspended as described above and centrifuged at 3000 g for 5 min.
4. Supernatant is removed, leaving the cell pellet intact, and cells washed again in 10 ml of SB
5. After centrifugation, a 1:4 dilution of secondary anti-mouse IgG (magnetic beads) in 100 μl of stain buffer is added to the cell pellet, and the resuspended pellet is incubated on ice for 25 min.
6. After incubation, 10 ml SB is added to wash the cell pellet, followed by centrifugation (3000 g) for 5 min. This process is repeated two more times in 5 ml SB.
7. Cells are finally resuspended in 500 μl of SB before being applied to a pre-washed magnetic bead column. Wash three times with 0.5 ml of SB and collect flow through from the column.
8. Remove the column from the magnet and elute with 1 ml of SB using plunger (*see* **Note 5**).
9. The flow through from the column is collected and saved for counting, and the retained eluate collected separately. Both flow through and eluate are brought up to 5 ml in SB after collection and counted to provide an estimate of the enrichment, prior to immunocytochemical and flow cytometry analysis.

3.3. Fixing Cells for Immunocytochemical and Flow Cytometry Analysis

1. Work in a fume hood. Wash cells obtained after completing the steps described in **Subheading 3.2**, 1 time in PBS^{++} (with Ca and Mg). Use aspirator to remove and a transfer pipette to add PBS (*see* **Note 6**).
2. Add enough PFA solution ($\sim$0.8 ml/1.8 cm^2) to cover bottom of well of chamber slides or dish.
3. Let sit at room temperature for 15–20 min.
4. Wash cells three times in 1 ml of PBS^{++}.
5. Store fixed cells at 4 °C until ready to stain with chosen markers.

3.4. SSEA-4 and Lectin Binding Profiling by Flow Cytometry

1. One milliliter of blocking solution is added to the fixed cells from dish (*see* **Subheading 3.3**.) for 30 min.
2. Cells are then placed in sterile 15-ml conical tubes in aliquots of 500,000 cells each and double stained with one of the 14 lectins at 5 μg/ml and SSEA-4 in a 1:100 dilution.
3. Cells are washed three times with 5 ml of PBS and then stained with secondary antibodies that include streptavidin–allophycocyanin (1:250, BD Biosciences, Franklin Lakes, NJ, USA) for recognition of biotinylated lectins and antigoat Mouse IgG conjugated Alexa 488 (1:2000) for recognition of SSEA-4. These secondary antibodies are chosen so that there is no overlap in the emission/excitation wavelengths and for double staining to be performed.
4. Unstained and stained enriched hESCs stained with secondary antibodies alone are used as controls (*see* **Note 7**).
5. Cytometry is performed using a Beckman Coulter Cytomics FC 500 Flow Cytometer, although obviously other cytometers will work well. Data analysis is performed using the RXP Analysis Software by Beckman Coulter and Windows Multi Document Interface for Flow Cytometry (WinMDI 2.8). Examples of flow cytometry histograms in hESCs are shown in **Fig. 1**.

3.5. Localization of Lectin and SSEA-4 Expression by Immunocytochemical Analysis

1. Cells fixed in wells of chamber slides (*see* **Subheading 3.3**.) are washed 1 time with 1 ml of PBS^{++}. Use aspirator to remove and transfer pipette to add PBS (*see* **Note 6**).
2. Add 0.8 ml of block solution to each well. Incubate at room temperature for 45 min.
3. Prepare cell surface marker (Lectin/SSEA4) in block solution at supplier recommended dilution. This is the primary antibody (1° antibody) solution.

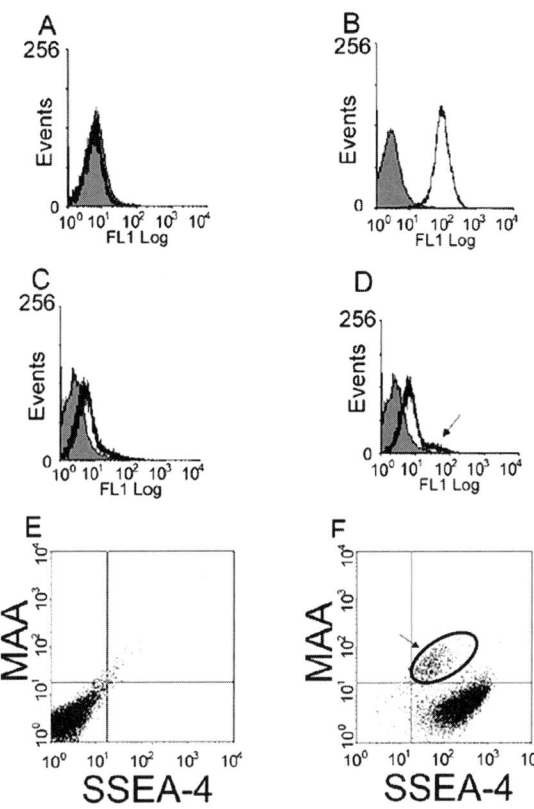

Fig. 1. Flow cytometry histograms of lectin binding in unstained and stained human embryonic stem cells (HESCs). (**A–D**) Histograms of stage-specific embryonic antigen-4 (SSEA-4) binding and representative lectins that were used in this study. To validate that double staining can be performed without signal interference, we determined that SSEA-4 expression was not found in the FL1 channel. (**A**) A histogram plot with the overlay image of SSEA-4 (black tracing) matching the histogram plot of unstained cells (gray fill). (**B**) a positive peak shift in the histogram overlay with Tomato lectin in black tracing and unstained cells (gray fill). (**C**) Lack of Lotus tetragonolobus lectin (black tracing) binding in the histogram overlay with unstained cells (gray fill). (**D**) a histogram overlay with Maackia amurensis (MAA) (black tracing) binding and two peaks, one representing a large population that overlays unstained cells and a smaller population denoted by arrow that shows a smaller population of $MAA^+/SSEA4^+$ cells. (**E**) plots of unstained hESCs. (**F**) subpopulations of cells characterized as $SSEA-4^+/MAA^-$ or as $SSEA-4^+/MAA^+$ cells (circle). (*Source*: Reproduced from **ref**. *12* with permission from BioMed Central Ltd, under the terms of the Creative Commons Attribution License.)

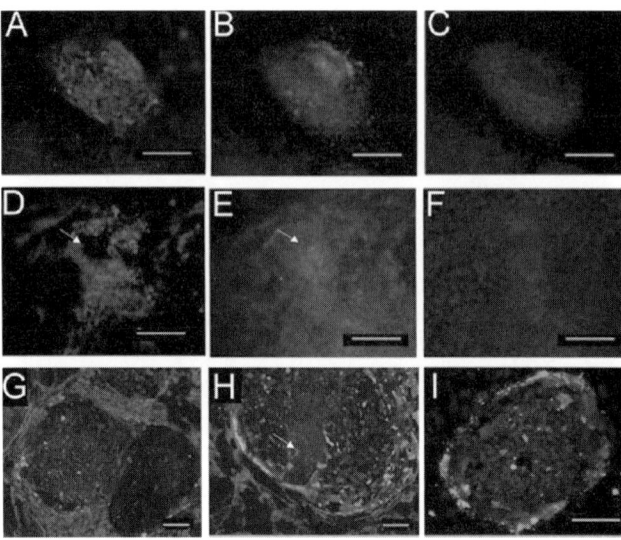

Fig. 2. Carbohydrate expression as determined by lectin binding using immuno-cytochemistry. (**A–C**) a human embryonic stem cell (hESC) colony that represents uniform lectin binding. Ricinus Communis agglutinin binding (**A**) is shown throughout this stage-specific embryonic antigen-4 (SSEA-4) positive colony (**B**). The 4,6-diaminodino-2-phenylindole (DAPI) nuclear stain image is also shown (**C**). Other lectins showed partial binding patterns, such as Vicia Villosa agglutinin binding, which is shown in a hESC colony (**D**) that has uniform SSEA-4 antibody binding (**E**). Arrows denote distinct SSEA-4-positive regions lacking VVA binding. DAPI nuclear staining is also shown (**F**). Phaseolus vulgaris erthyro-agglutinin (PHA-E) binding is shown in two separate images in (**G–H**). (**G**) There are two adjacent colonies, one that expresses strong binding of SSEA-4 antibody and weak to no binding of PHA-E and an adjacent colony showing binding of PHA-E without SSEA-4 antibody binding. (**H**) Another colony with a streak of stacked cells (as determined by high DAPI expression, see arrow) in the middle of the colony that are beginning to lose SSEA-4 expression but have strong PHA-E binding. However, the rest of the colony adjacent to this streak of cells is uniformly positive for SSEA-4 but is lacking PHA-E binding. (**2I**) lack of Dolichos biflorus agglutinin binding and presence of SSEA-4 and DAPI staining. Images and scale bars: (**A–G**) 20× magnification, 100 μm; (**H–I**) 10× magnification, 100 μm. (*Source*: Reproduced from **ref**. *12* with permission from BioMed Central Ltd, under the terms of the Creative Commons Attribution License.)

4. Aspirate off block solution from wells and add 300 μl of 1° antibody solution. Cover to prevent exposure to light and incubate for 1 h at room temperature. This can be extended to over night at 4 °C if necessary.

5. Wash cells 4 times in 1 ml of PBS^{++} for 5 min each wash.

6. Whilst completing washes in step 5, prepare 2° antibody in block solution at supplier recommended dilution. This solution contains secondary antibodies specific to both the lectin and SSEA-4.

7. Aspirate off last wash from wells and add 300 μl of 2° antibody solution. Cover and incubate for 1 h at room temperature. During incubation cover sample with foil to prevent fluorescence bleaching.

8. Wash wells four times in 1 ml of PBS^{++} for 5 min each wash.

9. Add 0.8 ml of a 1:10,000 dilution of DAPI in distilled H$_2$O to each well. Incubate for 5 min at room temperature. Cover with foil during incubation to prevent exposure to light.

10. Wash cells three times in 1 ml of PBS^{++}.

11. Verify that staining has taken place under fluorescence microscope before mounting.

12. Gently remove sides of chamber and aspirate excess surrounding PBS.

13. Place one drop of mounting media directly in center of each well area.

14. At an angle gently lower a coverslip onto the slide trying to avoid air bubbles where possible. Remove excess mounting media from slide and seal with nail varnish on all four sides.

15. Keep in dark storage until results are observed and documented. It is recommended to document the same day. Examples of carbohydrate expression as determined by lectin binding are shown in **Fig. 2**.

4. Notes

1. Our experience with the BG01 and BG02 cell lines have shown that at least 3-day-old mitotically inactivated MEFs are best for sustaining the undifferentiated state of the cell lines. Mitotically inactivated MEFs that are less than 3 days in age result in increased differentiation of the cultures.

2. While obtaining cells for analysis, the best cells are those that come off first. Once feeder layer begins to roll or break it is time to stop. It is best not to try hard to collect everything on the feeder layer or those that remain attached to plate.

3. We have noticed that this step is extremely important for an even distribution of cells across the dish so as to produce uniform-sized colonies across the dish. This helps maintain the undifferentiated nature of the stem cells during routine passaging.

4. It is important to note that depending on culture conditions, the percentage of SSEA-4-expressing cells within a hESC colony can vary. It is generally agreed that there is a high correlation between SSEA-4 expression and the presence of markers correlated with undifferentiated morphology. In our study, pluripotent hESCs have

been defined as those populations with 98–99% expression of SSEA-4, obtained after magnetic bead sorting

5. This step needs to be done very delicately as increased shear can lead to loss of viability of the cells.

6. The PBS wash in this step needs to be performed very gently as the cells have a tendency to dislodge easily.

7. An extra stringency control could also utilize isotype control antibodies. However, these have shown negligible cross-reactivity with cell surface antigens on hESCs.

Acknowledgments

This work was supported in part by funding provided by the Georgia Tech/Emory Center (GTEC) for the Engineering of Living Tissues, an ERC Program of the National Science Foundation under award no. EEC- 9731643. This work was also supported in part by funding from Bresagen Inc. and from the Department of Chemical and Life Science Engineering at Virginia Commonwealth University.

References

1. Shevinsky, L.H., Knowles, B.B., Damjanov, I., and Solter, D. (1982) Monoclonal antibody to murine embryos defines a stage-specific embryonic antigen expressed on mouse embryos and human teratocarcinoma cells. *Cell*, **30**, 697–705.

2. Kannagi, R., Cochran, N.A., Ishigami, F., Hakomori, S., Andrews, P.W., Knowles, B.B., and Solter, D. (1983) Stage-specific embryonic antigens (SSEA-3 and -4) are epitopes of a unique globo-series ganglioside isolated from human teratocarcinoma cells. *EMBO J*, **2**, 2355–2361.

3. Draper, J.S., Pigott, C., Thomson, J.A., and Andrews, P.W. (2002) Surface antigens of human embryonic stem cells changes upon differentiation in culture. *J Anat*, **200**, 249–258.

4. Andrews, P.W., Goodfellow, P.N., Shevinsky, L.H., Bronson, D.L., and Knowles, B.B. (1982) Cell-surface antigens of a clonal human embryonal carcinoma cell line: morphological and antigenic differentiation in culture. *Int J Cancer*, **29**, 523–531.

5. Ruan, S., Raj, B.K., and Lloyd, K.O. (1999) Relationship of glycosyltransferases and mRNA levels to ganglioside expression in neuroblastoma and melanoma cells. *J Neurochem*, **72**, 514–521.

6. Belloni, P.N., Nicolson, G.L. (1988) Differential expression of cell surface glyco-proteins on various organ-derived microvascular endothelia and endothelial cell cultures. *J Cell Physiol*, **136**, 398–410.

7. Takagi, Y., Talbot, N.C., Rexroad, C.E.J., and Pursel, V.G. (1997) Identification of pig primordial germ cells by immunocytochemistry and lectin binding. *Mol Reprod Dev*, **46**, 567–580.

8. Brown, P.J., Stephenson, T.J. (1989) Ulex europaeus agglutinin 1 lectin histochemical staining of dog hepatocellular and bile duct carcinomas. *Res Vet Sci*, **46**, 421–423.

9. Chapman, S.A., Bonshek, R.E., Stoddart, R.W., Jones, C.J., Mackenzie, K.R., O'Donoghue, and E., McLeod, D. (1995) Glycoconjugates of the human trabecular meshwork: a lectin histochemical study. *Histochem J*, **27**, 869–881.

10. Laitinen, L., Juusela, H., and Virtanen, I. (1990) Binding of the blood group-reactive lectins to human adult kidney specimens. *Anat Rec*, **226**, 10–17.

11. Mitalipova, M., Calhoun, J., Shin, S., Wininger, D., Schulz, T., Noggle, S., Venable, A., Lyons, I., Robins, A., and Stice, S. (2003) Human embryonic stem cell lines derived from discarded embryos. *Stem Cells*, **21**, 521–526.

12. Venable, A., Mitalipova, M., Lyons, I., Jones, K., Shin, S., Pierce, M., and Stice, S.L. (2005) Lectin binding profiles of SSEA-4 enriched, pluripotent human embryonic stem cell surfaces. *BMC Dev Biol*, **5**, 15.

6

Generation of a Monoclonal Antibody Library Against Human Embryonic Stem Cells

Micha Drukker, Christina Muscat, and Irving L. Weissman

Summary

Differentiated cell types derived from human embryonic stem cells (hESCs) may serve in the future to treat various human diseases and to model early human embryonic development in vitro. Fulfilling this potential, however, requires extensive development of methods and reagents for studying hESCs self-renewal and differentiation. One of the most widely used experimental approaches in the field of stem cell research is the identification of cell surface markers that can be used to prospectively define and isolate specific populations of stem cells and their progenitors. Here, we review an efficient method for generating monoclonal antibodies against cell surface antigens expressed by hESCs and stem cells at different stages of differentiation. This method may have profound implications for many aspects of hESC research and therapeutics.

Key Words: Human embryonic stem cells; Monoclonal antibodies.

1. Introduction

Human embryonic stem cell (hESC) lines are excellent candidates to serve as a valuable source of cells in transplantation medicine as they have the capacity to grow indefinitely in culture conditions without losing pluripotency and to differentiate to all cell types of the body upon induction of differentiation *(1–3)*. Moreover, owing to their broad differentiation potential, they may serve as an excellent tool to study commitment of pluripotent human cells to different cell types. For example, detailed differentiation protocols are available today for the derivation of neurons *(4–8)*, cardiomyocytes *(9–14)*, endothelial cells *(15)*,

From: *Methods in Molecular Biology, vol. 407: Stem Cell Assays*
Edited by: M. C. Vemuri © Humana Press, Totowa, NJ

hematopoietic precursors *(16,17)*, keratinocytes *(18)*, osteoblasts *(19)*, hepatocytes *(20,21)* and others.

Central to our understanding of how hESCs commit to different fates is the ongoing effort to identify and purify subpopulations of cells emerging during differentiation. Prospective isolation is also critical for therapeutic purposes; one must be able to sort out undifferentiated cells with the hazardous potential to form teratomas following transplantation *(22)*. Isolations of specific cell types are mainly based on either genetic tagging or the expression of cell surface markers. Genetically tagged cells are manipulated to express a reporter protein when the cells commit to a certain lineage. For example, Lavon et al. *(23)* used the pancreatic and duodenal homeobox factor-1 promoter fused to the enhanced green fluorescent protein encoding gene to study differentiation of hESCs to the pancreatic lineage. On the other hand, the cell surface marker identification method relies on immunoselection by antibodies recognizing lineage-specific or population-specific surface antigens. Cell surface marker-based isolation offers several advantages over gene manipulation techniques. First, antibody staining is simple and rapid whereas genetic manipulation of hESCs involves laborious plasmid construction, transfection and clone screening. Second, several antibodies can be used at once to examine the expression of multiple markers, whereas a similar genetic-based isolation would require multiple rounds of reporter construction and transfections. Finally, gene manipulation is only applicable in cases where the relevant genes are known in advance as this method requires detailed information about the gene promoter and the expression pattern. Thus, this method is rarely used as a screening tool for new types of cells, in contrast to the use of antibody combinations that can be utilized to identify and purify subsets of cells without any prior knowledge about their gene expression.

Here, we describe in detail a protocol for the generation of a hESC specific antibody library. This method may apply to any given type of cell if the number of cells available for immunizations exceeds 10 million. However, this procedure is not easily applied for rare populations of cells such as hematopoietic stem cells because they constitute only a small fraction of the peripheral blood and the bone marrow. Screening of the newly generated antibodies is simple, rapid and often yields couple of hundred monoclonal antibody clones. Antibody generation can be followed by studies to identify the target antigens and uncover their role in the biology of hESCs.

Monoclonal antibodies against human cell types are commonly generated by multiple rounds of injections of the target cells into the peritoneum of immunocompetent mice followed by harvesting the sensitized lymphocytes from the spleen and lymph nodes. The cells are fused to mouse myeloma cells and each hybridoma clone secretes monoclonal antibodies encoded by a specific B-cell genome *(24)*. Most of the antibody clones, however, are directed at human antigens that are broadly expressed by many cell types and are not unique for the cell type of interest. To enrich for hESC-specific antigens, we adopted a modified immunization technique, termed the Decoy Footpad Immunization *(25)*. It takes advantage of the fact that circulating T and B cells meet their cognate antigens in secondary lymph nodes and mature there for 4–6 days before their progeny cells are mobilized *(26)*. Decoy Footpad Immunization is initiated by repeated injections of human peripheral blood mononuclear cells (PBMCs) as "decoy" into the left hind footpad. This leads to trapping of B cells specific for generic human epitopes in the lymph node of the left hind limb. hESCs are then injected into the right hind footpad as the immunogen (target antigen). Most of the B cells trapped in the draining inguinal and popliteal right lymph nodes are hESC-specific as the non-specific B cells were already trapped in the draining lymph nodes of left hind limb. To boost the immune response against hESCs and to improve antibody diversification and specificity, repeated injections of the decoy and the target cells are carried out for a month, prior to harvesting of the popliteal lymph node from the right hind limb. Following hybridoma generation, hybrid-derived clones are selected and screened for their capacity to secrete hESC-specific antibodies. Resulting clones may also recognize antigens expressed on spontaneously differentiating hESCs that are present in the immunization mixture.

2. Materials

2.1. Cell Culture

1. hESC culture medium: 400 mL Dulbecco's Modified Eagle Medium: Nutrient Mix F-12 (DMEM/F-12) (1×), liquid, 1:1, with L-glutamine and 4-(2-hydroxyethyl)-1-piperazineethanesulfonic acid (HEPES) buffer (cat. no. 11330-032) supplemented with 100 mL Knockout Serum Replacement (cat. no. 10828-028), 5 mL MEM non-essential amino acids—100× solution (cat. no. 11140-050), 5 mL GlutaMAX™-I Supplement (cat. no. 35050-061), 5 mL penicillin–streptomycin (cat. no. 15140-122), 0.5 mL 2-mercaptoethanl (cat. no. 21985-023) (all from Invitrogen, Carlsbad, CA, USA) and one vial of 200 μL fibroblast growth factor FGF-basic (*see* **step 2**). Filter the medium using 0.22-μm filter unit and store refrigerated in the dark.

2. Recombinant human FGF-basic (Peprotech Inc, Rocky Hill, NJ, USA) is dissolved in 5 mM Tris–HCl at pH 7.6 to a concentration of 25 μg/mL and stored as 200 μL aliquots in −80 °C. Add one aliquot per 500 mL hESC medium, to reach a final concentration of 10 ng/mL. Avoid freezing and thawing.

3. Mouse embryonic fibroblast (MEF) medium: 500 mL DMEM High Glucose (cat. no. 11965-092, Invitrogen) supplemented with 50 mL characterized fetal bovine serum (FBS, cat. no. SH30071, HyClone, Ogden, UT, USA) and 5 mL penicillin–streptomycin (cat. no. 15140-122, Invitrogen). Filter the medium using $0.22-\mu m$ filter unit and store in refrigerator.

4. Basic RPMI 1640 medium: 400 mL RPMI 1640 medium without L-glutamine (cat. no. 21870-076, Invitrogen) supplemented with 50 mL characterized FBS (cat. no. SH30071, HyClone), 5 mL MEM non-essential amino acids—$100\times$ solution (cat. no. 11140-050, Invitrogen), 5 mL GlutaMAX™-I Supplement (cat. no. 35050-061, Invitrogen), 5 mL penicillin–streptomycin (cat. no. 15140-122, Invitrogen), 5 mL of 100 mM sodium pyruvate solution (cat. no. 11360-070, Invitrogen) and 0.5 mL 2-mercaptoethanl (cat. no. 21985-023, Invitrogen). Filter the medium using $0.22-\mu m$ filter unit.

5. RPMI 1640 complete medium: Same as described above, with 50 mL controlled process serum replacement (CPSR) heat inactivated Hybri-Max serum (cat. no. C0786-500 mL, Sigma, St Louis, MO, USA) instead of FBS.

6. RPMI 1640 complete fusion medium: Same as RPMI 1640 complete medium, supplemented with 50 mL P388D1 conditioned medium (described in **step 7**) and 10 mL hypoxantrine aminopterin and thymidine (HAT) medium supplement.

7. RPMI 1640 complete post-selection medium: Same as RPMI 1640 complete medium supplemented with 50 mL P3881 conditioned medium (described below) and 10 mL hypoxanthine and thymidine (HT) medium supplement.

8. HAT medium supplement, $50\times$ (cat. no. H-0262, Sigma). Prepare by adding 10 mL RPMI 1640 medium directly to the vial.

9. HT medium supplement, $50\times$ (cat. no. H-0137, Sigma). Prepare by adding 10 mL RPMI 1640 medium directly to the vial.

10. Polyethylene glycol 1500 (PEG 1500) (cat. no. 783641, Roche, Mannheim, Germany).

11. Gelatin solution: Prepare 1% autoclaved Gelatin Type A solution (cat. no. G-1890, Sigma) in distilled water.

12. Dulbecco's phosphate-buffered saline (PBS, $1\times$) without calcium/magnesium (cat. no. 14190-144, Invitrogen).

13. Sterile 1 mg/mL Collagenase Type IV (cat. no. 17104-019, Invitrogen) solution in DMEM/F-12.

14. Solution of trypsin (0.25%) with 1mM ethylenediaminetetraacetic acid (EDTA) (cat. no. 25200-056, Invitrogen).

15. Basement Membrane Matrix (Matrigel), Phenol Red-free, 10 mL (cat. no. 356237, BD, San Jose, CA, USA). Thaw the vial overnight on ice in a cold room ($4\,^{\circ}C$). In the hood, while the vial remains on ice, transfer 1 mL aliquots into chilled 1.5-mL microtubes. Store in $-20\,^{\circ}C$.

16. Basement Membrane Matrix, diluted solution. Thaw 1 mL aliquot overnight on ice in a cold room (4 °C). In the hood on ice, transfer the 1 mL aliquot to a 50-mL conical tube containing 19 mL DMEM/F-12 medium. Mix briefly and keep on ice. Store in −20 °C.
17. Dimethylsulphoxide (DMSO) (cat. no. D2650, Sigma).

2.2. Peripheral Blood Mononuclear Cell Isolation

1. Ficoll-Paque PLUS, 1.077 g/mL (cat. no. 17-1440-02, GE Healthcare, Uppsala, Sweden). Store at 4 °C in the dark.
2. Human peripheral blood, 30 mL, anticoagulated with heparin.

2.3. Animals and Injections

1. Four-week-old to eight-week-old Balb/c mice. Five animals per group. Use one gender to avoid mating.
2. Avertin stock solution: In 20-mL glass vial dissolve 25 g of 2,2,2-tribromoethanol (cat. no. T48402, Sigma) with 15.5 mL of tert-amyl alcohol (cat. no. 152463, Sigma). Store at room temperature (RT) wrapped with aluminum foil in a glass jar for up to 6 months.
3. Avertin working solution: Add 1 mL of stock solution to 49 mL of PBS (1.25% solution). Maintain constant stirring of PBS that is warmed to 37 °C while adding Avertin to avoid formation of crystals. Aliquot and store protected from light at 4 °C.

2.4. Immunization

1. Tuberculin syringe, 0.5 mL (cat. no.305620, BD).
2. Ethanol, 70%.

2.5. Surgery

1. Ethanol, 70%, surgical scissors, sharp dissecting scissors and surgical forceps.
2. Nylon Cell Strainer, 70 μm (cat. no. 352350, BD Falcon, Bedford, MA, USA).

2.6. Fluorescent-Activated Cell Sorter

1. Cell dissociation solution (cat. no. C5914, Sigma).
2. Staining medium. Filtered 2% FBS in PBS. Keep refrigerated.

3. Methods

Efficient generation of monoclonal antibodies against target cells relies on two major factors: the capacity to elicit strong immune response toward the injected cells and the efficiency of lymphocyte fusion with myeloma cells. For best results, it is advisable to immunize each animal with approximately

$1-2 \times 10^6$ hESCs at 3–4 days intervals for 1 month, meaning that a total of $40-80 \times 10^6$ hESCs are required for a group of five animals. Thus, it is advisable to prepare for continuous tissue culture work for a period of 1 month.

3.1. hESC Culture

This protocol assumes that the hESC culture starts from one plate of 1×10^6 cells (see Chapter 2 for detailed protocols for hESCs propagation).

1. Transfer approximately 1×10^6 4000 rad irradiated MEFs (*see* **Note 1**) in 10 mL MEF medium to 10-cm culture dish that was pre-coated with 6 mL gelatin solution for 10 min. Incubate overnight.
2. Aspirate MEF medium. Transfer approximately 1×10^6 hESC cell suspension to the MEF-coated plate in a total volume of 10 mL hESC culture medium. Incubate overnight.
3. Change culture medium every day for 3 to 5 days. The cells should form medium size colonies.
4. One day before splitting the cells, prepare three fresh MEF-coated plates.
5. To split hESCs, aspirate the medium, wash the plate with 6 mL PBS and aspirate again. Add 3 mL Collagenase Type IV solution to the cells. Incubate at 37 °C for 15 min (*see* **Note 2**).
6. Aspirate the solution and add 2 mL of fresh Collagenase Type IV solution. Incubate for additional 10 min.
7. The edges of the colonies should become loose. Gently spray 5 mL of medium on the colonies, until they become loose, and transfer to 15-mL conical tube. Spin 5 min at $400\,g$ at RT. Discard the medium and resuspend in 3 mL of hESC medium.
8. Aspirate MEF medium from the new plates. Transfer 1 mL hESC suspension to each MEF-coated plate and add 9 mL hESC culture medium. Incubate overnight.
9. For each immunization set (five animals) trypsinize one or two hESC plates, depending on the cell density: aspirate the medium, wash with 6 mL PBS, aspirate the PBS and add 1 mL 0.25% trypsin/EDTA solution to each plate (*see* **Note 2**). Incubate for 5 min (RT) and wash the cells with 4 mL hESC medium to quench trypsin.
10. Transfer to 15-mL conical tube and spin 5 min at $400\,g$ (RT). Aspirate the supernatant, resuspend in 1 mL PBS and count the cells using hemocytometer. There should be a total of $5-10 \times 10^6$ cells. Spin again under the same conditions. Aspirate the PBS carefully, leaving the cell pellet in a total volume of 100 μL. Tap gently on the tube to loosen the cells. Incubate on ice until injected to the animal (*see* **Note 3**).
11. Repeat **steps 4–11** to prepare cells for the next set of immunizations. Use the extra plate that was prepared in **step 8**.

3.2. Preparation of Peripheral Blood Mononuclear Cells

Human PBMCs are prepared to serve as the decoy immunogen. They can be prepared once and then frozen as aliquots of $5–10 \times 10^6$ PBMCs for a group of five animals.

1. PBMCs are prepared by density gradient centrifugation. Place 15 mL Ficoll-Paque in 50-mL conical tube, tilt the tube and overlay very slowly with 30 mL of human blood diluted 1:1 with PBS. Do not disrupt the surface tension of the density gradient material.
2. Centrifuge at $800\,g$ for 30 min (RT). Centrifuge brake should be off. Using 10-mL pipette carefully collect the mononuclear cell band at the Ficoll-Paque and serum interface and transfer the cell suspension to a new 50-mL tube. Make sure that you take as little of the density gradient as possible. Add 1 volume of PBS and centrifuge at $400\,g$ for 5 min (RT).
3. Discard the supernatant, resuspend in 50 mL PBS and spin under the same conditions. Repeat this step twice.
4. Discard the supernatant, resuspend in 10 mL MEF medium and count cells using hemocytometer. There should be a total of $50–150 \times 10^6$ cells if starting with 30 mL of blood.
5. Adjust the cell suspension to a concentration of $5–10 \times 10^6$ cells/mL with MEF medium. Add 10% DMSO to the cell suspension and freeze aliquots of 1 mL in cryofreezing containers.

3.3. Immunization

1. The first two rounds of immunizations are carried out with decoy cells alone. Each time thaw one vial of PBMCs in 37 °C bath, add 5 mL of MEF medium and spin at $400\,g$ for 5 min (RT). Discard the medium and resuspend in 5 mL PBS and spin again.
2. Aspirate the PBS carefully, leaving the PBMC pellet in a total volume of 100 μL. Loosen the pellet by gently tapping on the tube. Incubate on ice until administrated to the animal.
3. From the third injection onwards, inject hESCs and PBMCs in parallel. Prepare hESCs as described in **Subheading 3.1**.
4. Day 6, PBMCs only, left hind footpad; Day 3, PBMCs only, left hind footpad; Days 0, 3, 6, 9, 12, 15, 18, 21: Immunization schedule should be as follows: PBMCs (in left hind footpad) and hESCs (in right hind footpad).
5. Mouse anesthesia: Inject 15 μL Avertin (working solution) per 1 g of mouse body weight into the peritoneum. After about 1 min, mouse will be immobilized and will remain so for about half an hour.
6. Syringe loading: Use one syringe for each cell type and mark the syringe. Resuspend the cells briefly and pull the plunger to load the cells into the syringe.

Avoid making bubbles. There should be a total of 100 μL in each syringe (*see* **Note 3**). Place the syringes on ice.

7. Pinch the mouse hind limb to verify that the mouse is fully anesthetized. If the mouse can feel this, it will try to withdraw its leg from your grasp. Do not commence immunization until there is no reflex to this test.

8. Wipe the footpad with 70% ethanol.

9. Inject the "decoy cells" (PBMCs) into the left footpad: The injection site should be in between the "knuckles" of the mouse footpad (**Fig. 1A** and **B**). Insert the needle until it reaches the underside of the cushion footpad and slowly inject 20 μL of the cell suspension (**Fig. 1C**). Pull out the needle slowly to prevent leaks. Immediately after the needle is withdrawn, apply pressure (with thumb) for a few seconds to prevent backflow of the cells from injection site (**Fig. 1D**). To inject hESCs repeat the exact procedure in the right footpad immediately following left footpad decoy injection.

10. Once immunization is complete, wrap the mouse in a tissue to help keep it warm. A heating pad or indirect heat can be used to warm the animal until it regains consciousness (*see* **Note 4**). Return the animal to its cage.

3.4. P388D1 Conditioned Medium

1. Interleukin-16 (IL-6) conditioned medium supports hybridoma growth. To prepare conditioned medium, thaw $1-4 \times 10^6$ P388D1 cells in 37 °C bath and transfer together with 5 mL of basic RPMI 1640 medium to 15 mL-conical tube. Spin at 400 *g* for 5 min (RT). Discard the supernatant, resuspend in 10 mL basic RPMI 1640 medium and transfer to a 10-cm plate. Incubate for 24 h.

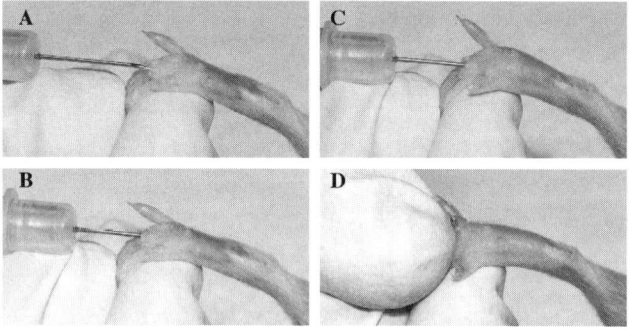

Fig. 1. Mouse footpad injection. (**A**) The injection site is between the "knuckles" of the mouse footpad. (**B**) The needle reaches the underside of the cushion footpad and (**C**) then the area swells when 20 μL of cell suspension is injected. (**D**) To prevent backflow apply pressure at the injection site for a few seconds.

2. Transfer the cells to the hood, resuspend and count the cells using hemocytometer. If the cells reached a density of $4–6 \times 10^5$ cells/mL then they should be split to a final concentration of 2×10^5 cells/mL in 10 mL aliquots of the same medium in T-75 flasks.

3. Split the cells again 2–4 days later when the cells reach a concentration of $4–6 \times 10^5$ cells/mL. Transfer the cells in 50mL aliquots of the same medium into T-175 flasks.

4. Grow the cells until they reach a concentration of 1×10^6 cells/mL. Transfer the medium together with the cells into 50-mL conical tubes and spin at $400\,g$ for 5 min at RT (do not discard the flasks). Aspirate the supernatant and resuspend 1×10^6 cells/mL in basic RPMI 1640 without FBS to maximize IL-6 secretion. Add the cells to the original flasks and return to the incubator.

5. At this point, it is recommended to freeze few cell aliquots for future use. Adjust cell suspension to a concentration of $1–4 \times 10^6$ cells/mL, add 10% DMSO to the cell suspension and freeze aliquots of 1 mL in cryofreezing containers.

6. Grow the cells for an additional 7–10 days. Transfer the medium together with the cells into 50-mL conical tubes and spin at $400\,g$ for 5 min at RT. Transfer 50 mL supernatant aliquots to fresh 50-mL conical tubes and freeze in $-20\,°C$ refrigerator. These aliquots contain high concentration of IL-6 and will be used to support hybridoma growth. Discard the cells.

3.5. Myeloma Cell Preparation

1. Seven to ten days prior to fusion, thaw $1–4 \times 10^6$ SP2/0 mouse myeloma cells in $37\,°C$ bath and transfer together with 5 mL of RPMI 1640 complete medium to a 15-mL conical tube. Spin at $400\,g$ for 5 min in RT. Discard the supernatant and resuspend in 10 mL RPMI 1640 complete medium. Transfer to 10-cm plate and incubate for 48 h.

2. Transfer the cells to the hood, resuspend and count the cells using hemocytometer. If the cells reach a density of $4–6 \times 10^5$ cells/mL, they should be split to a final concentration of 2×10^5 cells/mL in total volume of 10 mL RPMI 1640 complete medium in T-75 flasks. Incubate for 48 h.

3. As before, count the cells, adjust to a concentration of 2×10^5 cells/mL in RPMI 1640 complete medium and transfer 50 mL aliquots into T175 flasks. There should be approximately 3–4 flasks.

4. It is estimated that approximately $2–4 \times 10^7$ myeloma cells are required for every fusion of popliteal lymph node ($1–2 \times 10^8$ cells are harvested from every node). Thus, if five animals are immunized, then a total of $1–2 \times 10^8$ myeloma cells are required. It is important to note that for successful fusion, the cells should be

cultured in log phase, therefore they should not be allowed to reach a density greater than 8×10^5 cells/mL.

5. On the fusion day, collect the SP2/0 cells from T-175 flasks to 50-mL tubes, spin the cells at $400\,g$ for 5 min (RT), discard the supernatant, resuspend all the cells in total volume of 10 mL RPMI 1640 complete medium and count the cells using hemocytometer.

3.6. Lymph Node Isolation and Leukocyte Preparation

On injection day 25, or 4 days after the final injection, harvest the popliteal lymph node from the right hind limb. The isolated leukocytes from the lymph nodes would serve as fusion partners to the myeloma cells. Have the following setup ready before operating:

- Pre-warm the PEG 1500 tube in tissue culture incubator. It is important that the fusion is carried out at 37 °C. Leave the tube in the incubator until used.
- Pre-warm the RPMI 1640 complete fusion medium in 37 °C water bath.

1. Humanely euthanize the mice by terminal CO_2 inhalation.
2. Place the mouse in a supine position and pin down the forelimbs and the hind limbs with needles to a dissecting board.
3. Swab the right hind limb with 70% ethanol.
4. Make a small incision in the skin above the ankle (**Fig. 2A**) and loosen skin by gently inserting forceps jaw underneath the skin and sliding it left and right (**Fig. 2B**).
5. Enlarge the incision upward toward the leg. Grab the skin, stretch it down and pin it to the board. That will expose the area where the popliteal node is located or hidden within a fat pad (**Fig. 2C**).
6. Separate the fat pad from the muscle, exposing the lymph node, and gently dissect out the lymph node (**Fig. 2D**).
7. Transfer the lymph nodes to a 50-mL conical tube. For best results, it is better to carry out two independent fusions. Thus, it is advisable to separate the lymph nodes into two tubes, each containing 1–3 lymph nodes (*see* **Note 5**).
8. In the hood, transfer the nodes to a sterile cell strainer atop a 50-mL conical tube and homogenize the nodes by mincing them with the sterile thumb rest end of a syringe plunger until most of the tissue is dispersed. Then, wash the strainer thoroughly with 50 mL of basic RPMI 1640 medium in order to collect the cells in the suspension.
9. Spin for 5 min, $400\,g$, at RT. Remove the supernatant carefully (with a pipette, do not aspirate) and resuspend in 5 mL basic RPMI 1640 medium. Count the cells using hemocytometer.

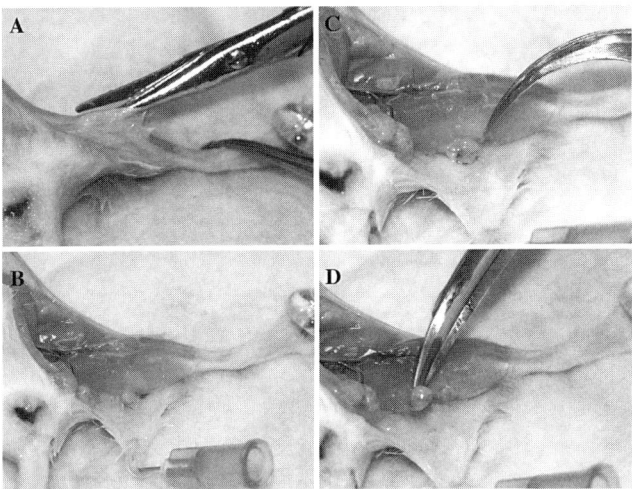

Fig. 2. Isolation of mouse popliteal lymph node. (**A**) Mouse is pined down in supine position and a small incision is made in the skin above the ankle. (**B**) Skin is stretched down and pinned to the board. (**C**) The popliteal lymph node is located within a fat pad. (**D**) Fat pad is separated from the muscle, exposing the lymph node and the node is removed by fine forceps.

3.7. Fusion with Myeloma Cells

Successful fusion between the lymph node cells and the myeloma cells is critical for the generation of monoclonal antibody library (*see* **Note 6**). Carry out the following steps carefully.

1. In a fresh 50-mL conical tube combine 2×10^7 myeloma cells per 1×10^8 lymph node cells. Centrifuge for 5 min at $400\,g$ (RT). Remove the supernatant carefully using a pipetman, do not aspirate.
2. Resuspend the cells with 10 mL basic RPMI 1640 medium. Spin again and discard supernatant. Repeat this step twice.
3. Remove the PEG 1500 from the incubator, swab with 70% ethanol and place in the hood.
4. Gently tap the cell pellet to loosen it.
5. Add 4 mL of PEG 1500 down the side of the tube at a rate of 1 mL/min while gently stirring (be very gentle).
6. Inactive the PEG 1500 by adding RPMI 1640 complete medium in the following manner: 1 mL/min while gently stirring; 3 mL/3min gently; 3 mL/3min directly to cell suspension; 5 mL/min; 10 mL/min and 25 mL/min.

7. Let the cells sit for 5 min in RT or 37 °C incubator. Spin for 5 min at 400 g (RT). Resuspend gently in RPMI 1640 complete fusion medium.

8. For each lymph node harvested, seed the cells in five 96-well plates in RPMI 1640 complete fusion medium (i.e., for 5 lymph nodes plate 25 96-well plates). The total volume of each 96 plate is approximately 18.5 mL. Transfer the cells together with the calculated volume of RPMI 1640 complete fusion medium into square medium bottle and swirl gently.

9. Pour 40 mL cell suspension into 50-mL cell reservoir well and using 12-channel or 8-channel micropipette dispense 190 μL into flat-bottom 96-well plates.

10. Place the plates in the incubator at 7% CO_2.

3.8. Propagating and Screening Clones

Following 4–5 days of culture, it is recommended to carry out partial medium change due to massive cell death. Propagate the clones in HAT-containing medium until they are expanded and screened. Later the cells are transferred to HT-containing medium instead of HAT-containing medium (*see* below). It is important to add the HT supplement as hypoxanthine and thymidine are necessary for the salvage nucleoside synthesis pathway (*see* **Note 6**).

1. Using multi-channel micropipette, draw 100 μL/well from the 96-well plates (do not resuspend). Use barrier tips once and immediately discard them to ensure that cells are not passed between wells.

2. Pour RPMI 1640 complete fusion medium into 50-mL cell reservoir well and using 12-channel or 8-channel micropipette dispense 150 μL into each well of the 96-well plates.

3. Return the plates to the incubator.

4. Positive clones should appear within 2–3 weeks. You can identify the clones by examining the plates under the microscope. Massive clusters containing thousands of dividing cells should form. When the clones expand further, the medium turns yellow and cell aggregates become apparent to the naked eye.

5. Positive clones should be expanded in the following manner: count the number of the positive clones. Prepare 500 μL RPMI 1640 complete fusion medium in each well of a 24-well plate for all the clones and 96-well plates containing 100 μL of the same medium. Using 200 μL fresh barrier tip, resuspend each expanded clone once and transfer 100 μL to a 24-well plate. Resuspend twice and transfer 100 μL to a 96-well plate. Mark both plates in the same manner, so you will be able to know which clones were transferred to the 24-well plate and matched 96-well plate.

6. You may want to add 100 μL RPMI 1640 complete fusion medium to each of the original 96-well plate and to keep it for additional few days as a backup.

7. Return all the plates to the incubator for additional 2–3 days. It is essential that during these days, preparations for clone analysis would be carried out.

8. Following clone screening (*see* **Subheading 3.9.**) expand the positive clones from the 24-well plates to six-well plates by transferring each clone to one six-well plate containing RPMI 1640 complete post-selection medium. Grow the cells for additional 2–3 days and then freeze aliquots as described in **Subheading 3.11**. If you choose to keep growing the clones maintain them in RPMI 1640 complete post-selection medium.

3.9. Clone Screening

There are several methods that one may use to test antibody secretion from hybridoma clones. In our opinion the best method to test secretion and at the same time to examine the staining pattern is by FACS (*see* **Note 7**).

1. As described in **Subheading 3.1** carry out **steps 1–9** for the analysis of 100 hybridoma clones. The only difference is that the cells are not seeded on MEF-coated plates as given in **step 8**. Instead, plate the cells on three Matrigel-coated plates.
2. Add 2 mL diluted Matrigel to each 10-cm tissue culture dish. Swirl the plate until it is uniformly covered with Matrigel.
3. Incubate in the hood for 3 h.
4. Aspirate the remaining solution and pass the hESC suspension to these plates. Typically 3–5 days later, the cells will form medium large colonies.
5. At the day of clone analysis, dissociate the culture by cell dissociation solution (*see* **Note 2**). Aspirate the medium, wash with 6 mL PBS and add 2 mL cell dissociation solution to each plate. Return cells to the incubator for 15 min. Occasionally tap on the plates to break cell aggregates.
6. Take the plates to the hood and using 1-mL micropipette dissociate the cells by sucking the buffer and spraying it on the attached colonies. Then, pass the cells through a 70-μm nylon cell strainer. Add 10 mL staining medium and spin down for 5 min at 400 g (4 °C). Aspirate the supernatant, resuspend in 1 mL staining medium and incubate on ice.
7. Count the cells using hemocytometer. You will need between 8–12 × 10^4 cells per analysis. Therefore, if you plan to test approximately 100 clones a total of 8–12 × 10^6 cells are required. Adjust the cell-suspension to 1.6–2.4 × 10^6 cells/mL in staining medium. One milliliter is sufficient for 20 stains.
8. Transfer the cells to a reservoir well and dispense 50 μl aliquots into 96-well U-bottom plates. Place the plate on ice.
9. Spin the 96-well clone-containing plates, which were prepared as described in **subheading 3.7** (**step 5**), for 10 min at 400 g (RT). Using a multichannel micropipette, transfer 50 μL from the supernatant of each clone into the 96-well U-bottom plates that contain the hESCs and resuspend once (*see* **Note 8**). Change tips between clones.

10. Incubate on ice for 1 h. Spin the 96-well plate for 10 min at 500 g (4 °C). Discard the supernatant by briefly overturning the plate over a sink and immediately return the plate to upright position.

11. Using multichannel micropipette, resuspend the clones in a 100 μL of FACS medium containing anti-mouse IgG secondary antibody. Incubate for 1 h on ice. We recommend using anti-mouse IgG polyclonal phycoerythrin-conjugated antibody such as eBioscience (cat. no. 12-4012, San Diego, CA, USA).

12. Spin again under the same conditions and discard supernatant. Resuspend in 200 μL FACS medium and then spin again. Discard supernatant.

13. At the last step you will need to transfer the stained cells to FACS tubes and to analyze the cells by FACS. This step depends on the machine used and on the adequate volume of cell suspension needed. We recommend to resuspend the cells in 250 μL FACS medium and to acquire 3×10^4 events per sample.

3.10. Analysis of FACS Data

Due to the heterogeneous nature of hESC culture (some differentiation always takes place) generating a monoclonal antibody library against these cells may yield antibodies that are not only specific to undifferentiated cells but also to differentiated derivatives. **Figure 3** illustrates four FACS analyses of expression patterns that were generated by staining hESCs with monoclonal antibodies that were produced using this protocol. The analysis in panel A shows that only a minority of the cells expresses the antigen that is recognized by this antibody clone. It indicates that this antibody may identify a marker for differentiated cells. In panel B there is a clear staining of all the cells. Thus, it is likely that this antibody recognizes a marker for undifferentiated hESCs. The variation in expression may indicate that the cells are not uniform with regard to the expression of the marker. On the other hand, in panel C, the expression pattern indicates that most of the cells express very high levels of the antigen. But still, a small population is not stained, and therefore, it may indicate that some differentiated cells down-regulate this marker. In panel D, there is almost uniform expression pattern, indicating that this antibody stains all the cells.

Before freezing the positive clones, it is important to establish that the antibodies, which uniformly stain hESCs, do not stain other human cells. Those that do should not be studied further as they recognize human antigens which are not hESC specific. We recommend staining at least one additional human cell line, such as Hela cells, in order to rule out this option.

3.11. Freezing and Thawing Clones

Ideally you may expect between 100–300 monoclonal hybridoma clones from a typical lymph node fusion. Usually about half of the clones do not

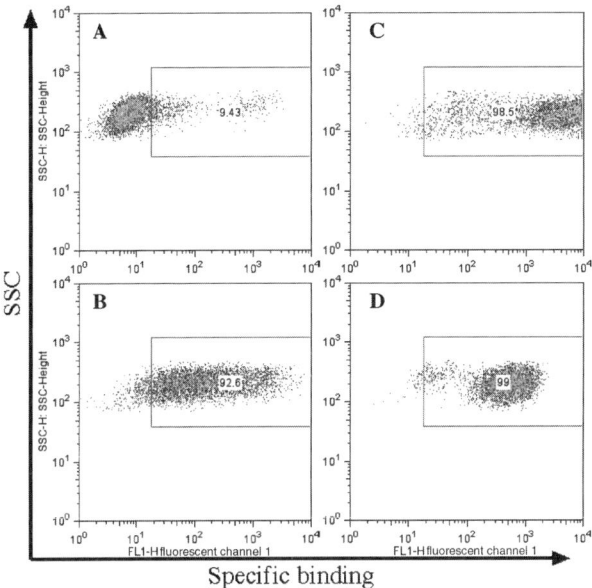

Specific binding

Fig. 3. Fluorescent-activated cell-sorter analysis of four monoclonal antibodies that were raised against human embryonic stem cells (hESCs). Note the different expression patterns. (**A**) Clone that recognize only a minority of the cells. (**B**) Monoclonal antibody that heterogeneously stains the whole population. (**C**) All of the cells are stained by this antibody and the majority of them express very high levels of the recognized antigen. (**D**) This antibody detects almost uniform expression of an antigen that is expressed in hESCs. The x-axis shows the expression level of the antigens that are recognized in hESCs by the monoclonal clones. The y-axis shows the side-scatter of the analyzed cells. Events inside the box indicate cells that are recognized by the antibody.

secrete antibodies or have anti-human specificity. After eliminating these, you are likely to have between 100 and 200 clones. As soon as screening is done, it is crucial to freeze at least two or even three aliquots of all of them due to their unstable genome (*see* **Note 9**).

1. Centrifuge 1×10^6–1×10^7 cells for 5 min at $400\,g$ ($4\,°C$). Resuspend the pellet in 1 mL cold sterile freezing medium (FBS containing 10% DMSO) and aliquot in cryofreezing containers.
2. Put the vials in freezing rack and place in $-70\,°C$ freezer. After several days remove containers from the freezer and place then in liquid nitrogen tank.
3. Reviving hybridoma cell lines: The frozen cultures are thawed quickly at $37\,°C$ by placing the cryofreezing vials in a water bath. Then, transfer the cells with 5 mL

RPMI 1640 complete medium to 15-mL conical tube. Centrifuge 5 min at 400 g (RT) and resuspend the pellet in 10 ml RPMI 1640 complete medium.

4. Notes

1. It is recommended that the same strain of mouse that is used to produce the MEFs would also be used as the immunized strain in order to prevent immune response toward the MEFs. Alternatively, one may grow the hESCs prior to immunization on Matrigel (described in **Subheading 3.9.**), for one passage, to dilute the MEF content.

2. Three methods of cell dissociation are used in this protocol. The reason is that passage by Collagenase IV yields excellent hESC culture as the cells are passed as small clumps. However, this method is not ideal for injecting the cells due to clogging of the needle and hence trypsin is preferred. The cell dissociation solution is necessary to dissociate cells prior to FACS as trypsin might cleave some of the antigens that are expressed by the cells.

3. To ensure sterility, a good practice would be to use a fresh syringe for every animal. If you choose to do so, load 20 μL of each cell type into five different syringes.

4. It is recommended to inject penicillin–streptomycin solution to the animals undergoing immunization, in order to prevent contaminations that might occur due to injection through the footpad. We inject 5 μL of penicillin–streptomycin tissue culture stock solution per 1 g body weight while the mice are still immobilized following immunization.

5. It is recommended that at least two separate fusions be carried out as described in **Subheading 3.6** as fusion efficiency vary greatly We routinely separate the lymph nodes to two groups and carry out two fusions in parallel.

6. HAT are used to select hybrid cells as myeloma cells are defective in either the hypoxanthine–guanine phosphoribosyl transferase (HGPRT) or the thymidine kinase (*TK*) gene. Therefore, both the salvage (by the mutation) and the de novo (by aminopterin) nucleoside pathways are blocked in myeloma cells under selection. Only B-cell myeloma hybrids are HGPRT+TK+(B cells have normal gene copies) and can survive in the presence of aminopterin as HPRT and TK supply nucleosides from hypoxanthine and thymidine.

7. If not enough cells are available for clone screening by FACS, the first round screen may be carried out by ELISA test to detect mouse IgG secretion (ELISA kit cat. no. 11333151001, Roche).

8. The supernatant from the 96-well plates that contain the propagated clones can be used for future screening by storing it in 4 °C. Following centrifugation step that is described in **step 9 Subheading 3.9**, the rest of the supernatant can be transferred into new 96-well plates. Do not disrupt the cell pellet while transferring the supernatant. To prevent evaporation, seal the microwells with adhesive plastic covers.

9. Due to the unstable genome of cell hybrids, hybridomas tend to lose multiple chromosomes while they divide. Thus, it is critical to sub-clone clones of interest. This is carried out by a limited dilution (0.5 cells/well) of the clone cells into 96-well plates. When subclones reappear, you should repeat the FACS analysis in order to pick the best subclones. It is important to note that occasionally subclones might also lose antibody secretion and therefore a second round of subcloning may be required. Propagate the subclones in RPMI 1640 complete medium. However, if the cells grow slowly or do not secrete the antibody effectively, grow the cells in RPMI 1640 complete post-selection medium to support cell growth.

References

1. Thomson, J. A., Itskovitz-Eldor, J., Shapiro, S. S., Waknitz, M. A., Swiergiel, J. J., Marshall, V. S., et al. (1998) Embryonic stem cell lines derived from human blastocysts. *Science* **282**, 1145–1147.
2. Reubinoff, B. E., Pera, M. F., Fong, C. Y., Trounson, A. and Bongso, A. (2000) Embryonic stem cell lines from human blastocysts: somatic differentiation in vitro. *Nat. Biotechnol.* **18**, 399–404.
3. Itskovitz-Eldor, J., Schuldiner, M., Karsenti, D., Eden, A., Yanuka, O., Amit, M., et al. (2000) Differentiation of human embryonic stem cells into embryoid bodies comprising the three embryonic germ layers. *Mol. Med.* **6**, 88–95.
4. Carpenter, M. K., Inokuma, M. S., Denham, J., Mujtaba, T., Chiu, C. P. and Rao, M. S. (2001) Enrichment of neurons and neural precursors from human embryonic stem cells. *Exp. Neurol.* **172**, 383–397.
5. Reubinoff, B. E., Itsykson, P., Turetsky, T., Pera, M. F., Reinhartz, E., Itzik, A., et al. (2001) Neural progenitors from human embryonic stem cells. *Nat. Biotechnol.* **19**, 1134–1140.
6. Schuldiner, M., Eiges, R., Eden, A., Yanuka, O., Itskovitz-Eldor, J., Goldstein, R. S., et al. (2001) Induced neuronal differentiation of human embryonic stem cells. *Brain Res.* **913**, 201–205.
7. Zhang, S. C., Wernig, M., Duncan, I. D., Brustle, O. and Thomson, J. A. (2001) In vitro differentiation of transplantable neural precursors from human embryonic stem cells. *Nat. Biotechnol.* **19**, 1129–1133.
8. Schulz, T. C., Palmarini, G. M., Noggle, S. A., Weiler, D. A., Mitalipova, M. M. and Condie, B. G. (2003) Directed neuronal differentiation of human embryonic stem cells. *BMC Neurosci.* **4**, 27.
9. Kehat, I., Kenyagin-Karsenti, D., Snir, M., Segev, H., Amit, M., Gepstein, A., et al. (2001) Human embryonic stem cells can differentiate into myocytes with structural and functional properties of cardiomyocytes. *J. Clin. Invest.* **108**, 407–414.
10. Kehat, I., Gepstein, A., Spira, A., Itskovitz-Eldor, J. and Gepstein, L. (2002) High-resolution electrophysiological assessment of human embryonic stem cell-derived cardiomyocytes: a novel in vitro model for the study of conduction. *Circ. Res.* **91**, 659–661.

11. Mummery, C., Ward, D., van den Brink, C. E., Bird, S. D., Doevendans, P. A., Opthof, T., et al. (2002) Cardiomyocyte differentiation of mouse and human embryonic stem cells. *J. Anat.* **200**, 233–242.

12. Xu, C., Police, S., Rao, N. and Carpenter, M. K. (2002) Characterization and enrichment of cardiomyocytes derived from human embryonic stem cells. *Circ. Res.* **91**, 501–508.

13. Mummery, C., Ward-van Oostwaard, D., Doevendans, P., Spijker, R., van den Brink, S., Hassink, R., et al. (2003) Differentiation of human embryonic stem cells to cardiomyocytes: role of coculture with visceral endoderm-like cells. *Circulation* **107**, 2733–2740.

14. He, J. Q., Ma, Y., Lee, Y., Thomson, J. A. and Kamp, T. J. (2003) Human embryonic stem cells develop into multiple types of cardiac myocytes: action potential characterization. *Circ. Res.* **93**, 32–39.

15. Levenberg, S., Golub, J. S., Amit, M., Itskovitz-Eldor, J. and Langer, R. (2002) Endothelial cells derived from human embryonic stem cells. *Proc. Natl. Acad. Sci. U. S. A.* **99**, 4391–4396.

16. Kaufman, D. S., Hanson, E. T., Lewis, R. L., Auerbach, R. and Thomson, J. A. (2001) Hematopoietic colony-forming cells derived from human embryonic stem cells. *Proc. Natl. Acad. Sci. U. S. A.* **98**, 10716–10721.

17. Chadwick, K., Wang, L., Li, L., Menendez, P., Murdoch, B., Rouleau, A., et al. (2003) Cytokines and BMP-4 promote hematopoietic differentiation of human embryonic stem cells. *Blood* **102**, 906–915.

18. Green, H., Easley, K. and Iuchi, S. (2003) Marker succession during the development of keratinocytes from cultured human embryonic stem cells. *Proc. Natl. Acad. Sci. U. S. A.* **100**, 15625–15630.

19. Sottile, V., Thomson, A. and McWhir, J. (2003) In vitro osteogenic differentiation of human ES cells. *Cloning Stem Cells* **5**, 149–155.

20. Rambhatla, L., Chiu, C. P., Kundu, P., Peng, Y. and Carpenter, M. K. (2003) Generation of hepatocyte-like cells from human embryonic stem cells. *Cell Transplant.* **12**, 1–11.

21. Lavon, N., Yanuka, O. and Benvenisty, N. (2004) Differentiation and isolation of hepatic-like cells from human embryonic stem cells. *Differentiation* **72**, 230–238.

22. Schuldiner, M., Itskovitz-Eldor, J. and Benvenisty, N. (2003) Selective ablation of human embryonic stem cells expressing a "suicide" gene. *Stem Cells* **21**, 257–265.

23. Lavon, N., Yanuka, O. and Benvenisty, N. (2006) The effect of over expression of Pdx1 and Foxa2 on the differentiation of human embryonic stem cells into pancreatic cells. *Stem Cells.* **24**, 1923–1930.

24. Kohler, G. and Milstein, C. (1975) Continuous cultures of fused cells secreting antibody of predefined specificity. *Nature* **256**, 495–497.

25. Yin, A. H., Miraglia, S., Zanjani, E. D., Almeida-Porada, G., Ogawa, M., Leary, A. G., et al. (1997) AC133, a novel marker for human hematopoietic stem and progenitor cells. *Blood* **90**, 5002–5012.
26. Weissman, I. L., Peacock, M. and Eltringham, J. R. (1973) Regional lymph node irradiation: effect on local and distant generation of antibody forming cells. *J. Immunol.* **110**, 1300–1306.

7

Analytical Methods for Cancer Stem Cells

Vinagolu K. Rajasekhar

Summary

The primary characteristics of adult stem cells are maintaining prolonged quiescence, ability to self-renew and plasticity to differentiate into multiple cell types. These properties are evolutionarily conserved from fruit fly to humans. Similar to normal tissue repair in organs, the stem cell concept is inherently impregnated in the etiology of cancer. Tumors contain a minor population of tumor-initiating cells, called "cancer stem cells." The cancer stem cells maintain some similarities in self-renewal and differentiation features of normal adult stem cells. Therefore, various methods developed originally for the analysis and characterization of adult stem cells are being extended to evaluate cancer stem cells. Relevant methods that are used generally across normal stem cells as well as cancer stem cells are summarized. Combination of two or more of these methods for validation of cancer stem cells appears to be a promising approach for the precise isolation and analysis of cancer stem cells.

Key Words: Cancer stem cells; Adult stem cells; Side-population assay; Sphere formation assay; Lineage labeling assay; Organ/tissue repopulation assay; Slow-cycling population assay; Self-renewal; Pluripotency.

1. Introduction

Several standardized assay protocols exist for the enrichment and analysis of adult stem cells. Many of these assays are extended to characterize cancer stem cells from lung *(1)*, brain *(2)*, prostate *(3)*, myeloma *(4)*, acute myeloid leukemia *(5)*, chronic myeloid leukemia *(6)* and acute lymphoblastic leukemia *(7)*, as these cells appear to share several of the primary characteristics of adult stem cells. In principle, adult stem cells are characterized by adopting the methods employed for analyzing embryonic stem (ES) cells. But, in contrast

From: *Methods in Molecular Biology, vol. 407: Stem Cell Assays*
Edited by: M. C. Vemuri © Humana Press, Totowa, NJ

to the ES cells, the isolation and purification of these cells are hampered by lack of universal adult stem cell phenotypic markers *(8)*. Another technically challenging problem is the scarcity of stem cell number in the pool of their origin, although there appears to be more than one such pool in a given organ.

2. Current Protocols for Enrichment and Analysis of Cancer Stem Cells

The following is a brief review of assays that are currently in use to prospectively identify, isolate, characterize, and validate the adult stem cells and cancer stem cells: (i) side-population (SP) assay, (ii) clonal sphere formation assay, (iii) slow-cycling population assay, and (iv) organ/tissue repopulation assay.

2.1. Side-Population Assay

Membrane pumps in stem cells effectively prevent influx of potentially harmful chemicals. For example, many stem cells are known to efficiently pump out DNA-binding dyes such as Hoechst 33342, which preferentially binds A–T-rich regions of DNA. The Hoechst 33342 is excited by ultraviolet light (395 nm) and emits blue light (450 nm). By exploiting this principle, the SP assay, a method developed first by Margaret Goodell *(9)* and now utilized for not only identifying the presence of putative stem cells and their analytical separation from rest of the neighboring cells, utilizes fluorescence-activated cell sorter (FACS). This analytical cell sorting procedure is popularly called flow cytometry as the assay relies on selective ability of stem cells to effectively efflux Hoechst 33342, a fluorescent and cell permeant DNA-binding dye. By virtue of this property, it is possible to distinguish putative stem cells as a sortable, Hoechst 33342 negative population of cells in a particular peripheral location on the FACS scan *(9)*. Therefore, this distinct pool of cells is referred to as "SP", which has been identified in several tissues of mammalian species *(10,11)*.

An SP is reported in a number of other organ tissues and respective tumors, including mammary gland *(12)*, skeletal muscle *(13)*, prostate *(14)*, pancreas *(15)*, lung *(16)*, liver *(17)*, skin *(18)*, testis *(19)*, heart *(20)*, epidermis *(18)*, gastrointestinal tissues *(21)*, retina *(22)*, and soon. Cancer cell lines also contained an SP *(23)*, which is considered to be crucial for tumor malignancy and drug resistance. Therefore, the SP phenotype is largely considered a universal stem cell biochemical marker.

The transmembrane adenosine 5′-triphosphate (ATP)-binding cassette (ABC) transporter MDR1 (P-glycoprotein/ABCB1) was originally speculated as responsible for this phenotype *(9)*. Later, another ABC transporter, namely Breast Cancer Resistance Protein (Bcrp1, which is also known as Abcg2 murine/ABCG2 human), is identified as largely responsible for the SP phenotype *(24)*. Physiologically both the transporters appear to protect stem cells against the cytotoxic actions of xenotoxins or endogenous compounds by transporting them out *(23)*. Overexpression of ABCG2 directly conferred the SP phenotype to bone marrow cells and caused a reduction in maturing progeny in both in vitro and transplantation assays. The hematopoietic stem cells (HSCs) are a subset of bone marrow cells that are capable of self-renewal and also multipotent in forming all types of blood cells. Following a loss of hematopoietic cells, the HSC that is assayable in the SP phenotype is mobilized to cycling, become, non-SP, and can reconstitute hematopoiesis when transplanted into lethally irradiated mice. Thus, the SP has been the most suitable marker to detect quiescent HSC in combination with some of the assayable cell surface markers *(25)*.

2.2. Clonal Sphere Formation Assay

The clonal sphere formation assay is another method utilized for isolating putative stem or progenitor cells, expanding and thereby purifying them from rest of their neighboring cells. This assay involves dissociation into single cell suspension of stem cell containing tissues and subsequent culture on non-adherent substrata in the presence of media supporting growth of stem cells till they form organized floating aggregates called spheres. Each sphere normally contains a minimum about six cells. However, there remain interesting queries like what is the exact nature of dividing cells within these spheres, how a certain number of cells within a sphere decide to shut down their further proliferation and maintain the sphere homeostasis, do these sphere-derived cells form secondary and tertiary spheres and if so what is the efficiency of re-sphere formation, if all the cells in spheres form re-spheres and can the sphere cells differentiate in situ, and so on are some of the questions that need to be addressed in detail.

Although the assay relies on the unique ability of stem cells to form spheres, it does not prove sphere clonality, which can be established by proving that one primary sphere forming cell gives rise to a single clonal sphere. The following types of evidences can establish the clonality of the spheres with the increasing order of stringency: (i) Increasing the primary cell number (1000–500,000 cells/ml) should increase the number of growth factor-dependent spheres linearly, (ii) when cells labeled with enhanced green fluorescent or yellow fluorescent (eGFP or YFP) reporter plasmids, and unlabeled control cells are mixed, the resulting spheres should be mutually exclusive (a situation that can be regarded as competitive clonality assay), (iii) when immobilized as individual cells on matrigel or methylcellulose, the primary cells should generate the same growth factor-dependent spheres, (iv) when re-plated on any of the above adherent or non-adherent substrata, single cells from the primary spheres should generate secondary spheres that express the same functional cell markers that could be evaluated by flow cytometry or immunohistochemistry, and 5) if the primary and secondary spheres can be induced to differentiate to specific lineages *(26)*.

The growth capacity of the sphere-generating cells can be estimated by measuring the size of the spheres *(27)*. But, little is known with regard to the developmental signaling pathways impinging on the clonogenic sphere formation. Present day availability of potent small molecular inhibitors specifically intersecting different steps in the crucial signaling pathways can be exploited to set up a careful drug screening strategy with this clonal sphere formation assays. Lastly, the practical advantage of this method is the ability to cryopreserve the spheres for future analysis, which facilitates the development of sphere banks. The sphere formation assay is utilized to identify adult stem cells and cancer stem cells from a number of tissues, including mammary gland *(28)*, brain *(29)*, skeletal muscle *(30)*, pancreas *(31)*, retina *(32)*, skin *(33)*, and human melanoma *(34)*.

2.3. Slow-Cycling Population Assay

Detection of proliferating cells based on a nucleotide label such as 5-bromo-2-deoxyuridine (BrdU) or tritiated (^{3}H) thymidine incorporation and determination of phenotype by immunofluorescence labeling are conventional methods for recognizing stem and progenitor cell populations in adult tissues *(35)*. Because of their characteristic feature of slow-cycling nature of organ-specific adult stem cells, this assay has been popularly called as slow-cycling population assay method. The principle behind this assay is that cells with slow-cycling time can be distinguished by retention of the label, which is incorporated into the DNA of cells during DNA synthesis from the cells with rapid cycling time. Following administration of a pulse of a label, the cells are monitored for long periods of chase. Thus, only the slow-cycling cells are expected to retain a concentration of label sufficiently high to allow their detection with the anti-BrdU antibody staining or radioactive label. Therefore, adult organ-specific stem cells can be identified as "label-retaining cells or slow-cycling cells." Almost 75% of HSCs are known to be quiescent in normal bone marrow and can retain BrdU for relatively long periods, indicating that

HSC are slow-cycling cells *(36)*. As the BrdU labeling is specific for S-phase of cell cycle, additional labeling with other antibodies such as phosphorylated histone H3 which is a marker for cells in M phase or T187-phospho-p27(KIP1), can be supplemented for distinction of stem cells from the proliferating cells *(37)*. The slow-cycling population assay is utilized to identify adult stem cells and cancer stem cells from mammary gland *(38)*, skeletal muscle *(39)*, pancreas *(40)*, lung *(41)*, heart *(42)*, epidermis *(43)*, epithelium *(44)*, prostate *(45)*, and human hair follicle bulge cells *(46)*.

However, various technical issues impede the accuracy in the incorporation of BrdU or its detection in the cells that retain it as it is (i) difficult to demonstrate that a pluripotent cell in vitro is a slow cycling one in vivo, because of the destructive nature of BrdU detection procedure; (ii) because the BrdU incorporation occurs only during the S phase of cell cycle, unless all cells are synchronous it is not expected that all the stem cells will be equally labeled; and (iii) the label retaining is dependent on the length of the cell cycle, and the length of cell cycle in stem cells appear to alter as the organ matures *(47)*. Thus, quantification of stem cells through this assay will require their identification, isolation, and characterization using non-invasive methods such as employing specific surface markers of stem cells.

2.4. Lineage Labeling Assay

Lineage labeling assay was originally developed for understanding the development at the cellular level such as identification of cellular origin, lineage relationships, division patterns, migration streams, and differentiation *(48)*. In short, cells will be labeled and their fate is traced in vivo. Labeling of cells is carried out through various approaches such as genetic fate mapping, development of genetic mosaics by mitotic recombination *(49)* or Cre-recombinase-mediated transgenesis *(50)*, chromosome loss *(51)*, interspecies bio-assays *(52)*, replication-defective retroviral vectors *(53)*, and/or involvement of biochemical approaches such as injecting activatable lineage marker fluorescent dyes *(54)* or enzymes directly into target cells and transplantation of labeled donor cells into unlabeled hosts *(55)*. Mapping of destination of embryonic neural crest and mesodermal stem cells utilizing a Cre-recombinase-mediated transgenesis *(56)* and migration of myoblasts across basal lamina during skeletal muscle development *(57)* are representative examples of this assay. Furthermore, by combining with whole-mount labeling, lineage labeling assay revealed hither to unidentified spatial relationships between stem cells and its transit amplifying progeny in the basal layer of human epidermis *(58)*. Due to several inherent limitations with these techniques such as injury to the cells, cell separation from native environment, weak and/or transient expression of markers, unrestricted clonality, and so on a modified lineage tracing method was developed to follow the fate of central nervous system midline progenitors *(48)*. In this case, target cells are non-invasively labeled in their own natural environment with a lipophilic fluorescent tracer (diI) and the development of a progenitor cell as well as composition of their lineages can be followed. It remains to be seen if the combination of these techniques can be extended to cancer stem cells and also to metastasis initiating stem/progenitor cells.

2.5. Organ/tissue Repopulation Assay

Stem cells can be limitless sources for tissue or organ regeneration. The organ/tissue repopulation assay is a method originally developed based on intrinsic self-renewal property of adult stem cells and on expectation that given a feasible cellular context such stem cells would functionally repopulate tissue of their origin. These properties make stem cell transplantation therapies feasible. A single HSC transplanted into an irradiated or immunocompromised mice could regenerate entire hematopoietic and lymphoid systems *(59)*. This long-term multi-lineage repopulating activity was the most consistent marker for adult stem cells such as (HSCs). Based on competitive in vivo prolif-eration of fetal and adult HSCs in lethally irradiated mice for repopulation *(60)*, repopulating units (RUs) denoting the amount of repopulating activity *(61)* and competitive repopulating units (CRUs) denoting the number of stem cells *(62)* have been utilized to quantitatively express stem cell activity *(63)*. A mean activity of stem cells (MAS) can therefore be determined (MAS = RU/CRU), when different stem cell population qualities are compared *(64)*. Recently, the role of fibroblast growth factors (FGF)-1, FGF-2, and FGF-4 have been realized to preserve the long-term repopulating ability of HSC in serum-free cultures, although stem cell niche-dependent cell–cell contact and associated short-range interactions appear to play a significant role during in vivo repopulation *(65)*. The long-term repopulating HSCs give rise to short-term repopulating HSC, whose normal function appears to depend on stem cell leukemia factor gene *(66)*.

3. Future Outlook

It has been a practice to utilize more than one assay to identify and charac-terize a stem cell population. For example, label-retaining undifferentiated large light cells (distinguishable from that of small light cells) in breast epithelium expresses the putative stem cell markers, $p21^{CIP1}$, Musashi-1, and Keratin 19, displays a "SP" phenotype and is enriched for steroid receptor expression

(67). Recently, mouse ES cells were stably transfected with the enhanced green fluorescent protein expressed under the control of an Oct4 promoter and successfully isolated residual undifferentiated stem cells from the embryoid body cultures by fluorescent tracking *(68)*. The molecular basis for this experimental design was that different pluripotent stem cells including the ES cells, embryonal carcinoma cells, and primordial germ cells express the Oct4, which is a POU-domain transcription factor family member as the pluripotency marker *(69)*. Such approaches may be extended to isolate the residual undifferentiated cancer stem cells from the differentiated tumor tissues. Also, retroviral gene transduction of circulating progenitor cells in metastatic breast cancer patients has long been realized *(70)*. Thus, identifying and cataloging distinct marker proteins in various adult stem cells with more than one of these assays and contrasting them with the respective tumor cell marker patterns will be an invaluable asset for deriving pure populations of cancer stem cells by the above-envisioned approaches.

Recently, Tomishima et al. *(71)*, demonstrated a powerful method to identify and purify stem cells and their differentiated progeny based on gene transcription. The approach included modification of bacterial artificial chromosomes (BACs) from a previously constructed green fluorescent protein (GFP) transcriptional fusion library (Gene Expression Nervous System Atlas [GENSAT]) for use in ES cells, and generation of multiple BAC transgenic ES cell lines that fluoresce upon activation of target genes. GFP expressing BACs can be selected from the library of Gene Expression Nervous System Atlas [GENSAT] that consists of over 400 genes and continues to grow. Through this approach, time-lapse imaging can be utilized to observe differentiation, and cells that express the gene of interest can be separated for more precise gene expression or further transplantation studies. Another application for these reporter stem cells is in high throughput screens. Such screens could identify small molecule inhibitors or genes that affect signaling pathways. This method has a potential to be utilized for isolation and characterization of cancer stem cells. Furthermore, by virtue of the in-built reporter system, this method can be exploited in high throughput screens for identifying small molecule inhibitors or genes that affect various signaling pathways in cancer stem cells.

Acknowledgments

I thank Drs. Martin Begemann, Julie Cerrato, and Yvette Chin for helpful discussions. Birgit L Baur is appreciated for help during preparation of the manuscript.

References

1. Kim, C. F., Jackson, E. L., Woolfenden, A. E., Lawrence, S., Babar, I., Vogel, S., Crowley, D., Bronson, R. T. and Jacks, T. (2005) Identification of bronchioalveolar stem cells in normal lung and lung cancer. *Cell* **121**, 823–35.
2. Singh, S. K., Clarke, I. D., Terasaki, M., Bonn, V. E., Hawkins, C., Squire, J. and Dirks, P. B. (2003) Identification of a cancer stem cell in human brain tumors. *Cancer Res.* **63**, 5821–8.
3. Collins, A. T., Berry, P. A., Hyde, C., Stower, M. J. and Maitland, N. J. (2005) Prospective identification of tumorigenic prostate cancer stem cells. *Cancer Res.* **65**, 10946–51.
4. Matsui, W., Huff, C. A., Wang, Q., Malehorn, M. T., Barber, J., Tanhehco, Y., Smith, B. D., Civin, C. I. and Jones, R. J. (2004) Characterization of clonogenic multiple myeloma cells. *Blood* **103**, 2332–6.
5. Lapidot, T., Sirard, C., Vormoor, J., Murdoch, B., Hoang, T., Caceres-Cortes, J., Minden, M., Paterson, B., Caligiuri, M. A. and Dick, J. E. (1994) A cell initiating human acute myeloid leukaemia after transplantation into SCID mice. *Nature* **367**, 645–8.
6. Bedi, A., Zehnbauer, B. A., Collector, M. I., Barber, J. P., Zicha, M. S., Sharkis, S. J. and Jones, R. J. (1993) BCR-ABL gene rearrangement and expression of primitive hematopoietic progenitors in chronic myeloid leukemia. *Blood* **81**, 2898–902.
7. George, A. A., Franklin, J., Kerkof, K., Shah, A. J., Price, M., Tsark, E., Bockstoce, D., Yao, D., Hart, N., Carcich, S., Parkman, R., Crooks, G. M. and Weinberg, K. (2001) Detection of leukemic cells in the CD34(+)CD38(−) bone marrow progenitor population in children with acute lymphoblastic leukemia. *Blood* **97**, 3925–30.
8. Blau, H. M., Brazelton, T. R. and Weimann, J. M. (2001) The evolving concept of a stem cell: entity or function?. *Cell* **105**, 829–41.
9. Goodell, M. A., Brose, K., Paradis, G., Conner, A. S. and Mulligan, R. C. (1996) Isolation and functional properties of murine hematopoietic stem cells that are replicating in vivo. *J Exp. Med* **183**, 1797–806.
10. Hirschmann-Jax, C., Foster, A. E., Wulf, G. G., Goodell, M. A. and Brenner, M. K. (2005) A distinct "side population" of cells in human tumor cells: implications for tumor biology and therapy. *Cell Cycle 4* 203–5.
11. Hirschmann-Jax, C., Foster, A. E., Wulf, G. G., Nuchtern, J. G., Jax, T. W., Gobel, U., Goodell, M. A. and Brenner, M. K. (2004) A distinct "side population" of cells with high drug efflux capacity in human tumor cells. *Proc. Natl. Acad. Sci. U.S.A.* **101**, 14228–33.
12. Lynch, M. D., Cariati, M. and Purushotham, A. D. (2006) Breast cancer, stem cells and prospects for therapy. *Breast Cancer Res* **8**, 211.
13. Kawiak, J., Brzoska, E., Grabowska, I., Hoser, G., Streminska, W., Wasilewska, D., Machaj, E. K., Pojda, Z. and Moraczewski, J. (2006) Contribution of stem cells to skeletal muscle regeneration. *Folia Histochem Cytobiol* **44**, 75–9.

14. Bhatt, R. I., Brown, M. D., Hart, C. A., Gilmore, P., Ramani, V. A., George, N. J. and Clarke, N. W. (2003) Novel method for the isolation and characterisation of the putative prostatic stem cell. *Cytometry A* **54**, 89–99.

15. Zhang, L., Hu, J., Hong, T. P., Liu, Y. N., Wu, Y. H. and Li, L. S. (2005) Monoclonal side population progenitors isolated from human fetal pancreas. *Biochem Biophys Res Commun* **333**, 603–8.

16. Majka, S. M., Beutz, M. A., Hagen, M., Izzo, A. A., Voelkel, N. and Helm, K. M. (2005) Identification of novel resident pulmonary stem cells: form and function of the lung side population. *Stem Cells* **23**, 1073–81.

17. Chiba, T., Kita, K., Zheng, Y. W., Yokosuka, O., Saisho, H., Iwama, A., Nakauchi, H. and Taniguchi, H. (2006) Side population purified from hepatocellular carcinoma cells harbors cancer stem cell-like properties. *Hepatology* **44**, 240–51.

18. Yano, S., Ito, Y., Fujimoto, M., Hamazaki, T. S., Tamaki, K. and Okochi, H. (2005) Characterization and localization of side population cells in mouse skin. *Stem Cells* **23**, 834–41.

19. Kubota, H., Avarbock, M. R. and Brinster, R. L. (2003) Spermatogonial stem cells share some, but not all, phenotypic and functional characteristics with other stem cells. *Proc Natl Acad Sci USA*. **100**, 6487–92.

20. Tomita, Y., Matsumura, K., Wakamatsu, Y., Matsuzaki, Y., Shibuya, I., Kawaguchi, H., Ieda, M., Kanakubo, S., Shimazaki, T., Ogawa, S., Osumi, N., Okano, H. and Fukuda, K. (2005) Cardiac neural crest cells contribute to the dormant multipotent stem cell in the mammalian heart. *J Cell Biol* **170**, 1135–46.

21. Haraguchi, N., Inoue, H., Tanaka, F., Mimori, K., Utsunomiya, T., Sasaki, A. and Mori, M. (2006) Cancer stem cells in human gastrointestinal cancers. *Hum Cell* **19**, 24–9.

22. Bhattacharya, S., Jackson, J. D., Das, A. V., Thoreson, W. B., Kuszynski, C., James, J., Joshi, S. and Ahmad, I. (2003) Direct identification and enrichment of retinal stem cells/progenitors by Hoechst dye efflux assay. *Invest Ophthalmol Vis Sci* **44**, 2764–73.

23. Jonker, J. W., Freeman, J., Bolscher, E., Musters, S., Alvi, A. J., Titley, I., Schinkel, A. H. and Dale, T. C. (2005) Contribution of the ABC transporters Bcrp1 and Mdr1a/1b to the side population phenotype in mammary gland and bone marrow of mice. *Stem Cells* **23**, 1059–65.

24. Zhou, S., Schuetz, J. D., Bunting, K. D., Colapietro, A. M., Sampath, J., Morris, J. J., Lagutina, I., Grosveld, G. C., Osawa, M., Nakauchi, H. and Sorrentino, B. P. (2001) The ABC transporter Bcrp1/ABCG2 is expressed in a wide variety of stem cells and is a molecular determinant of the side-population phenotype. *Nat. Med* **7**, 1028–34.

25. Suda, T., Arai, F. and Shimmura, S. (2005) Regulation of stem cells in the niche. *Cornea* **24**, S12–S17.

26. Dontu, G. and Wicha, M. S. (2005) Survival of mammary stem cells in suspension culture: implications for stem cell biology and neoplasia. *J Mammary Gland Biol Neoplasia* **10**, 75–86.

27. Charrier, C., Coronas, V., Fombonne, J., Roger, M., Jean, A., Krantic, S. and Moyse, E. (2006) Characterization of neural stem cells in the dorsal vagal complex of adult rat by in vivo proliferation labeling and in vitro neurosphere assay. *Neuroscience* **138**, 5–16.

28. Ayyanan, A., Civenni, G., Ciarloni, L., Morel, C., Mueller, N., Lefort, K., Mandinova, A., Raffoul, W., Fiche, M., Dotto, G. P. and Brisken, C. (2006) Increased Wnt signaling triggers oncogenic conversion of human breast epithelial cells by a Notch-dependent mechanism. *Proc Natl Acad Sci USA*. **103**, 3799–804.

29. Singh, S. K., Hawkins, C., Clarke, I. D., Squire, J. A., Bayani, J., Hide, T., Henkelman, R. M., Cusimano, M. D. and Dirks, P. B. (2004) Identification of human brain tumour initiating cells. *Nature* **432**, 396–401.

30. Vourc'h, P., Lacar, B., Mignon, L., Lucas, P. A., Young, H. E. and Chesselet, M. F. (2005) Effect of neurturin on multipotent cells isolated from the adult skeletal muscle. *Biochem Biophys Res Commun* **332**, 215–23.

31. Lehnert, L., Lerch, M. M., Hirai, Y., Kruse, M. L., Schmiegel, W. and Kalthoff, H. (2001) Autocrine stimulation of human pancreatic duct-like development by soluble isoforms of epimorphin in vitro, *J Cell Biol* **152**, 911–22.

32. Sun, G., Asami, M., Ohta, H., Kosaka, J. and Kosaka, M. (2006) Retinal stem/progenitor properties of iris pigment epithelial cells. *Dev Biol* **289**, 243–52.

33. Toma, J. G., McKenzie, I. A., Bagli, D. and Miller, F. D. (2005) Isolation and characterization of multipotent skin-derived precursors from human skin. *Stem Cells* **23**, 727–37.

34. Fang, D., Nguyen, T. K., Leishear, K., Finko, R., Kulp, A. N., Hotz, S., Van Belle, P. A., Xu, X., Elder, D. E. and Herlyn, M. (2005) A tumorigenic subpopulation with stem cell properties in melanomas. *Cancer Res* **65**, 9328–37.

35. Braun, K. M. and Watt, F. M. (2004) Epidermal label-retaining cells: background and recent applications. *J Investig Dermatol Symp Proc* **9**, 196–201.

36. Paterson, R. F., Ulbright, T. M., MacLennan, G. T., Zhang, S., Pan, C. X., Sweeney, C. J., Moore, C. R., Foster, R. S., Koch, M. O., Eble, J. N. and Cheng, L. (2003) Molecular genetic alterations in the laser-capture-microdissected stroma adjacent to bladder carcinoma. *Cancer* **98**, 1830–6.

37. Kase, S., Yoshida, K., Ohgami, K., Shiratori, K., Ohno, S. and Nakayama, K. I. (2006) Phosphorylation of p27(KIP1) in the mitotic cells of the corneal epithelium. *Curr Eye Res* **31**, 307–12.

38. Smith, G. H. (2005) Label-retaining epithelial cells in mouse mammary gland divide asymmetrically and retain their template DNA strands. *Development* **132**, 681–7.

39. Shinin, V., Gayraud-Morel, B., Gomes, D. and Tajbakhsh, S. (2006) Asymmetric division and cosegregation of template DNA strands in adult muscle satellite cells. *Nat Cell Biol* **8**, 677–82.

40. Duvillie, B., Attali, M., Aiello, V., Quemeneur, E. and Scharfmann, R. (2003) Label-retaining cells in the rat pancreas: location and differentiation potential in vitro. *Diabetes* **52**, 2035–42.

41. Giangreco, A., Reynolds, S. D. and Stripp, B. R. (2002) Terminal bronchioles harbor a unique airway stem cell population that localizes to the bronchoalveolar duct junction. *Am J Pathol* **161**, 173–82.
42. Urbanek, K., Cesselli, D., Rota, M., Nascimbene, A., De Angelis, A., Hosoda, T., Bearzi, C., Boni, A., Bolli, R., Kajstura, J., Anversa, P. and Leri, A. (2006) From the cover: stem cell niches in the adult mouse heart. *Proc Natl Acad Sci USA.* **103**, 9226–31.
43. Wu, W. Y. and Morris, R. J. (2005) In vivo labeling and analysis of epidermal stem cells. *Methods Mol Biol* **289**, 73–8.
44. Bickenbach, J. R. (2005) Isolation, characterization, and culture of epithelial stem cells. *Methods Mol Biol* **289**, 97–102.
45. Patrawala, L., Calhoun, T., Schneider-Broussard, R., Li, H., Bhatia, B., Tang, S., Reilly, J. G., Chandra, D., Zhou, J., Claypool, K., Coghlan, L. and Tang, D. G. (2006) Highly purified CD44+ prostate cancer cells from xenograft human tumors are enriched in tumorigenic and metastatic progenitor cells. *Oncogene* **25**, 1696–708.
46. Ohyama, M., Terunuma, A., Tock, C. L., Radonovich, M. F., Pise-Masison, C. A., Hopping, S. B., Brady, J. N., Udey, M. C. and Vogel, J. C. (2006) Characterization and isolation of stem cell-enriched human hair follicle bulge cells. *J. Clin. Invest.* **116**, 249–60.
47. Oliver, J. A., Maarouf, O., Cheema, F. H., Martens, T. P. and Al-Awqati, Q. (2004) The renal papilla is a niche for adult kidney stem cells. *J. Clin. Invest.* **114**, 795–804.
48. Bossing, T. and Technau, G. M. (1994) The fate of the CNS midline progenitors in Drosophila as revealed by a new method for single cell labelling. *Development* **120**, 1895–906.
49. Garcia-Bellido, A., Ripoll, P. and Morata, G. (1973) Developmental compartmentalisation of the wing disk of Drosophila. *Nat New Biol* **245**, 251–3.
50. Ganat, Y. M., Silbereis, J., Cave, C., Ngu, H., Anderson, G. M., Ohkubo, Y., Ment, L. R. and Vaccarino, F. M. (2006) Early postnatal astroglial cells produce multilineage precursors and neural stem cells in vivo. *J Neurosci* **26**, 8609–21.
51. Janning, W. (1978) Gynandromorph fate maps in Drosophila. *Results Probl Cell Differ* **9**, 1–28.
52. Illmensee, K., Levanduski, M. and Zavos, P. M. (2006) Evaluation of the embryonic preimplantation potential of human adult somatic cells via an embryo interspecies bioassay using bovine oocytes. *Fertil Steril* **85 Suppl 1**, 1248–60.
53. Blau, H. M. and Hughes, S. M. (1990) Retroviral lineage markers for assessing myoblast fate in vivo. *Adv Exp Med Biol* **280**, 201–3.
54. Vincent, J. P. and O'Farrell, P. H. (1992) The state of engrailed expression is not clonally transmitted during early Drosophila development. *Cell* **68**, 923–31.
55. Technau, K., Renkl, A., Norgauer, J. and Ziemer, M. (2005) Necrolytic migratory erythema with myelodysplastic syndrome without glucagonoma. *Eur J Dermatol* **15**, 110–2.

56. Matsuoka, T., Ahlberg, P. E., Kessaris, N., Iannarelli, P., Dennehy, U., Richardson, W. D., McMahon, A. P. and Koentges, G. (2005) Neural crest origins of the neck and shoulder. *Nature* **436**, 347–55.

57. Hughes, S. M. and Blau, H. M. (1990) Migration of myoblasts across basal lamina during skeletal muscle development. *Nature* **345**, 350–3.

58. Jensen, U. B., Lowell, S. and Watt, F. M. (1999) The spatial relationship between stem cells and their progeny in the basal layer of human epidermis: a new view based on whole-mount labelling and lineage analysis. *Development* **126**, 2409–18.

59. Osawa, M., Hanada, K., Hamada, H. and Nakauchi, H. (1996) Long-term lympho-hematopoietic reconstitution by a single CD34-low/negative hematopoietic stem cell. *Science* **273**, 242–5.

60. Micklem, H. S., Ford, C. E., Evans, E. P., Ogden, D. A. and Papworth, D. S. (1972) Competitive in vivo proliferation of foetal and adult haematopoietic cells in lethally irradiated mice. *J Cell Physiol* **79**, 293–8.

61. Harrison, D. E., Jordan, C. T., Zhong, R. K. and Astle, C. M. (1993) Primitive hemopoietic stem cells: direct assay of most productive populations by competitive repopulation with simple binomial, correlation and covariance calculations. *Exp Hematol* **21**, 206–19.

62. Szilvassy, S. J., Humphries, R. K., Lansdorp, P. M., Eaves, A. C. and Eaves, C. J. (1990) Quantitative assay for totipotent reconstituting hematopoietic stem cells by a competitive repopulation strategy. *Proc Natl Acad Sci USA.* **87**, 8736–40.

63. Ema, H., Sudo, K., Seita, J., Matsubara, A., Morita, Y., Osawa, M., Takatsu, K., Takaki, S. and Nakauchi, H. (2005) Quantification of self-renewal capacity in single hematopoietic stem cells from normal and Lnk-deficient mice. *Dev Cell* **8** 907–14.

64. Ema, H. and Nakauchi, H. (2000) Expansion of hematopoietic stem cells in the developing liver of a mouse embryo. *Blood* **95**, 2284–8.

65. Yeoh, J. S., van Os, R., Weersing, E., Ausema, A., Dontje, B., Vellenga, E. and de Haan, G. (2006) Fibroblast growth factor-1 and -2 preserve long-term repopulating ability of hematopoietic stem cells in serum-free cultures. *Stem Cells* **24**, 1564–72.

66. Curtis, D. J., Hall, M. A., Van Stekelenburg, L. J., Robb, L., Jane, S. M. and Begley, C. G. (2004) SCL is required for normal function of short-term repopulating hematopoietic stem cells. *Blood* **103**, 3342–8.

67. Clarke, R. B., Spence, K., Anderson, E., Howell, A., Okano, H. and Potten, C. S (2005) A putative human breast stem cell population is enriched for steroid receptor-positive cells. *Dev Biol* **277**, 443–56.

68. Ensenat-Waser, R., Santana, A., Vicente-Salar, N., Cigudosa, J. C., Roche, E., Soria, B. and Reig, J. A. (2006) Isolation and characterization of residual undiffer-entiated mouse embryonic stem cells from embryoid body cultures by fluorescence tracking. *In Vitro Cell Dev Biol Anim* **42**, 115–23.

69. Nichols, J., Zevnik, B., Anastassiadis, K., Niwa, H., Klewe-Nebenius, D., Chambers, I., Scholer, H. and Smith, A. (1998) Formation of pluripotent stem cells

in the mammalian embryo depends on the POU transcription factor Oct4. *Cell* **95**, 379–91.

70. Coles, R. E., Boyle, T. J., Kurtzberg, J., Stewart, A., Peters, W. P. and Lyerly, H. K. (1993) Retroviral gene transduction of circulating progenitor cells in patients with metastatic breast cancer. *Surg Oncol* **2**, 1–6.
71. Tonishima, M.J., Hadjantonakis, A.K., Gong, S. and Studer, L. (2007) *Stem Cells* **25**, 39–45.

8

Micro RNA Profiling

An Easy and Rapid Method to Screen and Characterize Stem Cell Populations

Uma Lakshmipathy, Brad Love, Christopher Adams, Bhaskar Thyagarajan, and Jonathan D. Chesnut

Summary

MicroRNAs (miRNAs) are small regulatory RNAs varying in length between 20 and 24 nucleotides. They are thought to play a key role during development by negative gene regulation at the post-transcriptional level. Recent studies using quantitative polymerase chain reaction (QPCR) and northern blot analysis have reported the presence of several miRNA unique to specific cell types. The NCode™ multispecies miRNA array provides a means for simultaneously profiling the expression patterns of hundreds of known miRNAs in a given cell type or biological sample. Using this method, miRNA expression patterns in embryonic and adult stem cell lines can be characterized and compared with each other. The accuracy of NCode™ miRNA array data can be further confirmed by QPCR analysis of putative array hits. This array-based screening platform is a fast and easy to use analytical tool that allows one to asses the state of stem cell lines following multiple passages in culture as well as a discovery tool that eliminates the need to screen large numbers of candidate regulatory miRNAs by northern blot or PCR. In this chapter, we describe in detail the method to carry out miRNA array analysis in human embryonal carcinoma cells and confirm the array results using QPCR.

Key Words: MicroRNA; Stem cell populations; Characterization; Array.

From: *Methods in Molecular Biology, vol. 407: Stem Cell Assays*
Edited by: M. C. Vemuri © Humana Press, Totowa, NJ

1. Introduction

MicroRNAs (miRNAs) are small non-coding RNA molecules that regulate gene expression across a wide range of cell types and species *(1)*. They were first discovered in the nematode *Caenorhabaitis elegans (2,3)* and have since been shown in flies, plants, and mammals *(4–6)*. They are approximately 21–23 bases in length and are derived from precursors that are typically 70–110 bases long *(1)*. miRNAs regulate gene expression by binding to sites within the 3'-UTR of mRNAs by either inducing degradation of the mRNA or blocking its translation *(1,7)*. They are distinct from endogenous small interfering RNAs (siRNAs) as they are derived from genomic loci that are spatially separated from the gene they target, whereas siRNAs are derived from mRNAs, transposons, or viruses *(8)*.

MicroRNAs have been shown to play an important role in mammalian development as evidenced by studies in mammalian stem cells. It has been shown that they are expressed in a tissue-specific and developmental stage-specific manner *(6,9–12)*. It was therefore not surprising that Houbaviy et al. *(13)* discovered a set of miRNAs that were specifically expressed in mouse embryonic stem cells. Although the targets for these miRNAs have not been identified, they are believed to play a role in differentiation of these cells into various lineages. More recently, human embryonic stem cell-specific miRNAs have been identified *(14)* and shown to be highly related to each other in that many are expressed as polycistronic primary transcripts. This group also showed that differentiation of stem cells leads to rapid down regulation of embryonic stem cell (ESC)-specific miRNAs, suggesting that they play an important role in the maintenance of the pluripotent state of human ESCs (hESCs). Although it is a general notion that miRNAs play an important role in the development and differentiation, there have been only a few published studies that document such function. A direct involvement of miRNAs in the development of mouse neural lineages *(15)* and in the differentiation of hematopoietic stem cells into skeletal muscle and skin cells *(16–18)* has recently been reported. It is therefore crucial that the role of these molecules be elucidated in greater detail.

To further our understanding and delineate the role of miRNAs in stem cells, it is imperative that we understand their expression patterns in stem cells and in cells derived at different stages of differentiation. In this study, we describe a protocol designed to determine global expression patterns of miRNAs and compare them across different conditions of growth and differentiation by using miRNA arrays. These arrays allow the monitoring of the expression of all known miRNAs and thereby determine species that are differentially expressed between different conditions. Knowledge gained from these experiments can be validated using more conventional means, and candidate miRNA species could be studied in further detail using either gain or loss of function studies.

2. Materials

2.1. Cell Culture

1. Culture media for human embryonal carcinoma (hEC) cells: Dulbecco's Modified Eagle's Medium (DMEM) high glucose (cat. no. 10310-021, Invitrogen, Carlsbad, CA, USA) constituted with 10% ESC qualified Fetal bovine serum (FBS) (cat. no. 10439-024, Invitrogen) and Glutamax (cat. no. 35050-061, Invitrogen).
2. Trypsin–ethylenediamine tetra acetic acid (EDTA) (0.25%) (cat. no. 25200-056, Invitrogen) for harvesting and dissociating 2102Ep hESCs (kindly provided by Dr Peter Andrews, University of Sheffield, UK).
3. Trypan blue (cat. no. 15250-061, Invitrogen).
4. Sterile cell scrapers (cat. no. 15621-005, VWR, Westchesler, PA, USA) for harvesting Ntera2 cl.D1 hESCs (cat. no. CRL1973, ATCC, Manassas, VA, USA).

2.2. RNA and miRNA Isolation

1. TRIzol reagent (cat. no. 15596-026, Invitrogen).
2. Micro-to Midi total RNA purification system (cat. no. 12183-018, Invitrogen).
3. Purelink™ miRNA isolation kit (cat. no. K1570-01, Invitrogen).
4. DNase I, Amplification grade (cat. no. 18068-015, Invitrogen).
5. Phenol: chloroform: isoamylalcohol, St. Louis, MO, USA (cat. no. 15593-031, Invitrogen).
6. Chloroform (cat. no. C2432, Sigma (25:24:1, v/v)).
7. Absolute Ethanol, 75% ethanol.
8. Diethyl pyrocarbonate (DEPC)-treated water (cat. no. 750023, Invitrogen).
9. NovexR15% TBE-Urea denaturing gel (cat. no. EC6885BOX, Invitrogen).

2.3. miRNA Array/QPCR

1. Ncode™ miRNA labeling system (cat. no. MIRLS-20, Invitrogen): Kit has components for 20 reaction or 10 dye swap experiments. Each kit has all the materials required for the procedures described in **subheadings 3.4.1** to **3.4.3**. Additional required materials are listed below.
2. Ncode™ multispecies miRNA micro array (cat. no. MIRA-05, Invitrogen): Contains 5 microarray slides. Each slide has probes targeting all the known miRNAs—human, mouse, rat, drosophila, *C. elegans*, and zebra fish. In addition, several human predicted as well as NCode controls are present on each array slide *(19)*.
3. NCode™ miRNA QPCR kit (cat. no. MIRQ-100 and MIRC-10, Invitrogen).
4. Hybridization chamber (cat. no. 2551, Corning, Corning, NY, USA).
5. Hybridization ovens set to 55 °C and 62 °C.
6. Ultrapure™ EDTA (0.5 M), pH 8.0 (cat. no. 15575-020, Invitrogen).

7. Glass Lifter cover slips (24 × 60 mm, Erie Scientific, Portsmouth, NH, USA).
8. Microarray wash solution 1: Prepare 2× sodium dodecyl sulfate (SSC), 0.2% SDS-sodium Chloride (SDS) solution by diluting Ultrapure™ 20× SSC (cat. no. 15557-044, Invitrogen), and Ultrapure™ 10% SDS solution (cat. no. 15553-027, Invitrogen) and store at room temperature. Just before start of the experiment, prewarm to 42 °C for the wash step after the first hybridization and 55 °C for the wash after the second hybridization step.
9. Microarray wash solution 2: Prepare 2× SSC solution and store at room temperature.
10. Microarray wash solution 3: Prepare 0.2× SSC and store at room temperature.
11. Glass Coplin jars with lid (cat. no. 100500-230, VWR).
12. Microcentrifuge.
13. Microarray reader.

3. Methods

3.1. Experimental Design

Comparison of two samples can be carried out by differential labeling using the NCode miRNA labeling kit using different methods:

1. Homotypic labeling: In this method, sample A is labeled with both green (Alexa 3) and red (Alexa 5) dyes and are hybridized to a single miRNA array. The data are then compared with data obtained from another miRNA array of sample B hybridized with dual color in a similar way. This method is not optimal, because chip to chip variability exists leading to potentially inconsistent results.
2. Heterotypic labeling: This method relies on two samples A and B labeled with two different dyes and hybridized on the same chip. For example, sample A labeled with Alexa 3 is co-hybridized with sample B labeled with Alexa 5. Such a direct comparison of two samples on the same chip minimizes any chip to chip variability. This method however does not rule out dye bias or tendency of labeling by one particular dye to be more efficient than the other in a given experiment.
3. Heterotypic dye swap: This is the ideal method for comparing two samples. Here, both the chip variation and dye bias are addressed and results obtained can be normalized to yield results with more statistical validity. In this method, sample A is labeled with the first dye (Alexa 3) and sample B with the second dye (Alexa 5) on the first chip and on the second chip sample A is labeled with Alexa 5 and sample B with Alexa 3.

3.2. Collection of Cells

1. Culture hEC cells, 2102Ep and Ntera2, in DMEM containing 10% FBS and Glutamax at moderate to high density.
2. Remove growth media and wash cells twice with 1× PBS.

3. Add 2 ml TyrpLE and incubate at room temperature for 3–5 min. Gently tap the culture dish to loosen attachment and harvest cells into a 15-ml conical tube.

4. Spin at 800 g and discard supernatant. Resuspend cells in appropriate amount of 1× PBS or serum-free media and count the number of cells using a hemacytometer in the presence of Trypan blue to exclude dead cells.

5. Aliquot 2–3 × 10^6 cells per 15-ml tube and spin cells and discard supernatant.

6. Cell pellets can be directly used for miRNA enrichment or stored at –80 °C for future use (*see Note, cat.no.1*)

3.3. RNA and miRNA Isolation

To carry out 2–3 miRNA arrays and corresponding quantitative reverse transcriptase polymerase chain reaction (qRTPCR) validation, it is ideal to start with a cell pellet of 2–3 × 10^6 cells and isolate the total RNA using TRIzol reagent or any other organic extraction method (*see Note, cat.no.2*) The general work flow for miRNA array and qRTPCR is shown in **Fig. 1**.

3.3.1. Isolation of Total RNA From Cell Pellet

1. Lyse cells in 1 ml of TRIzol reagent per 5–10 × 10^6 mammalian cells by repeated pipeting. Incubate the samples for 5 min at room temperature to ensure complete dissociation of nucleoprotein complexes.

2. Add 0.2 ml of chloroform per milliliter of TRIzol reagent. Shake tubes vigorously for 15 s and incubate at room temperature for 2–3 min. Centrifuge samples at 12,000 g for 15 min at 4 °C.

3. Collect the upper aqueous phase into a fresh tube and add 0.5 ml isopropyl alcohol per milliliter of TRIzol reagent. Incubate samples at room temperature for 10 min and centrifuge at 12,000 g for 10 min at 4 °C.

4. Remove supernatant and wash pellet with 1 ml of 75% ethanol. Mix by vortexing and centrifuge at 7500 g for 5 min at 4 °C.

5. Air dry the pellet and resuspend in 50–100 μl of RNase-free water, and to ensure complete dissolution of RNA, incubate at 55–60 °C for 10 min.

6. Determine the quality of isolated total RNA by measuring absorbance at A260 nm and by running on a 15% Urea-TBE gel.

3.3.2. miRNA Enrichment From Cell Pellet or Total RNA

1. Resuspend cell pellet or 20 μg of total RNA in 300 μl of Binding buffer (L3) provided in the Purelink miRNA isolation kit. Mix well by pipeting up and down until the cells are uniformly lysed. If the cells are not uniformly lysed, giving a cloudy viscous solution, add an additional 300 μl of the binding buffer and distribute into two tubes equally.

2. Add 300 μl of 70% ethanol to the cell lysate and mix by vortexing.

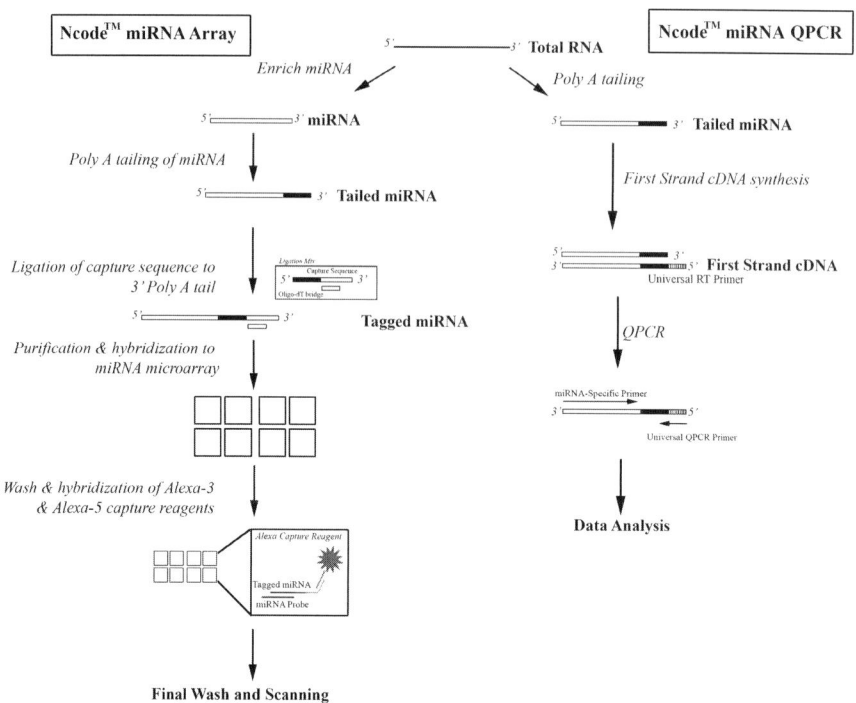

Fig. 1. Work flow scheme for Ncode™ microRNA (miRNA) profiling and subsequent Ncode QPCR validation. It is important to start with total RNA population, preferably isolated using TRIzol or any other organic extraction method and not through column-based methods which removes the miRNA fraction.

3. Transfer the 600 μl of the above cell lysate with binding buffer and ethanol to a spin cartridge provided in the Purelink™ miRNA isolation kit.

4. Centrifuge at 12, 000 g for 1 minute at room temperature. The column contains large RNA fraction and the flow through contains miRNA. (Transfer the spin cartridge with the bound large RNA fraction into another collection tube for subsequent wash with Wash solutions 1 and 2 and subsequent elution with 70 μl water using the Micro-to-midi total RNA purification system.)

5. The flow through contains miRNA. Add 700 μl of 100% ethanol to 600 μl of the flow through to yield a final concentration of 70% ethanol. Mix by vortexing.

6. Add 600 μl of the above mix to a fresh spin cartridge and centrifuge at 12,000 g for 1 min at room temperature. Repeat the process with the remaining sample.

7. Discard the flow through and place the spin cartridge into a fresh collection tube and proceed to the washing step.

8. Wash spin cartridge with 500 μl of wash buffer (W5) with ethanol and centrifuge for 12,000 g for 1 min at room temperature.

9. Repeat the above step wash with 500 μl of wash buffer and discard the flow through. Centrifuge the spin cartridge at maximum speed for 2–3 min to remove residual wash buffer and proceed to the elution step.

10. Transfer the spin cartridge to a clean 1.7-ml recovery tube. Add 70 μl of sterile, RNase-free water and incubate for 1 min at room temperature.

11. Centrifuge the spin cartridge at maximum speed for 1 min at room temperature.

12. Discard the cartridge, measure the concentration of miRNA, and check the integrity of the sample by running on a 15% Urea-TBE gel. Typically, enriched miRNAs have a molecular weight of less than 100 bp (**Fig. 2**).

13. The eluted miRNA can be directly used for downstream application or stored at −80 °C for future use. (*see Note, cat.no.*3)

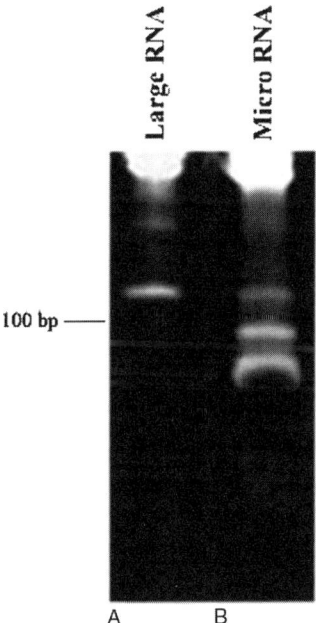

Fig. 2. Simultaneous purification of large and microRNA (miRNA)-enriched RNA using the micro-to-midi total RNA purification system and Purelink miRNA isolation kit. Large RNA (**A**) and miRNA-enriched fraction (B) separated on a 15% Novex TBE-Urea gel and stained with SYBR gold. Enriched miRNA has a molecular weight of less than 100 bp.

3.4. miRNA Array

The procedure described below is for two cell types (A and B) in a heterotypic dye swap experiment. For the first array, sample A is labeled with Alexa 3 (1) and sample B with Alexa 5 (2) while on the second slide the dye labels are swapped with sample A labeled Alexa 5 (2) and sample B labeled Alexa 3 (1).

3.4.1. Poly A Tailing

1. Label four eppendorf tubes as 1A, 2A, 1B and 2B.
2. Dilute 500 ng of miRNA to 14.5 µl with DEPC water.
3. Add the following components to the above 14.5 µl of diluted miRNA.

   a. 5 µl of 5× miRNA reaction buffer.
   b. 2.5 µl of $MnCl_2$.
   c. 1 µl of diluted ATP (Dilute ATP 1:50-1 µl of ATP + 49 µl of 1 mM Tris, pH 8).
   d. 1 µl of Poly A Polymerase (PAP).
   e. 1 µl of Ncode control.

4. Mix gently and centrifuge. Incubate for 15 min at 37 °C.

3.4.2. Ligation of Capture Sequence

1. Briefly centrifuge the tailed miRNA to collect the sample at the bottom of the tube.
2. Separate the tubes into two categories, A1 and B1 on one side and A2 and B2 on the other.
3. Add 6 µl of 6× Alexa3 ligation mix to A1 and B1 tubes and 6µl of Alexa5 ligation mix to A2 and B2 tubes.
4. To all the tubes add 3 µl of DEPC water and 2 µl of T4 DNA ligase.
5. Mix gently by flicking the tubes and briefly microfuge.
6. Incubate at room temperature for 30 min.
7. Stop the reaction by adding 4 µl of 0.5 M EDTA. Briefly vortex and centrifuge. Add 60 µl of 1× Tris, EDTA (TE) buffer to yield a final volume of 100 µl. Vortex and centrifuge the tagged miRNA.

3.4.3. Purification of PolyA Tailed/Ligated miRNA

1. Add 700 µl of binding buffer to the tagged DNA. Mix well and spin down to collect the contents at the bottom of the tube.
2. When comparing two samples (A and B) with two color labeling, load samples A1 and B2 on the same column and samples B1 and A2 on the same column.
3. Load the entire amount from one tube to the column (A1 into column 1 and B1 into column 2) and spin at 4000 g for 1 min. Discard the flow through and add the second sample into the same column (B2 into column1 and A2 into column 2). Spin at 4000 g or 1 min and discard the flow through.

4. Add 600 µl of wash buffer to the column and spin for 30–60 s at 15,000 g. Discard the flow through and spin the column for an additional minute at 15,000 g to remove residual wash buffer.
5. Place the column in a 1.5-ml collection tube and apply 25 µl of DEPC water directly to the membrane.
6. Let stand at room temperature for 1 min and centrifuge for 2 min at full speed to elute the tagged miRNA ready for hybridization.

3.4.4. First Hybridization and Wash

1. Thaw and resuspend the 2× SDS-based hybridization buffer by heating at 70 °C for 10 min. Vortex well to resuspend the buffer evenly.
2. For each array add equal volume (25 µl) of thawed 2× SDS-based hybridization buffer.
3. Gently vortex and briefly microfuge the hybridization mix. Incubate at 75–80 °C for 10 min and then transfer to hybridization temperature (52 °C) until the samples are loaded on the array.
4. Place an array slide in the hybridization chamber. Place 25 µl of water in the slots at the ends of the hybridization chamber to keep the chamber hydrated during hybridization.
5. Apply a 24-cm × 60-cm glass lifter cover slip to the array. Add the hybridization mix along the edge of the cover slip taking care not to introduce air bubbles.
6. Record the slide number and the corresponding samples applied on it.
7. Seal the slide hybridization chamber and place the array overnight (16–20 h) in a hybridization oven with the temperature set to 52 °C.
8. Remove the slide from the hybridization chamber and gently immerse and shake the slide into a container with 2× SSC, 0.2% SDS solution prewarmed to 42 °C until the lifter slip floats away.
9. Place the slide in a staining jar with 2× SSC, 0.2% SDS solution prewarmed to 42 °C and wash for 10 min with moderate agitation on a rocking platform.
10. Transfer the slides to a container with 2× SSC and wash for 10–15 min at room temperature with moderate agitation.
11. Transfer the slides for the final wash in a container with 0.2× SSC for 10–15 min with moderate agitation.
12. Remove the slides and immediately spin-dry them at 800 g for 3–4 min using a table top centrifuge. To minimize background, it is important to avoid drying of the slide prior to the spin or any direct physical contact to the array surface.

3.4.5. Second Hybridization With Probe and Wash

1. Prepare Alexa3 and Alexa5 capture reagents prior to setting up the second hybridization.

   (a) Thaw the Alexa3 and Alexa5 capture reagents in the dark and vortex at maximum setting for 3 s.

(b) Microfuge briefly and incubate at 50–55 °C for 10 min.

(c) Vortex for an additional 3–5 min and microfuge.

2. Thaw and resuspend 2× SDS-based hybridization buffers by heating at 70 °C for 20 min. Vortex vigorously to ensure a uniform homogenous solution and microfuge briefly.

3. For each two-color array prepare an alexa hybridization mix with the following components:

(a) Alexa3 Capture reagent (2.5 µl).

(b) Alexa5 Capture reagent (2.5 µl).

(c) DEPC water (20 µl).

(d) 2× SDS-based hybridization solution (25 µl).

4. Vortex, briefly microfuge and incubate the hybridization mix first at 75–80 °C for 10 min and then at 62 °C until ready to load the array.

5. Gently vortex, microfuge briefly and load the hybridization mix on the microarray covered with a lifter cover slip along the edges carefully avoiding air bubbles.

6. From here the slides must be kept away from light as much as possible.

7. Seal the hybridization chamber and incubate overnight in a hybridization oven set at 62 °C.

8. After overnight incubation, shrug the lifter slip off the array by gently shaking in a container containing 2× SSC, 0.2% SDS prewarmed to 60 °C.

9. Transfer the slides to a staining container with 2× SSC, 0.2% SDS prewarmed to 60 °C and wash for 15 min at room temperature with moderate agitation.

10. Wash for 10–15 min in 2× SSC at room temperature with moderate agitation.

11. Wash for 10–15 min in 0.2× SSC at room temperature with moderate agitation.

12. Dry the array immediately by using a table top centrifuge carefully avoiding any physical contact to the array surface or drying of the wash media on the slide.

13. The slides are now ready to be scanned.

3.4.6. Slide Read

1. Scan the slides and collect data using an array reader. While performing dye swap experiments ensure that the signal intensity for both the green and red channels are comparable by adjusting the PM gain (sensitivity of the photomultiplier tube) of each channel under the instrument settings. It is important that the resulting spots have balanced color in order to extract reliable differential expression data from the hybridized spots.

2. Open the "read array data" using Genepix Pro or a similar software. To extract data from each spot and their corresponding miRNA identity, align the read miRNA slide to the spots on the miRNA array to the Ncode Mutispecies GAL file using the white "landing lights" at the top of the printed array (*see Note, cat.no.***4**).

3.5. Statistical Analysis of Data

1. Data shown were background corrected and normalized (*see Note, cat.no.*5) using the generalized Latin Squares Dye Swap/Loop Design Model *(20)* given by

$$\log(t_{ijkg}) = \mu + A_i + D_j + V_k + G_g + AG_{ig} + VG_{kg} + \varepsilon_{ijkg}$$

where $i = 1, 2\ldots, I$, $j = 1, 2, \ldots, J$, $k = 1, 2, \ldots, K$ and $g = 1, 2, \ldots, n$, additionally here we will assume that log is base 2. Note that these observations are confounded, meaning that on each array with the corresponding dye combination there are a limited number of samples that can be observed. t_{ijkg} is the signal from the ith array, j^{th} dye, k^{th} tissue and g^{th} gene. μ is the overall average signal in the experiment. A_i is the average signal correlating with the ith array. Note that there are a total of I arrays, with the same probes. D_j is the average signal correlating with the jth dye. Note that there are a total of J dyes, typically $J = 2$. V_k is the average signal correlating with the kth sample or variety. Note that there are a total of K samples. G_g is the average signal correlating with the gth gene/probe on the array. Note that there are a total of n probes on each array.

AG_{ig} is the interaction between the array and the probe, this is looking to control for geographical issues on a single array that would effect a probe in all dyes.

VG_{kg} is the interaction between the kth sample and the gth probe on the array, this is in general the term of interest in the model, as it is the term that deals with finding differential gene expression. ε_{ijkg} is the error term of the model, that is, what cannot be explained by the model. It is assumed the average of this term is 0 and that they are identically and independently distributed.

2. For a classical Latin Squares Dye Swap experiment $I = J = K = 2$, meaning that the experiment is typically using two samples, two arrays, with two dyes. With this minimal experiment, differential probes can be found, and statistical information can be found. Generally for classical loop designs, $I = K$ and $J = 2$, though it can be done if I does not equal K.

3. The experimental design for the typical Loop Design model, where suppose $I = K$ and $J = 2$, would look like the following:

	Dye 1	Dye 2
Array 1	Sample 1	Sample 2
Array 2	Sample 2	Sample 3
Array 3	Sample 3	Sample 4
...	...	...
Array $I-1$	Sample $K-1$	Sample K
Array I	Sample K	Sample 1

4. To estimate the various parameters in the model (here we will assume that $I = K$ and $J = 2$ for ease of notation, moving away from this format can be generalized from the given equations), the following equations are used:

$$\mu = \frac{\sum_{ijkg} \log(t_{ijkg})}{nJK}$$

$$A_i = \frac{\sum_{jkg} \log(t_{ijkg})}{nJ} - \mu$$

$$D_j = \frac{\sum_{ikg} \log(t_{ijkg})}{nI} - \mu$$

$$V_k = \frac{\sum_{ijg} \log(t_{ijkg})}{nJ} - \mu$$

$$G_g = \frac{\sum_{ijk} \log(t_{ijkg})}{IJ} - \mu$$

$$AG_{ig} = \frac{\sum_{jk} \log(t_{ijkg})}{J} - A_i - G_g - \mu$$

$$VG_{kg} = \frac{\sum_{ij} \log(t_{ijkg})}{J} - V_k - G_g - \mu$$

5. From the proceeding equations, differential markers can then be identified by looking at 2 raised to the power of $|VG_{kg} - VG_{k'g}|$ which will give you fold change of the gth gene that is between tissues k and k'. The sign of $VG_{kg} - VG_{k'g}$ will designate what direction the fold changes. This validity of this model has been confirmed by comparing the fold change obtained by array results with qRTPCR (*see* **Table 1**).
6. Calculation of a p-value for the difference between the tissue k and k', requires bootstrapping the error distribution. This can be a time-consuming calculation and should be implemented carefully *(21,22)*.
7. To rank genes, rank the genes within an array, dye and sample combination based on signal intensity, and then take the median rank for each gene across all samples. The median ranks can then be reranked for each sample to report the observed gene ranks (*see* **Table 2**).

3.6. NCode™ miRNA qRTPCR

Results obtained on NCode™ miRNA arrays can be further confirmed using Ncode™ miRNA qRTPCR. Because a housekeeping miRNA abundantly expressed at constant levels in all cell types has not yet been identified, it is best to carry out the miRNA QPCR with total RNA, and subsequent normalization can be done with housekeeping genes such as GAPDH or β-2 microglobulin.

Table 1
List of microRNAs expressed in two human embryonal carcinoma cell lines

Name	Ntera2 vs. 2102 average Fold Change		Known function
	Array	qPCR	
miR_368	−6.09	−1.32	hESC
miR_302d	−4.34	−4	hESC/EC
miR_302c	−14.65	−25	hESC/EC
miR_302a	−27.33	−11.11	hESC/EC
miR_200c	−105.12	−100	hESC/EC
miR_125b	03.12	22.01	hESC differentiation
miR_21	17.71	1.52	Anti-apoptotic
miR_222	2.07	1.09	Basic cellular function
miR_103	4.48	2.27	Basic cellular function
miR_106b	4.42	2.48	Unknown
miR_214	2.3	3.41	Unknown
miR_93	5.86	2.71	Unknown

EC, embryonal Carcinoma; hESC, human embryonic stem cell.
Some of the microRNAs have been shown to be associated with human ESC/EC cells. MicroRNA expression was analyzed by Ncode™ miRNA array and results validated using Ncode™ QPCR. Data obtained from Ncode™ miRNA array using heterotypic dye swap experiment were normalized and fold change calculated as described (p < 0.002). Ct values obtained by qRTPCR was used to calculate ΔΔCt values as described. Fold change in expression was calculated as $2^{-\Delta\Delta Ct}$ and values less than 1 expressed as its reciprocal with a negative value.

3.6.1. DNase Treatment of TRIZOL® Extracted Total RNA Samples

1. Treat isolated 10 μg of total RNA with 1 μg of DNase for 2–3 h at 37 °C in a total volume of 100 μl.
2. Add an equal volume of phenol : chloroform and vortex vigorously. Centrifuge the sample for 5 min at room temperature.
3. Transfer the top aqueous phase to a fresh tube and add 2–3 volumes of cold ethanol and 1 μl of glycogen (enables visualization of RNA pellet). Chill on ice for 15–30 min and spin down for 5–10 min at 12,000 g for 15 min at 4 °C.
4. Discard supernatant and wash pellet with 100 μl of 70% ethanol.
5. Resuspend the pellet in 20 μl of RNase-free water. Measure absorbance at 260 nm and assess the quality of DNased RNA by electrophoresis on a 15% Urea-TBE gel.

Table 2
**The top 20 miRNAs expressed in 2102Ep cells and
their relative rank number in Ntera2 cells**

miRNA	2102	Ntera 2
miR_373	1	35
miR_302a	2	8
miR_302d	3	2
miR_302b	4	4
miR_372	5	34
miR_21	6	5
miR_371	7	70
miR_423	8	10
miR_296	9	13
miR_302a[a]	10	18
miR_302c	11	39
miR_422b	12	80
miR_320	13	9
miR_106a	14	1
miR_302c[a]	15	55
miR_93	16	11
miR_200c	17	236
miR_25	18	32
miR_337	19	52
miR_198	20	111

miRNA, microRNA.
Ranking was determined based on the normalized signal
intensity obtained on the Ncode™ miRNA array. miRNAs in
bold re species known to be associated with pluripotent cells.
[a] derived from the same precursor

3.6.2. Poly A Tailing Reaction

Start with 500 ng of the DNased total RNA and set up the polyA tailing
reaction by adding the following components:

1. 5× reaction buffer (5 μl).
2. $MnCl_2$ (2.5 μl).
3. Diluted ATP [1 μl (1 μl ATP + 9 μl of 1mM Tris, pH 8)].
4. PAP enzyme (1 μl).
5. RNA/H_2O (29 μl).

Mix gently and incubate at 37 °C for 15 min. Each Poly A tailing reaction can be used for six subsequent RT-PCRs.

3.6.3. RT Reaction

1. Add the following components to RNA from the above reaction.
   - (a) PolyA tailed RNA (2 μl).
   - (b) Anneal buffer (1 μl).
   - (c) Universal RT Primer (25 μM) (3 μl).
2. Incubate at 65 °C for 5 min and keep on ice at 4 °C for 1 min.
3. Add the following to the above 8 μl reaction.
   - a. 2× RT reaction mix (10 μl)
   - b. RT enzyme mix (2 μl).
4. Incubate at 50 °C for 50 min and at 85 °C for 5 min.

The resulting cDNA can be stored at −20 °C or directly used for QPCR. Just before use dilute the resulting cDNA 1:10 times as $MnCl_2$ can interfere in subsequent PCR.

3.6.4. qPCR

1. Set up 20 μl volume qPCR by adding the following components.
   - (a) 2× SYBR SupermixUDG with ROX (10 μl).
   - (b) miRNA primer (10 μM) (0.4 μl).
   - (c) Universal QPCR primer (included in kit, 10 μM) (0.4 μl).
   - (d) H_2O [7.2 μl (excluding cDNA)].
   - (e) cDNA template (2 μl).

miRNA primers are DNA version of the exact sequence as the corresponding miRNA sequence with the U is replaced with T.

2. Cycling conditions: 50 °C for 2 min, 95 °C for 3 min; 95 °C for 15 min, 60 °C for 30 min ×45 cycles; 45 °C for 1 min, Melt curve analysis.
3. The resulting threshold cycle Ct values can be compared between two different samples using the relative quantification method. Here, the Ct values of the samples are normalized to a house keeping mRNA such as GAPDH or β-2 microglobulin (Note, cat.no.5) and the normalized Delta Ct values of the sample ($Ct_{GeneofInterest} - Ct_{Housekeepinggene}$) compared to a second cell type to yield Delta Delta Ct ($DeltaCt_{Sample1} - DeltaCt_{Sample2}$). Fold difference in gene expression of the sample 1 from the sample 2 can be calculated using the equation $2^{-DeltaDeltaCt}$. In conclusion, the Ncode™ QPCR therefore offers a more sensitive and quantitative method to further evaluate and validate the markers identified by the NCode™ miRNA array (*see* **Table 1**) (Note, cat.no.**6**) (Note, cat.no.**7**).

4. Notes

1. Several of the miRNA sequences are conserved between species and it is therefore recommended that in the case of cells grown on feeder layer care be taken to separate the cell of interest from the feeders. Some of the commonly used methods to achieve this are listed below:

   (a) FACS cell sorting based on presence or absence of surface marker on cell type of interest.
   (b) Physical removal of feeder cells by scrapping and mechanical dissection of cells colonies based on morphology.
   (c) Alternately, cells can be maintained under feeder-free conditions for a generation prior to miRNA analysis under conditions that would not alter its stem cell characteristics.

2. RNA isolation should be carried out using TRIzol or any other organic extraction methods. Use of column-based methods generally excludes miRNA and other small RNAs and are eluted out at the very first step of RNA binding to the column.

3. The in vitro stability of miRNA cannot be easily assessed and it is therefore prudent to save the total RNA at $-80\,^{\circ}$C and use the enriched miRNA within a few days of enrichment.

4. One of the major error in analyzing samples is the use of incorrect GAL files. Please match the slide number with its corresponding GAL file. This information can be obtained on http:www.invitrogen.com.

5. Statistical analysis of data is crucial in obtaining reliable data. Described here is one such analysis based on Martin, Kerr, and Churchill equation. Other known array statistical can also be applied to the results.

6. β-Actin is not recommended as a house keeping gene for Ncode miRNA qRTPCR because a portion of the universal QPCR primer for miRNA amplification contains this sequence and can lead to linear amplification of the β-actin gene.

7. qRTPCR results can also be reported as copy number using a standard curve generated from a pure synthetic template diluted over several logs. The Ct values of the reference cell line can then be converted to copy number, and based on fold difference obtained using the relative quantification method, copy number of candidate miRNAs in the sample is deduced. The copy number so generated will still be a relative quantification and cannot be assumed to be absolute numbers.

References

1. Bartel, D. P. MicroRNAs: genomics, biogenesis, mechanism, and function (2004) *Cell*, **116**, 281–97.
2. Lee, R. C., Feinbaum, R. L., and Ambros, V. The C. elegans heterochronic gene lin-4 encodes small RNAs with antisense complementarity to lin-14 (1993) *Cell*, **75**, 843–54.

3. Wightman, B., Ha, I., and Ruvkun, G. Posttranscriptional regulation of the heterochronic gene lin-14 by lin-4 mediates temporal pattern formation in C. elegans (1993) *Cell*, **75**, 855–62.

4. Lagos-Quintana, M., Rauhut, R., Lendeckel, W., and Tuschl, T. Identification of novel genes coding for small expressed RNAs (2001) *Science*, **294**, 853–8.

5. Moss, E. G. MicroRNAs: hidden in the genome (2002) *Curr Biol*, **12**, R138–40.

6. Pasquinelli, A. E., Reinhart, B. J., Slack, F., Martindale, M. Q., Kuroda, M. I., Maller, B., Hayward, D. C., Ball, E. E., Degnan, B., Muller, P., Spring, J., Srinivasan, A., Fishman, M., Finnerty, J., Corbo, J., Levine, M., Leahy, P., Davidson, E., and Ruvkun, G. Conservation of the sequence and temporal expression of let-7 heterochronic regulatory RNA (2000) *Nature*, **408**, 86–9.

7. Olsen, P. H., and Ambros, V. The lin-4 regulatory RNA controls developmental timing in Caenorhabditis elegans by blocking LIN-14 protein synthesis after the initiation of translation (1999) *Dev Biol*, **216**, 671–80.

8. Bartel, B., and Bartel, D. P. MicroRNAs: at the root of plant development? (2003) *Plant Physiol*, **132**, 709–17.

9. Aravin, A. A., Lagos-Quintana, M., Yalcin, A., Zavolan, M., Marks, D., Snyder, B., Gaasterland, T., Meyer, J., and Tuschl, T. The small RNA profile during Drosophila melanogaster development (2003) *Dev Cell*, **5**, 337–50.

10. Lagos-Quintana, M., Rauhut, R., Yalcin, A., Meyer, J., Lendeckel, W., and Tuschl, T. Identification of tissue-specific microRNAs from mouse (2002) *Curr Biol*, **12**, 735–9.

11. Krichevsky, A. M., King, K. S., Donahue, C. P., Khrapko, K., and Kosik, K. S. A microRNA array reveals extensive regulation of microRNAs during brain development (2003) *RNA*, **9**, 1274–81.

12. Sempere, L. F., Sokol, N. S., Dubrovsky, E. B., Berger, E. M., and Ambros, V. Temporal regulation of microRNA expression in Drosophila melanogaster mediated by hormonal signals and broad-complex gene activity (2003) *Dev Biol*, **259**, 9–18.

13. Houbaviy, H. B., Murray, M. F., and Sharp, P. A. Embryonic stem cell-specific MicroRNAs (2003) *Dev Cell*, **5**, 351–8.

14. Suh, M. R., Lee, Y., Kim, J. Y., Kim, S. K., Moon, S. H., Lee, J. Y., Cha, K. Y., Chung, H. M., Yoon, H. S., Moon, S. Y., Kim, V. N., and Kim, K. S. Human embryonic stem cells express a unique set of microRNAs (2004) *Dev Biol*, **270**, 488–98.

15. Krichevsky, A. M., Sonntag, K. C., Isacson, O., and Kosik, K. S. Specific microRNAs modulate embryonic stem cell-derived neurogenesis (2006) *Stem Cells*, **24**, 857–64.

16. Chen, C. Z., Li, L., Lodish, H. F., and Bartel, D. P. MicroRNAs modulate hematopoietic lineage differentiation (2004) *Science*, **303**, 83–6.

17. Chen, J. F., Mandel, E. M., Thomson, J. M., Wu, Q., Callis, T. E., Hammond, S. M., Conlon, F. L., and Wang, D. Z. The role of microRNA-1 and microRNA-133 in skeletal muscle proliferation and differentiation (2006) *Nat Genet*, **38**, 228–33.

18. Yi, R., O'Carroll, D., Pasolli, H. A., Zhang, Z., Dietrich, F. S., Tarakhovsky, A., and Fuchs, E. Morphogenesis in skin is governed by discrete sets of differentially expressed microRNAs (2006) *Nat Genet*, **38**, 356–62.
19. Goff, L. A., Yang, M., Bowers, J., Getts, R. C., Padgett, R. W., and Hart, R. P. Rational probe optimization and enhanced detection strategy for microRNAs using microarrays (2005) *RNA Biol*, **2**, 93–100.
20. Kerr, M. K., Martin, M., and Churchill, G. A. Analysis of variance for gene expression microarray data (2000) *J Comput Biol*, **7**, 819–37.
21. Kerr, M. K., and Churchill, G. A. Experimental design for gene expression microarrays (2001) *Biostatistics*, **2**, 183–201.
22. Manly, B. F. J. (1997) *Randomization, bootstrap* and *Monte Carlo Methods in Biology.* Second Edition. Chapman and Hall/CRC, Boca Raton.

9

Gene Transfer Via Nucleofection Into Adult and Embryonic Stem Cells

Uma Lakshmipathy, Shannon Buckley, and Catherine Verfaillie

Summary

The use of embryonic and adult stem cells as therapeutic agents is gaining momentum. A major impediment in the use of stem cells for genetic disorders is their ability to undergo genetic modification. The recognition of various site-specific integration methods open up a new avenue for gene therapy in stem cells. However, this necessitates efficient delivery of DNA molecule into cells. Most commercially used liposome-mediated transfection reagents are toxic or work poorly with stem cells. Electroporation, while effective in transfecting stem cells, is rather harsh and leads to excessive cell death. Nucleofection, a technology by Amaxa, uses a combination of electric pulse in an appropriate media, which decreases the toxicity and promotes efficient transfection of stem cells. Various types of adult and embryonic stem cells can be successfully transfected using this method, as described in this chapter.

Key Words: Mouse bone marrow; Embryonic stem cells; Adult stem cells; Transfection; Nucleofection.

1. Introduction

Efficient transfection of cells is a critical step in carrying out gene transfer experiments. While several methods exist for achieving high levels of transfection in cell lines and some types of primary cells *(1)*, this is not true with stem cells. Both embryonic stem cells and adult stem cells transfect poorly with most transfection methods. Electroporation has been the traditional method for efficient gene transfer in certain stem cells such as mouse embryonic stem cells (mESCs) but is toxic to most adult stem cells as well as human embryonic stem cells (hESCs) *(2,3)*. Several lipid-based methods have been reported to yield

From: *Methods in Molecular Biology, vol. 407: Stem Cell Assays*
Edited by: M. C. Vemuri © Humana Press, Totowa, NJ

success in transfecting on both adult stem cells and ESCs to varying degrees *(4–6)*. In adult stem cells such as bone marrow (BM)-derived multipotent adult progenitor cells (MAPCs), lipid-mediated transfection cannot be used as these cells differentiate when seeded at high density *(7)*.

Recently, the nucleofection technology from Amaxa has shown promising results with primary cells that do not transfect easily. Nucleofection is an electroporation-based method where combinations of proprietary solution together with defined nucleofection programs that have set electrical parameters give rise to low toxicity and efficient transfection of cells. We and others have shown that both adult stem cells and hESCs are efficiently transfected by this method *(2,3,8–13)*. Here, we describe in detail the experimental protocol for transfecting stem cells via nucleofection and evaluate transfection efficiency.

2. Materials

All media components were obtained from Invitrogen, Carlsbad, CA, USA unless otherwise stated.

2.1. Mouse BM

1. BM Flushing media: Modified Iscove's media (cat. no. 10-016-CV, Mediatech, Herndon, VA, USA) or 1× (phosphate-buffered saline) (cat. no. 21-040-CV, Mediatech, Herndon, VA, USA) is used to collect BM cells.
2. Mouse BM growth media: Stem Pro-SFM media (cat. no. 10639-011, Invitrogen) with supplement and 10 µg/ml of mouse stem cell factor (cat. no. 455-MC, R&D Systems, Minneapolis, MN, USA), 5 µg/ml mouse thromobopoeintin (cat. no. 488-TO, R&D systems, Minneapolis, MN, USA), and L-glutamine (cat. no. 25030–081, Invitrogen).
3. Ammonium chloride (cat. no. 07850, Stem Cell Technologies, Vancouver, BC, Canada) is aliquoted in 5–10 mls of ammonium chloride is aliquoted and stored at −20 °C. Freshly thawed aliquot is used for red blood cell (RBC) lysis.
4. Cell strainer, 70 µm (cat. no. 352350, BD Biosciences San Diego, CA, USA)
5. Cell counting is carried out in the presence of TURK (cat. no. 885016 Ricca Chemical Company, Arlington, TX, USA)

2.2. Mouse MAPC

1. Mouse MAPC medium: DMEM (low glucose, cat. no. 11885-084, Invitrogen), 1× MCDB (cat. no. M-6770, Sigma), insulin–transferrin–selenium (cat. no. I-3146, Sigma), 1mg/ml linoleic acid–(bovine serum albumin) (cat. no. L-9530, Sigma), 0.1 nM L-ascorbic acid (cat. no. A-8960, Sigma), 0.5 µM dexamathasone (cat. no. D-2915, Sigma), 10 ng/ml PDGF (cat. no. 220-BB, R&D Systems, Minneapolis, MN, USA), 10 ng/ml Epidermal Growth Factor (EGF) (cat. no. E-9644, Sigma),

10^3 U LIF (cat. no. ESG1107, Chemicon, Temecula, CA, USA), and β-mercaptoethanol (cat. no. 21985-023, Invitrogen).

2. Plates were coated with 10 ng/ml fibronectin (cat. no. F0895, Sigma) for at least 30 min at room temperature prior to seeding cells.

3. MACS columns, CD45 and Terr119 depletion beads were obtained from Miltenyi Biotec Inc., Auburn, CA, USA.

4. MACS buffer: PBS containing 0.5% BSA, and 2 mM ethylendiamine tetra acetic acid (EDTA).

2.3. Mouse ESC

1. Mouse ESC culture medium: DMEM (cat. no. 11995-065, Invitrogen) supplemented with 15% fetal calf serum (FCS) (Hyclone, Logan, UT, USA ES certified), β-mercaptoethanol (cat. no. 21985-023 Invitrogen), nonessential amino acids (cat. no. 11140-050, Invitrogen), L-glutamine (cat. no. 25030-081, Invitrogen) and Luekemia inhibitory factor (LIF) (cat. no. ESG1107, Chemicon, Temecula, CA, USA).

2. gelatin solution (0.1%) was prepared from 2% gelatin (cat. no. G1393, Sigma)

2.4. Transfection

1. Nucleofector and Nucleofection kits (cat. no. VPA-1003, Amaxa, Gaithersburg, MD, USA Human CC34+ kit, cat. no. VPE-1001 Human MSC kit; cat. no. VPH-1001, Mouse ESC kit).

2. Endotoxin-free plasmid peGFP-N1 (cat. no. 6085-1, Clontech, Mountainview, CA) was prepared using the endo-free plasmid maxi kit (cat. no. 12362, Qiagen, Valencia, CA, USA).

2.5. Analysis of Transfection

1. Staining medium: 0.1% BSA or 3% FCS in PBS.

2. Propidium iodide (cat. no. P3566, Invitrogen).

3. Falcon tube with cell strainer cap for fluorescent-associated cell sorter (FACS) analysis (cat. no. 352335, BD Biosciences, San Diego, CA, USA).

4. All antibodies used for immunophenotyping of mouse BM cells were from BD Biosciences (cat. no. 557397l, Mac1-PE; cat. no. 553129, Gr1-APC; cat. no. 553672, Terr1-PE; cat. no. 553092 B220-APC; cat. no. 557308, CD4-PE; cat. no. 553035 CD8-APC; cat. no. 553771 CD45-BIO; cat. no. 554067 Secondary Ab-APC; cat. no. 55336 Sca1-PE; cat. no. 553356 cKit-APC).

3. Methods

3.1. Isolation of MAPCs From Mouse BM

1. Anesthetize and Fill adult normal mouse between 4 and 6 weeks according to approved protocols. Disinfect the entire body surface of the animal with 70% ethanol and place the animal on its back and through a small incision in the skin

close to the hind legs cut the skin carefully by separating it from the underlying muscle. Once the skin is removed, the muscles can be carefully teased away from the bone cautiously without cutting the bone. Under aseptic conditions, transfer the entire length of the femur and if needed tibia as well into a Petri dish containing cold PBS.

2. Flush out BM from 2–4 femurs by inserting a 23-gauge needle fitted into a 2-ml syringe with cold PBS in a tissue culture hood. Continue flushing till the bone appears white (pinkish exterior is due to muscles, ignore). Repeat the procedure for all isolated bones. Disperse red clumps of cells collected by passing through 23-gauge needle until no clumps are seen. The cells can also be passed through a 70-μm cell strainer to remove contaminating muscle and bone debris.

3. Transfer cell suspension in PBS to a 15-ml conical tube. Spin for 5 min, $250\,g$ (Brakes: high) at 4 °C in a Sorvall legend RT table top refrigerated centrifuge. Discard the supernatant carefully and retain the loose pellet of cells.

4. Add 5 ml of ammonium chloride to the red color cell pellet and pipette up and down. This step lyses RBCs (~ 1 ml per 10×10^6 cells). Immediately spin the cell and discard the supernatant. Wash cells once with $1\times$ PBS to get rid of residual ammonium chloride and resuspend the cell pellet in $1\times$ PBS (*see* **Note 1**).

5. Take a small aliquot of cells ($2\,\mu l$) and add $18\,\mu l$ Turk. Count number of cells (Note: Cells are diluted 1:10). Two femurs from 8-week-old to 12-week-old CB57Bl/6 BM yields around $50–70 \times 10^6$ cells.

6. Use approximately $10–15 \times 10^6$ total-BM cells per transfection.

3.2. Isolation and Characterization of Mouse MAPC

1. Flush out BM from 2–4 femurs, lyse RBC using ammonium chloride and prepare total BM cells pellet as described in **Subheading 3.1**.

2. Meanwhile, place Ficoll-paque in a fresh 15-ml tube. Carefully layer equal volume of the cells resuspended in $1\times$ PBS over Ficoll. Centrifuge for 20 min, $250\,g$, at room temperature (20–25 °C) with brakes off (*see* **Note 2**).

3. Monocytes form a buffy coat at the interface of Ficoll and PBS. Remove the top PBS layer carefully leaving behind 1 ml on top of the buffy coat.

4. Pipette 10 ml MAPC medium with FCS into a 10-ml pipette to coat the pipette. Carefully collect the buffy coat with smooth swirling motion. (*see* **Note 3**)

5. Transfer to a 15-ml tube and mix the contents by pipeting up and down. Add MAPC medium and centrifuge for 5 min, $250\,g$, at 4 °C (Brakes: high).

6. Discard supernatant and resuspend pellet in 10 ml MAPC medium and count the number of cells. Plate at $1–2 \times 10^6$ cells per well in a six-well plate coated with fibronectin (FN).

7. Change medium every 3rd day and observe for the appearance of adherent colonies (1–2 weeks). Colonies start appearing after 10–14 days. Trypsinize and combine all colonies. Replate cells in appropriate size flask at 75% confluency. Expand cells to $2–3 \times 10^6$.

8. To deplete residual CD45-positive and Ter119-positive cells, incubate cells with 20 μl of CD45 beads and 20 μl of Ter119 beads for 15 min at 4 °C. Wash with MACS buffer using 20× volume and resuspend the cells in 500 μl of MACS buffer. Meanwhile wash the negative selection column with MACS buffer. Layer cells in 500 μl of MACS buffer onto the column and add 5 ml (four times) of MACS buffer when the cells enter into the column. Collect the flow through, spin the cells and count.

9. Plate cells as bulk in 10-cm FN-coated dish or at 10 cells per well in FN-coated 96-well plates for clonal expansion. Change medium every 3 days. After colonies grow out, expand them in larger wells. Normally, 1% of the wells will yield long-term cultures.

10. Split and expand cells (usually 1–2 splits). Assess for multipotency of cells by setting up tri-lineage differentiations as described earlier *(7)*.

3.3. Maintenance of Mouse ESC

1. Feeder cells for mouse ESC are inactivated murine embryonic fibroblasts (iMEFS), mitotically inactivated using Mitomycin C (*see* **Note 4**)

2. Prepare feeder plates a day before starting ESC culture. Choose plate size depending on the number of cells per frozen vial. Add desired number of cells from the suspension (~ 1–2×10^6 cells per 6-cm dish).

3. Replace fresh ES culture medium everyday after plating, and when cells reach 40% confluence, the culture should be split.

4. Aspirate medium and wash cells with PBS. To split cells, add 1 ml of trypsin–EDTA (2.5 ml for 10-cm dish) and incubate for 2.5 min. Use a p1000 pipetman to resuspend the detached cells using sterile tips. This step is important to break up the cells into single cell suspension.

5. Transfer the trypsinized cells to a 15-ml tube containing 5–8 ml of ES culture medium. Spin at $3000\,g$ for 5 min. Discard supernatant. Flick the pellet and resuspend in fresh medium, count the number of cells using a hemacytometer or split at a defined ratio (e.g., 1:3 or 4 if split every 48 h or 1:5 or 6 if split every 72 h).

6. Prior to nucleofection, maintain the mESC under feeder-free conditions for one generation by seeding cells on 0.1% gelatin-coated plates.

7. If carrying out FACS after nucleofection, plate transfected cells on gelatin-coated dishes. If carrying out stable selections, plate the transfected cells on drug-resistant iMEFs. For example, if transfected with a plasmid with Neomycin resistant gene, grow mESC on NeoR MEFs.

3.4. Transfection

1. Start with 0.5–1×10^6 adherent cells or over 10×10^6 total BM cells.

2. Spin the cells and discard the supernatant. Remove residual medium using a pipetman. (*see* **Note 5**).

3. Add 100 μl of nucleofection solution (Amaxa Inc., Gaithersburg MD, USA) (*see* **Note 6**). Pipette cells up and down gently to form a uniform suspension. Add

5–10 μl of DNA (∼ 6–10 μg of DNA). Mix well and transfer to a 4-mm electrode gap cuvette (supplied by Amaxa Inc., Gaithersburg MD, USA) without introducing air bubbles.

4. Place the cuvette in the cuvette holder, set the nucleoporator to Program A-23 for mouse MAPC and mESC and Program U08 for total BM cells, and nucleoporate. Remove the cuvette from the holder and carefully retrieve the cells using plastic pipettes (provided in the kit) and transfer to Eppendorf tubes containing 1 ml of growth medium pre-warmed to 37 °C.

5. Centrifuge the transfected cells on a desktop centrifuge at 250 g for 5 min and remove the supernatant without disturbing the pellet (*see* **Note 7**). Allow the cell pellet to recover at 37 °C for 5–10 min.

6. Add fresh medium to the pellet and pipette up and down gently to resuspend the cells. Assuming 75% cell death, plate cells at appropriate density in Petri plate (*see* **Note 8**). Replace growth medium after 12–16 h and monitor for gene expression.

3.5. Analysis of Transfection

Transfection efficiency can be measured by various methods. If cells are transfected with a marker such as green fluorescent protein (GFP) or B-gal expressing construct, visual inspection of cells or FACS analysis is the most commonly employed method. In the absence of a visual marker, transfection efficiency is often measured by QPCR analysis to determine the expression of the transgene mRNA or by western blot analysis. These two methods are qualitative as the percent of transfected cells cannot be accurately assessed by these methods.

3.5.1. Fluorescence Microscopy

1. Remove medium and replace with fresh growth medium.

2. Observe the cells under bright field (see **Fig. 1A**) and count the number of cells. Switch to fluorescence mode and count the number of green cells (*see* **Fig. 1B**) under fluorescence light to quantify cells transfected with GFP vector.

3. Repeat the process for five to six random areas. Add the total number of cells and corresponding number of fluorescent cells to calculate the percentage of transfected cells. This method is appropriate for cells seeded at low density. For cells transfected at high density or a heterogeneous cell population such as total BM, a more accurate and sensitive method such as FACS analysis has to be utilized (*see* **Fig. 2**).

3.5.2. FACS Analysis

3.5.2.1. TITRATE ANTIBODY: PRIOR TO THEIR FIRST USAGE

1. Dilute antibodies (Abs) with staining media (dilute all Abs at 1×, 2×, 5×, 10×, 20×, 40×, 80× and 100×).

2. Prepare the cells that express the antigen to be analyzed.

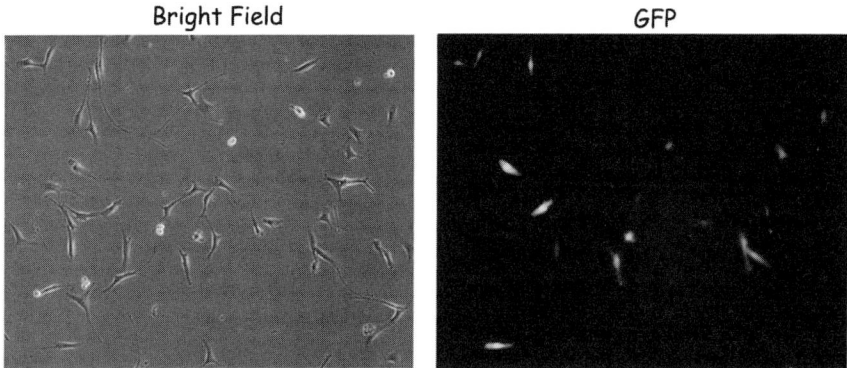

Fig. 1. Analysis of transfection by fluorescence microscopy. Twenty-four hours after nucleoporation of bone marrow-derived mouse multipotent adult progenitor cells with GFP, cells were examined under a fluorescence microscope. Because the cell density is low, transfection efficiency can be calculated by counting the number of cells in the bright field (left panel) vs. the number of GFP+ cells (right panel).

3. Include unstained cells and cells stained with isotype Ab as negative controls.
4. Count number of cells and use 1 million cells (50–100,000 cells work as well) per condition (Note: Use same cell number in every experiment.) Starting with higher number of cells is preferred as setting up parameters during FACS analysis takes time and 10,000 or greater events should be collected for reliable data.
5. Centrifuge cells for 5 min, 250 g, at 4 °C and discard supernatant.
6. Add 5 µl of Ab diluted to varying concentration into separate sample tubes containing cells.
7. Mix well and incubate cells on ice for 25–30 min.
8. If primary antibodies are not directly conjugated to fluorescent tag, carry out the second step incubation with secondary Ab tagged to fluorescent tag.
9. Wash with 10 ml staining media. Discard supernatant and resuspend cells in 0.5 ml staining media and add 3 µg/ml propidium iodide to detect dead cells.
10. Analyze cells by FACS.

3.5.2.2. ONE-STEP STAINING WITH FLUORESCENT-LABELED AB(S)

1. Trypsinize cells ($\sim 2 \times 10^5$ cells per condition) and dilute cells in staining medium. Centrifuge cells for 5 min, $250 \times g$, at 4 °C and discard supernatant.
2. To the cell pellet, add 5 µl of diluted primary Ab(s) conjugated to fluorescent tag.
3. Flick the tube to resuspend the cell pellet, mix well, and incubate on ice for 25–30 min.
4. Wash cells with 10 ml of cold staining media. Centrifuge cells for 5 min, $250 \times g$, at 4 °C.

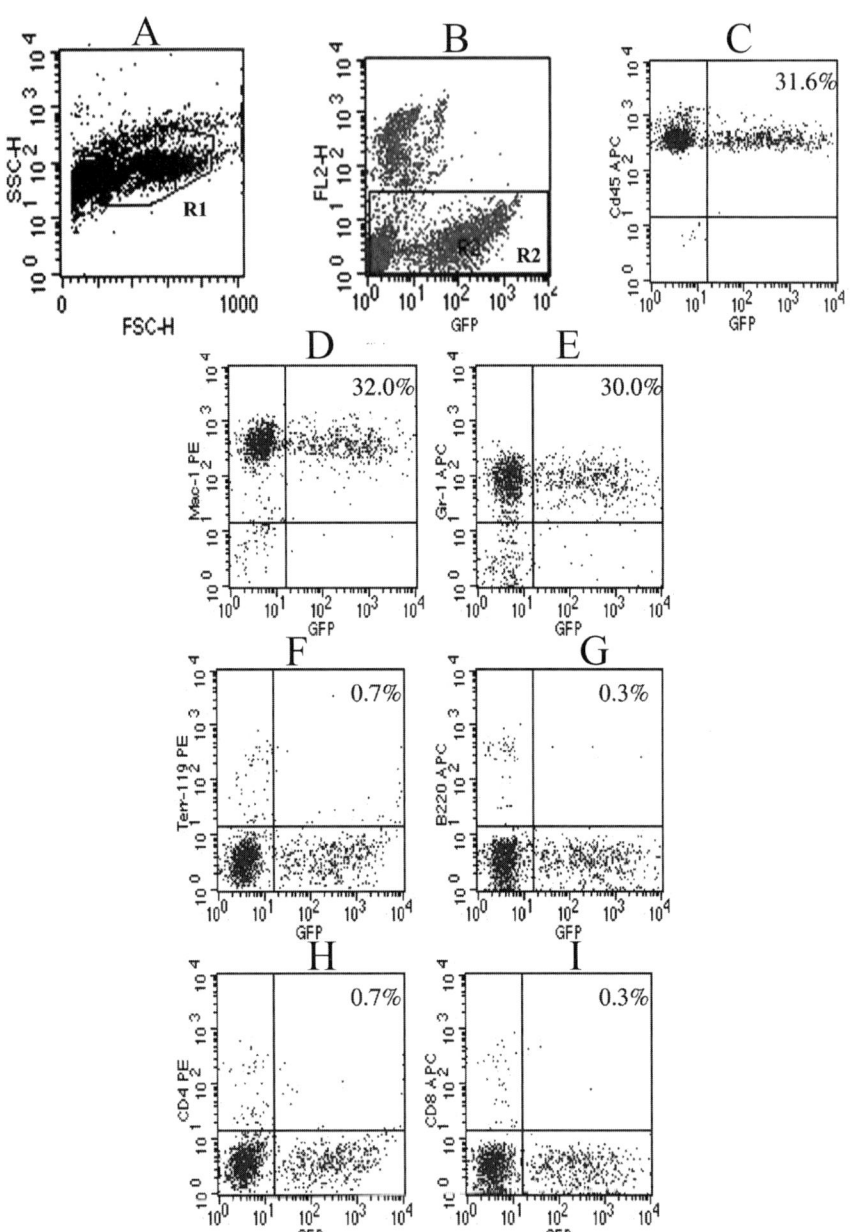

Fig. 2. Analysis of percent transfection of specific lineages in total bone marrow by flow cytometry. Twenty-four hours after nucleoporation of total bone marrow, samples were stained with propidium iodide and specific antibodies to assess percent of

5. Discard supernatant and re-suspend cells with 0.5 ml staining media.
6. Filter through FACS tubes fitted with filtered caps before analysis by FACS. Add 3 µg/ml propidium iodide to detect dead cells.

3.5.2.3. TWO-STEP STAINING WITH BIOTINYLATED ANTIBODY

1. Trypsinize cells and add staining media. Centrifuge cells for 45 min, $250 \times g$, at 4 °C.
2. Discard supernatant. Add 5 µl of appropriately diluted biotinylated primary Ab(s).
3. Include unstained cell, cells stained with isotype primary Ab, and cells stained with secondary Ab only as negative controls.
4. Flick the tube to resuspend cells and mix well. Incubate cells on ice for 25–30 min.
5. Wash cells with 10 ml of cold staining media. Centrifuge at $1000 \times g$, 4 °C for 5 min.
6. Discard supernatant. Add diluted streptavidin secondary Ab conjugated to fluorescent tag.
7. Mix well and incubate cells on ice for 25–30 min on ice.
8. Wash cells with 10 ml of cold staining media. Centrifuge cells at $1000 \times g$, 4 °C, for 5 min.
9. Discard supernatant and re-suspend cells with 0.5 ml staining media.
10. Filter through FACS filter tubes before analysis or sorting the cells by FACS.

Using FACS immunophenotyping analysis, the survival of GFP transfected cells positive for Sca-1 and c-Kit, a scanty population in the BM representing the hematopoietic stem cells *(14)*, can be measured and compared to appropriate controls (*see* **Fig. 3**) *see* **Note 9**.

3.5.3. Selection of Stable Clones

Mouse MAPC and mESC can be further cultured, and drug selection is carried out to derive stable clones expressing the gene of interest.

Fig. 2. transfection of specific hematopoietic lineages. (**A**) Fluorescent-associated cell sorter analysis showing SSC (side scatter) vs. FSC (forward scatter) and gated (R1) to exclude small dead cells and contaminating red blood cell. (**B**) The R1 gated cells were further analyzed by plotting FL1 (GFP) vs. FL2 (propidium iodide). Cells positive for propidium iodide were excluded, as they represent dead cells, using the gate R2. Live cells were further stained with antibodies labeled with the dye PE (FL2) or APC (FL3) and a general heamtopoietic lineage marker. Lineage-specific antibodies were used to assess the percent of transfection of hematopoeitic lineages. (**C**) CD45 (monocytes). (**D**) Mac-1: Macrophages. (**E**) Gr-1: Granulocyte. (**F**) Terr-1: Erythrocyte. (**G**) B220: B–cell. (**H**) CD4: T cell. (**I**) CD8: T cell.

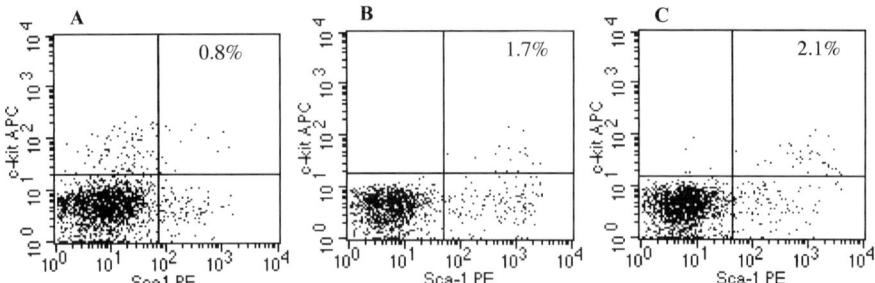

Fig. 3. Hematopoietic stem cell population (Lin–/Sca1+/c-kit+) is not lost with nucleoporation. Total bone marrow cells were gated for Lin– and then analyzed for Sca1+ and c-Kit+ cell populations. (**A**) Lin– population prior to nucleoporation. (**B**) Lin– bone marrow population 24 h after nucleoporation. (**C**) GFP+ and Lin– bone marrow population 24 h after nucleoporation.

3.5.3.1. mESC

1. Following transfection, seed mESC on iMEF and start drug selection. It is best to use a feeder layer that is resistant to the drug used for selection. Change medium containing appropriate drug every day.
2. At the end of first week, colonies begin to form. At this stage, drug-resistant colonies can be picked and seeded into individual wells of a 24-well or 48-well culture dish.
3. Continue drug selection for an additional week. Split and expand cells when necessary.
4. Screen colonies for expression of gene of interest either by QPCR or western blot analysis. If transfected with GFP construct, the colonies can be visually inspected using fluorescence microscopy.

3.5.3.2. mMAPC

1. In the case of MAPC, seed transfected cells at 10 cells per well in a 96-well plate prior to drug selection.
2. After 48–72 h, start drug selection. Change medium every alternate day.
3. Check all wells regularly to see any emerging clones. Split and expand cells when necessary.
4. Screen clones for expression of the gene of interest by western blot analysis and genomic integration of the gene by southern blot.

4. Notes

1. A small aliquot of the isolated total mouse BM should be kept aside for immunophenotyping by FACS.

2. Temperature is crucial for Ficoll-based density gradient, and samples should be centrifuged at room temperature. The brakes on the centrifuge need to be turned off to prevent disturbance of the buffy coat layer formed at the interface of aqueous and Ficoll layers during centrifugation.

3. During isolation of cells from BM, it is important to pipette media containing FBS or FBS alone into the pipette to prevent sticking of the cells to the walls of the pipette.

4. Depending on the mouse ESC cell line used, cells can be either maintained on feeder layers or on 0.1% gelatin-coated plates. Feeder cells (MEFs) can be mitotically inactivated either by irradiation or by treatment with Mitomycin C.

5. It is important to remove all medium during transfection because volumes can change if excess medium is left behind. Furthermore, alteration in composition of nucleofector solution could lead to suboptimal transfection or may affect delivery of set electrical parameters to the sample resulting in an error signal.

6. Kit varies for each cell type; hCD34+ kit for mouse total BM and hCD34+ cells, hMSC kit for human mouse and rat MAPC and mESC kit for mESC and hEC/ESC cells. Optimization for each cell type is therefore necessary for efficient transfection.

7. Because the composition of nucleofector solution is not known, it is important to dilute the cells with excess medium and spin the cells to remove any residual nucleofector solution prior to plating the transfected cells.

8. Cell survival of adult stem cells following nucleofection may decrease with increase in passage number. Depending on the type of adult stem cell, this could be largely due to heterogeneity in the cell population with passage.

9. FACS analysis is a quantitative assay for measuring not only the transfection efficiency but also monitoring cell survival and immunophenotypic characterization of the transfected cells. In the case of a heterogeneous sample such as total bone marrow, it is important to set gates to exclude cell debris and unlyzed RBC based on their side scatter and forward scatter, and dead cells based on propidium iodide staining.

References

1. Bonetta, L. (2005) The inside scoop-evaluating gene delivery methods. *Nat Methods*, **2**, 875–883.

2. Lakshmipathy, U., Pelacho, B., Sudo, K., Linehan, J.L., Coucouvanis, E., Kaufman, D.S. and Verfaillie, C.M. (2004) Efficient transfection of embryonic and adult stem cells. *Stem Cells*, **22**, 531–543.

3. Lakshmipathy U., Hammer, L. Verfaillie C. (2004) A nonviral gene transfer method for transfecting multipotent adult progenitor cells (MAPC). *Gene Therapy and Regulation*, **2**, 301–312.

4. Liu, Y.P., Dovzhenko, O.V., Garthwaite, M.A., Dambaeva, S.V., Durning, M., Pollastrini, L.M. and Golos, T.G. (2004) Maintenance of pluripotency in human

embryonic stem cells stably over-expressing enhanced green fluorescent protein. *Stem Cells Dev*, **13**, 636–645.

5. Nowling, T., Desler, M., Kuszynski, C. and Rizzino, A. (2002) Transfection of embryonal carcinoma cells at high efficiency using liposome-mediated transfection. *Mol Reprod Dev*, **63**, 309–317.

6. Ren, C.P., Zhao, M., Shan, W.J., Yang, X.Y., Yin, Z.H., Jiang, X.J., Zhang, H.B. and Yao, K.T. (2005) Establishment of human embryonic stem cell line stably expressing Epstein-Barr virus-encoded nuclear antigen 1. *Acta Biochim Biophys Sin (Shanghai)*, **37**, 68–73.

7. Jiang, Y., Jahagirdar, B.N., Reinhardt, R.L., Schwartz, R.E., Keene, C.D., Ortiz-Gonzalez, X.R., Reyes, M., Lenvik, T., Lund, T., Blackstad, M. et al. (2002) Pluripotency of mesenchymal stem cells derived from adult marrow. *Nature*, **418**, 41–49.

8. Aluigi, M., Fogli, M., Curti, A., Isidori, A., Gruppioni, E., Chiodoni, C., Colombo, M.P., Versura, P., D'Errico-Grigioni, A., Ferri, E. et al. (2006) Nucleofection is an efficient nonviral transfection technique for human bone marrow-derived mesenchymal stem cells. *Stem Cells*, **24**, 454–461.

9. Haleem-Smith, H., Derfoul, A., Okafor, C., Tuli, R., Olsen, D., Hall, D.J. and Tuan, R.S. (2005) Optimization of high-efficiency transfection of adult human mesenchymal stem cells in vitro. *Mol Biotechnol*, **30**, 9–20.

10. Kobayashi, N., Rivas-Carrillo, J.D., Soto-Gutierrez, A., Fukazawa, T., Chen, Y., Navarro-Alvarez, N. and Tanaka, N. (2005) Gene delivery to embryonic stem cells. *Birth Defects Res C Embryo Today*, **75**, 10–18.

11. Lorenz, P., Harnack, U. and Morgenstern, R. (2004) Efficient gene transfer into murine embryonic stem cells by nucleofection. *Biotechnol Lett*, **26**, 1589–92.

12. Quenneville, S.P., Chapdelaine, P., Rousseau, J., Beaulieu, J., Caron, N.J., Skuk, D., Mills, P., Olivares, E.C., Calos, M.P. and Tremblay, J.P. (2004) Nucleofection of muscle-derived stem cells and myoblasts with phiC31 integrase: stable expression of a full-length-dystrophin fusion gene by human myoblasts. *Mol Ther*, **10**, 679–87.

13. Siemen, H., Nix, M., Endl, E., Koch, P., Itskovitz-Eldor, J. and Brustle, O. (2005) Nucleofection of human embryonic stem cells. *Stem Cells Dev*, **14**, 378–83.

14. Spangrude, G.J., Klein, J., Heimfeld, S., Aihara, Y. and Weissman, I.L. (1989) Two monoclonal antibodies identify thymic-repopulating cells in mouse bone marrow. *J Immunol*, **142**, 425–30.

10

RNAi Knockdown of Transcription Factor Pu.1 in the Differentiation of Mouse Embryonic Stem Cells

Gang-Ming Zou, Meredith A. Thompson, and Mervin C. Yoder

Summary

Murine embryonic stem (mES) cells are pluripotent cells derived from the inner cell mass of the preimplantation blastocyst. These cells are primitive and undifferentiated and have the potential to become a wide variety of specialized cell types. Mouse ES cells can be regarded as a versatile biological tool that has led to major advances in our understanding of cell and developmental biology. To study specific gene function in early developmental events, gene knockout approaches have been traditionally used, however, this is a time-consuming and expensive approach. Recently, we have shown that small interfering RNA is an effective strategy to knockdown target gene expression, during ES cell differentiation, and consequently, one can alter cell fates in ES-derived differentiated cells. This method will be useful to test the function of a wide variety of gene products using the ES cell differentiation system.

Key Words: ES cell; RNAi; siRNA; Pu.1; Stem cells; Differentiation.

1. Introduction

Murine embryonic stem cell (mES) differentiation is a robust system that can be used to study the regulation of hematopoietic cell development *(1,2)*. As observed in developing murine embryos in vivo, differentiated mES cells express similar cell surface antigens and molecular expression patterns at the appropriate stages of progenitor cell development. Mature blood cells, such as red blood cells, platelets, neutrophils, eosinophils, mast cells, and dendritic and natural killer cells have been generated from mES cells *(3–6)*. ES cells are able to form embryoid bodies (EBs) in the absence of leukemia inhibitory factor

From: *Methods in Molecular Biology, vol. 407: Stem Cell Assays*
Edited by: M. C. Vemuri © Humana Press, Totowa, NJ

(LIF) in culture *(7)*. After dissociation of EBs to single cells by collagenase digestion, about 10% of cells express CD34 *(8,9)*. The isolated CD34$^+$ cells can differentiate into myeloid progenitor cells when induced with appropriate cytokines *(9)*.

RNA interference (RNAi), a term coined by Fire and his colleagues *(10)*, describes the inhibition of gene expression by double-stranded RNAs (dsRNAs) that have been introduced into worms. Guo and Kemphues, in the year 1995, first found that dsRNA was more effective at producing interference of gene expression than either strand individually. After injection into adult *Caenorthabditis elegans*, single-stranded anti-sense RNA had a modest effect in diminishing specific gene expression, whereas double-stranded mixtures caused potent and specific interference *(11)*. RNAi is a multi-step process involving the generation of small interfering RNAs (siRNAs) in vivo through the action of the RNase III endonuclease Dicer. The resulting 21-nucleotide (nt) to 23-nt siRNAs mediate degradation of their complementary RNA *(12)*.

Though the traditional gene knockout techniques play a principal role in analyzing gene function during normal murine development, it is an expensive and time-consuming technique. Recently, siRNA has been used successfully to knock down target gene expression in mammalian cells. ES cells are an attractive model for studying the molecular regulation of cell lineage commitment and cellular differentiation because ES cells give rise to cells derived from all three primary germ layers. Therefore, the ability to selectively knock down specific target genes using siRNA would aid in the understanding of multiple aspects of early murine development. Our approach to knock down Pu.1 gene expression in ES cell differentiation into hematopoietic cells is described in this chapter.

2. Materials

2.1. Cells, Medium, and Serum

1. The D3 ES cells were purchased at passage 7 from American Tissue Culture Collection (ATCC). D3 ES cells were derived from day 3 blastocysts of 129/SVJ mice and were maintained in culture with murine embryonic fibroblast feeder cells.
2. Iscove's Modified Dulbecco's Medium (IMDM) (cat. no. 12440-079, Invitrogen, Carlsbad, CA, USA).
3. Dulbecco's modified Eagle's medium (DMEM) (cat. no. 10569, Invitrogen, Carlsbad, CA, USA).
4. Fetal bovine serum (FBS) (cat. no. HCC6900, StemCell Technologies, Vancouver, Canada).

2.2. siRNA

1. Design Pu.1 siRNA sequence: The Pu.1 siRNA targeting sequence is AATGCAT-GACTACTACTCCTT (*see* **Notes 1** and **2**).
2. Synthesis of Pu.1 siRNA (*see* **Note 3**): Pu.1 siRNA was synthesized commercially from Dharmacon Research Inc (Lafayette, co, USA).
3. Control siRNA: Lamin A/C siRNA (cat. no. D-001620-03-05) was purchased from Dharmacon Research Inc. (*see* **Note 4**)

2.3. Antibodies

1. Rabbit anti-mouse Pu.1 antibody (cat. no. SC-352) and goat anti-human-actin antibodies (cat. no. SC-1615) were purchased from Santa Cruz Bio Technologies Inc (Santa Cruz, CA, USA).
2. Fluoroscein isothiocyanate (cat. no. 09434D) or biotin-labeled rat anti-mouse CD34 monoclonal antibody (cat. no. 03432D, were purchased from BD Pharmingen, San Diego, CA, USA).
3. Goat anti-human Lamin A/C polyclonal antibody (cat. no. SC-6214, Santa Cruz Biotechnologies Inc). This antibody reacts with murine Lamin A/C protein.

2.4. Cytokines

1. Murine LIF (mLIF) (cat. no. 02740, StemCell Technologies).
2. Murine interleukin-3 (mIL-3) (cat. no. 02733, StemCell Technologies).
3. Murine granulocyte-macrophage-colony stimulating factor (mGM-CSF), (cat. no. 02732, StemCell Technologies).
4. Murine stem cell factor (mSCF) (cat. no. 02731, StemCell Technologies).

2.5. Other Reagents

1. Anti-biotin beads (cat. no. 130-091-147) were purchased from Miltenyi Biotec (Auburn, CA, USA).
2. Oligofectamine 2000 (cat. no. 12252011) was purchased from Invitrogen.
3. Methylcellulose-based ES cell differentiation medium (cat. no. M312D) and collagenase (cat. no. 07902) were purchased from StemCell Technologies.
4. Gelatin was purchased from StemCell Technologies.
5. Mouse embryonic fibroblasts (cat. no. 00321) were purchased from StemCell Technologies.
6. Collagenase (cat. no. 07902) was purchased from StemCell Technologies.
7. Magnetic-associated cell sorting (MACS) buffer: 500 ml phosphate buffered saline (PBS) supplemented with 0.5 g BSA and 2 mM EDTA (pass the solution to 0.22-μm filter before use).

3. Methods

3.1. In Vitro Maintenance of ES Cells

1. Mouse D3 ES cells were maintained on murine embryonic fibroblast feeder cells or gelatinized tissue culture dishes (100 mm; Costar, Cambridge, MA, USA) in standard ES culture medium consisting of DMEM supplemented with 15% fetal calf serum (GIBCO, Grand Island, NY, USA), 0.1 mmol L-glutamine, 150 mmol monothioglycerol (MTG), 100 U/ml penicillin, 100 mg/ml streptomycin, and 1000 U/ml LIF (Stem Cell Technologies).
2. The culture medium was changed every day and the cells were passaged every 2 or 3 days *(13)*.

3.2. In Vitro Differentiation of ES Cells

1. D3 ES cells were added to 0.9% methylcellulose medium (StemCell Technologies), 15% FBS (StemCell Technologies), 100 ng/ml SCF (R&D System, Minneapolis, MN, USA), and 450 µM MTG (Sigma, St. Louis, MO, USA) at a cell concentration of 5000–10000 cells/ml plated in a 33-mm Petri dish. Efficient differentiation of ES cells to EBs occurred after 10 days of culture.
2. Harvesting EBs: The EBs were removed from methylcellulose by dilution with IMDM.
3. Isolating CD34$^+$ EB cell population by MACS: CD34$^+$ EB cell isolation was carried out using a MACS magnetic separation device as described previously *(14)*. In summary:

   a. Dissociate EB with Collagenase: add 3 ml Collagenase to the EBs in the Facon tube, and incubate at 37 °C for 1 h.
   b. Wash cells with IMDM.
   c. Label EB cells with anti-CD34-biotin at 2 µg per 1×10^6 cells.
   d. Incubate 15 min at 4 °C.
   e. Wash cells with MACS buffer.
   f. Ada anti-biotin bead (2 µl per 1×10^6 cells) to the cell suspension and incubate at 6–12 °C for 15 min.
   g. Wash cells with MACS buffer at 30× volume.
   h. Pass the cells to the LS/VS column.
   i. Wash the column completely with MACS buffer.
   j. Elute the cells.
   k. Collect eluted CD34$^+$ EB cell populations.

3.3. Preparation of dsRNA and Transfection of siRNA to EB Cells

1. The 21-nt Pu.1 dsRNA sequence and the protocol for transient transfection of siRNAs were reported previously *(15)*. CD34$^+$ EB cells were diluted with fresh medium without antibiotics and transferred to 6-well plates with 5×10^5 cells/well

(500 μl well). We performed a single transfection of siRNA duplex using Oligo-fectamine 2000 Reagent and assayed for gene expression silencing 2 days after transfection (*see* **Note 5**).

2. In a sterile 1.5-ml Eppendorf tube (tube A) add 50 nM Pu.1 siRNA or control siRNA in Opti-MEM medium to reach a final volume of 50 μl. In another sterile 1.5-ml Eppendorf tube (tube B) add 3 μl of Oligofectamine 2000 to 12 μl of Opti-MEM medium (total 15 μl). Incubate tubes A and B separately for 10 min at room temperature.

3. Transfer the contents of tube B to tube A and mix by inversion 5 to 10 times. Do not vortex the tube. Incubate the tube A (containing the mixture) for 25 min at room temperature. The solution will become turbid because siRNA binds to Oligofectamine 2000 to form a complex suspension.

4. Add 35 μl of fresh Opti-MEM medium to tube A to obtain a complex with a final volume of 100 μl.

5. Add the above formed complex (100 μl) to the wells containing cells. Cells are maintained in a low serum conditions (2% serum) according to manufacturer's instruction at 50% confluence. Incubate for about 4 h at 37 °C.

6. Add an equal volume of culture medium with 18% serum to quench the transfection. Allow the cells to continue to grow for 3 days in a 5% CO_2 incubator.

3.4. Culturing Conditions for siRNA Transfected CD34$^+$ EB Cells

1. CD34$^+$ EB cells were cultured with SCF or with SCF, mGM-CSF, and mIL-3 (10 ng/ml) to promote myeloid differentiation (*see* **Fig. 1**).

3.5. Western Blot Analysis for Pu.1

1. The protein was isolated from cultured cells after 72 h of siRNA treatment (*see* **Note 6**).

2. Pu.1 western blot analysis was performed as described previously *(15)*. In brief, 20 μg of protein was separated by Sodium dodecyl sulfate (SDS)-polyacrylamide electrophoresis using a 10% (w/v) polyacrylamide resolving gel and transferred electrophoretically to a nitrocellulose membrane.

3. The blots were blocked with 5% Tris-Buffered Saline (TBST) (TBS containing 0.2% Tween 20) buffer for 1 h. Immunoblotting was performed overnight at 4 °C using the Pu.1 primary antibody (Santa Cruz Biotechnologies Inc) at a 1:500 dilution, or the actin antibody at a 1:1000 dilution, and peroxidase-conjugated secondary antibodies were then used (Amersham Pharmacia Biotech Piscataway, NJ, USA). All immunoblots were visualized by the Amersham electrochemi-luminescence Advance Western Blotting Detection Kit according to the manufacturer's instructions (Amersham Pharmacia Biotech) (*see* **Fig. 2**).

A **B**

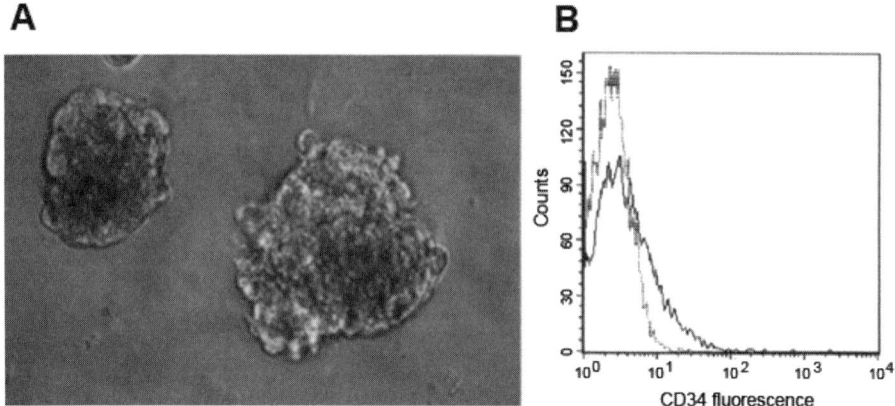

Fig. 1. Embryonic Stem (ES) cells were differentiated into embryoid bodies (EBs) in the absence of murine leukemia inhibitory factor. (**A**) The phenotype of two EBs on day 10 differentiated from D3 ES cells ($\times 400$). (**B**) CD34 antigen expression on EB cells (dark line) detected by flow cytometry (isotype control cells depicted with light line). CD34$^+$ cell populations were isolated from whole EB cell suspension by magnetic-associated cell sorting magnetic bead cell separation.

3.6. Immunocytochemistry for Pu.1

1. CD34$^+$ day 10 EB cells were cultured at 2×10^4 cells per well in a lab-Tek chamber slide system (Fisher Inc, Pittsburg, PA, USA) in the presence of 10 ng/ml SCF.
2. Cells were transfected with Pu.1 siRNA or Lamin A siRNA at 50 nM final concentration.
3. Cells were incubated for 72 h (*see* **Notes 7** and **8**).

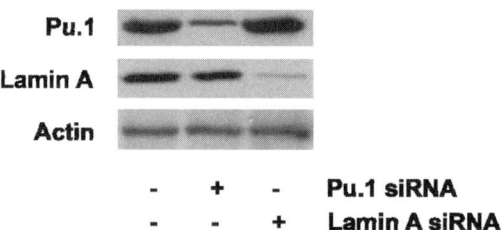

Fig. 2. Western blot analysis was performed to confirm the effect of Pu.1 small interfering RNA (siRNA) on the target gene's level of expression. As shown, Pu.1 expression was reduced when cells were treated by Pu.1 siRNA (lane 2) compared to siRNA untreated cells (lane 1) or control (Lamin A) siRNA-treated (lane 3) cells for 72 h. The data are representative of one of three individual experiments.

4. Fix the cells with 4% paraformaldehyde for 5 to 10 min.
5. The Pu.1 antibody (1:500) was added to the cell slide and incubated overnight at 4 °C with shaking.
6. The slide was washed with PBS for 10 min.
7. The anti-rabbit antibody was added and incubated for 30 min (1:250) (Sigma).
8. The slide was washed with PBS for 10 min.
9. The slide was exposed to Extravidin (1:250 dilution) for 3 h at room temperature.
10. The reaction product was visualized with 0.05 diaminobenzidine (DAB) and 0.1 M phosphatase buffer and 0.01% H_2O_2 (Sigma).
11. The positive cells were scored under a microscope at ×20 magnification microscope (*see* **Fig. 3**).

Control siRNA

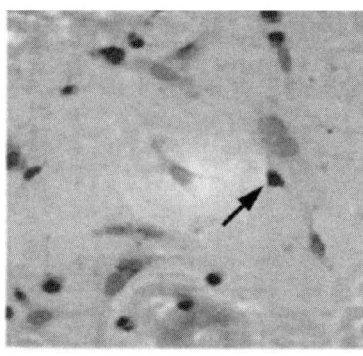

Pu.1 siRNA

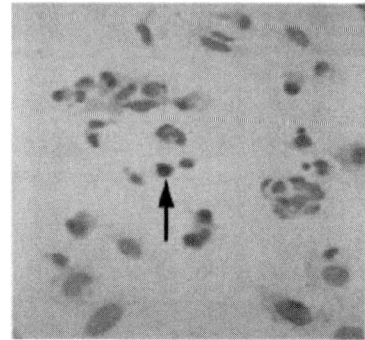

Fig. 3. Immunocytochemical staining of Pu.1 in both CD34$^+$ embryoid body cells treated with control small interfering (siRNA) or Pu.1 siRNA. Pu.1-positive stained cells significantly decreased in Pu.1 siRNA-treated cells compared with cells treated with control siRNA. The arrows show Pu.1 positive staining cells in nucleus.

4. Notes

1. Dharmacon Research Inc has been a valuable supplier of commercial RNAi synthesis in our experience.
2. Both Pu.1 dsRNA and Lamin A dsRNA (cat. no. D-100-5) were obtained from Dharmacon Research Inc. (To design the specifc siRNA sequence, refer the siRNA designer software at the Dharmacon website: www.dharmacon.com.) The key points to consider in selecting a siRNA sequence are as follows: (i) start 75 bases downstream from the start codon; (ii) locate the first AA dimer; (iii) record the next 19 nts following the AA dimer; (iv). Subject the chosen 21-base sequence to a BLAST search (NCBI) database to ensure that only one gene is targeted. A typical 0.2 μmol-scale RNA synthesis provides about 1 m of RNA, which is sufficient for 1000 transfections.
3. We recommend designing siRNA with symmetric 3'TT overhang as previously recommended by Elbashir et al. *(16)* to facilitate equal ratio of sense and antisense target RNA-cleaving siRNAs.
4. We used Lamin A/C siRNA as a control in this protocol. We also recommend using scrambled siRNA as a control for off-target non-specific effects *(17)*.
5. In siRNA transfection experiments, the efficiency of transfection may depend on the cell type, the passage number, and the confluency of the cells. Moreover, the time and the manner of formation of siRNA–liposome complexes (e.g., inversion versus vortexing) are also critical. (Note: Please follow the instructions provided by the manufacturers.) Low transfection efficiencies are the most frequent cause of unsuccessful silencing. Good transfection is a non-trivial issue and needs to be carefully examined for each new cell line to be used.
6. Depending on the abundance and the half-life (or turnover) of the target protein, a knockdown phenotype may become apparent after 1–3 days of siRNA transfection. If no phenotype is observed, depletion of the protein must be tested by immunofluorescence or western blot analysis.
7. siRNAs introduced here usually are effective to knock down the target gene expression and this effect is normally maintained 3–5 day. After that period, siRNA will be degraded and the knockdown effect will be lost and target gene expression will recover in the cells.
8. For longer period of target gene expression knockdown, generation and use of siRNA are discussed in detail by Hannon and Conklin *(18)*.

Acknowledgments

The authors thank Dr. Janet D Rowley at the University of Chicago for her collaboration in this work. This research was supported by CA 84405 and the Spastic Paralysis Foundation of the Illinois-Eastern Iowa Division of Kiwanis International.

References

1. Keller, G., Lacaud, G., and Robertson, S. (1999) Development of the hematopoietic system in the mouse, *Exp Hematol* **27**, 777–87.
2. Daley, G. Q. (2003) From embryos to embryoid bodies: generating blood from embryonic stem cells, *Ann N Y Acad Sci* **996**, 122–31.
3. Hamaguchi-Tsuru, E., Nobumoto, A., Hirose, N., Kataoka, S., Fujikawa-Adachi, K., Furuya, M., and Tominaga, A. (2004) Development and functional analysis of eosinophils from murine embryonic stem cells, *Br J Haematol* **124**, 819–27.
4. Lieber, J. G., Webb, S., Suratt, B. T., Young, S. K., Johnson, G. L., Keller, G. M., and Worthen, G. S. (2004) The in vitro production and characterization of neutrophils from embryonic stem cells, *Blood* **103**, 852–9.
5. Fujimoto, T. T., Kohata, S., Suzuki, H., Miyazaki, H., and Fujimura, K. (2003) Production of functional platelets by differentiated embryonic stem (ES) cells in vitro, *Blood* **102**, 4044–51.
6. Nakayama, N., Fang, I., and Elliott, G. (1998) Natural killer and B-lymphoid potential in CD34+ cells derived from embryonic stem cells differentiated in the presence of vascular endothelial growth factor, *Blood* **91**, 2283–95.
7. Potocnik, A. J., Kohler, H., and Eichmann, K. (1997) Hemato-lymphoid in vivo reconstitution potential of subpopulations derived from in vitro differentiated embryonic stem cells, *Proc Natl Acad Sci USA* **94**, 10295–300.
8. Lu, S. J., Li, F., Vida, L., and Honig, G. R. (2002) Comparative gene expression in hematopoietic progenitor cells derived from embryonic stem cells, *Exp Hematol* **30**, 58–66.
9. Nakayama, N., Lee, J., and Chiu, L. (2000) Vascular endothelial growth factor synergistically enhances bone morphogenetic protein-4-dependent lymphohematopoietic cell generation from embryonic stem cells in vitro, *Blood* **95**, 2275–83.
10. Fire, A., Xu, S., Montgomery, M. K., Kostas, S. A., Driver, S. E., and Mello, C. C. (1998) Potent and specific genetic interference by double-stranded RNA in Caenorhabditis elegans, *Nature* **391**, 806–11.
11. Guo, S., and Kemphues, K. J. (1995) par-1, a gene required for establishing polarity in C. elegans embryos, encodes a putative Ser/Thr kinase that is asymmetrically distributed, *Cell* **81**, 611–20.
12. Shi, Y. (2003) Mammalian RNAi for the masses, *Trends Genet* **19**, 9–12.
13. Zou, G. M., Reznikoff-Etievant, M. F., Hirsch, F., and Milliez, J. (2000) IFN-gamma induces apoptosis in mouse embryonic stem cells, a putative mechanism of its embryotoxicity, *Dev Growth Differ* **42**, 257–64.
14. Zou, G. M., Chen, J. J., Yoder, M. C., Wu, W., and Rowley, J. D. (2005) Knockdown of Pu.1 by small interfering RNA in CD34+ embryoid body cells derived from mouse ES cells turns cell fate determination to pro-B cells, *Proc Natl Acad Sci USA* **102**, 13236–41.

15. Zou, G. M., Wu, W., Chen, J., and Rowley, J. D. (2003) Duplexes of 21-nucleotide RNAs mediate RNA interference in differentiated mouse ES cells, *Biol Cell* **95**, 365–71.

16. Elbashir, S. M., Harborth, J., Lendeckel, W., Yalcin, A., Weber, K., and Tuschl, T. (2001) Duplexes of 21-nucleotide RNAs mediate RNA interference in cultured mammalian cells, *Nature* **411**, 494–8.

17. Zou, G. M., Chan, R. J., Shelley, W. C., and Yoder, M. C. (2006) Reduction of Shp-2 expression by siRNA reduces murine embryonic stem cell-derived in vitro hematopoietic differentiation, *Stem Cells* **24**, 587–94.

18. Hannon, G. J., and Conklin, D. S. (2004) RNA interference by short hairpin RNAs expressed in vertebrate cells, *Methods Mol Biol* **257**, 255–66.

11

StemBase
A Resource for the Analysis of Stem Cell Gene Expression Data

Christopher J. Porter, Gareth A. Palidwor, Reatha Sandie, Paul M. Krzyzanowski, Enrique M. Muro, Carolina Perez-Iratxeta, and Miguel A. Andrade-Navarro

Summary

StemBase is a database of gene expression data obtained from stem cells and derivatives mainly from mouse and human using DNA microarrays and Serial Analysis of Gene Expression. Here, we describe this database and indicate ways to use it for the study the expression of particular genes in stem cells or to search for genes with particular expression profiles in stem cells, which could be associated to stem cell function or used as stem cell markers.

Key Words: Databases; Bioinformatics; Gene expression; DNA microarrays; Serial Analysis of Gene Expression; Stem cells; Data mining.

1. Introduction

A number of recently developed methods for high-throughput analysis of gene expression allow profiling of gene expression in cell or tissue samples in a relatively simple and quick manner. For example, DNA microarrays were applied in the early 2000s for the characterization of genes expressed in stem cells *(1,2)*. However, it soon became evident that the comparison of heterogeneous gene expression information from multiple laboratories and samples was not a trivial problem *(3)*.

In order to profile gene expression in stem cells in an exhaustive and rational way, the Stem Cell Genomics Project (SCGP) was established in the 2003 by

From: *Methods in Molecular Biology, vol. 407: Stem Cell Assays*
Edited by: M. C. Vemuri © Humana Press, Totowa, NJ

the Canadian Stem Cell Network (http://www.stemcellnetwork.ca) as one of its first projects. The objective of the project is to analyze the gene expression in stem cells and their derivatives obtained from research groups across Canada. These are processed in a single facility at the Ontario Genomics Innovation Centre (OGIC) in Ottawa, ensuring a homogeneous experimental protocol and quality control. The data are being stored and made publicly accessible in StemBase *(4)*, a database of gene expression data obtained from stem cells and their differentiated derivatives, with associated query and analysis tools, which is also developed at the OGIC.

Currently, StemBase stores gene expression data derived from Affymetrix expression microarrays *(5)* and Serial Analysis of Gene Expression (SAGE) libraries *(6)*. All samples from which the data are derived were provided by SCGP members and include a variety of mouse embryonic and adult stem cells, human adult stem cells, and differentiated cell types. As of August 2006, the database contains data from 172 mouse samples, 44 human samples, and 3 rat samples. Data can be browsed and queried online or can be downloaded for analysis offline.

As this chapter describes a database, not an experimental method, the structure is slightly different to that of other chapters. **Subheading 2** describes the database structure and the data it contains. A combined Methods and Notes section (*see* **Subheading 3**.) describes how the data can be accessed and analyzed both on the StemBase web site and using standalone tools.

2. Materials

2.1. Data Source

Human Affymetrix microarray data in StemBase were generated using the HGU133 array set, and most mouse data were generated using the MOE430 array set, although a significant minority of the data uses the older MG_U74 array set. The few rat samples were processed using the RAE230 array set. The SAGE libraries created for the SCGP use the longSAGE variant *(7)*, which generates a 17-bp sequence tag from the 3′ end of each mRNA, and the abundance of each tag as an indication of the relative abundance of the mRNA in the source tissue. Most genes generate a unique sequence tag; thus, the tags observed in a library can be used to identify the genes expressed. Data are currently available from six longSAGE libraries generated from mouse cell lines; the libraries are Embryonic Stem Cell, Embryoid Body (9 days differentiation), Neurosphere, Committed Neuronal Progenitor, Undifferentiated Mammosphere, and Differentiated Mammosphere.

2.2. The Database

StemBase is a web application that runs on a Linux server under Apache (*see* database schema in **Fig. 1**). It is written in PHP and uses mySQL as the back end database. The StemBase schema organizes experimental data into "experiments"; these are groupings of "samples" which represent different conditions in a given experiment. Samples contain a set of "replicates" (usually 3). These are biological replicates made using the same sample under the same experimental conditions. Each replicate is associated with one or more "files" which contain the raw expression data, either the Affymetrix GeneChip files or the SAGE tag counts. These original files are stored in the filesystem so that they may be downloaded by users, while expression values from these files are generated and are stored in the database in the *GeneExpression* and *SAGETAG* tables respectively.

2.3. Data Quality and Curation

As most of the samples in StemBase were processed in a single lab, the core facility of OGIC, we are able to ensure uniformity of microarray and

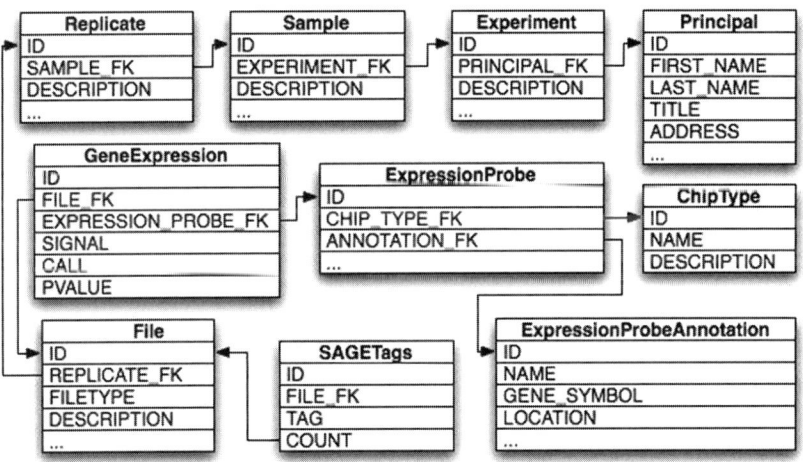

Fig. 1. Simplified StemBase database schema. This schema outlines the relationships between key tables in StemBase. Each box represents a table. The name of the table is at the top, followed by the list of the table fields. Ellipses indicate when fields were omitted for the sake of clarity. Arrows show connections between tables, for example, EXPERIMENT_FK in the sample table connects to ID field of the experiment table indicating which Experiment a given sample is a member of.

SAGE processing and provide consistently high quality data. The use of only Affymetrix microarrays means that we avoided the problems of comparison between microarray platforms, and the use of a limited number of array types means that most samples can be compared for any organism. With few exceptions samples were analyzed in triplicate, allowing more robust statistical analysis. A detailed set of information about the sample and experimental conditions were required from the researcher in order for a given data set to be accepted into StemBase.

2.4. Normalization

StemBase Affymetrix expression data are available as MAS5 *(8,9)* normalized values, p-values, and calls as provided by Affymetrix GCOS software. SAGE data are provided as raw tag counts.

2.5. Data Embargo

StemBase has a staged data release policy, proving preferential access to the originating researcher and SCGP members. For 4 months from the generation of the results, the data are available only to the original researcher. For the next 8 months, the data are available to the Stem Cell Network as a whole. After 1 year, the data are released to the public.

2.6. Export to Other Archives

Although StemBase data are freely available to the public, the database is not as well known as larger, more general purpose archives. In order to make StemBase data more readily accessible, there is an ongoing effort to export all StemBase data into the National Centre for Biotechnology Information's Gene Expression Omnibus (GEO) *(10)*. The presence of the StemBase data in GEO makes them available to a wider scientific community in the context of a central data repository. GEO as a whole, however, does not have the same levels of data consistency as StemBase and does not offer the interactive query and analysis tools detailed below. As of August 2, 2006, the Ontario Genomics Innovation Centre, on behalf of StemBase, was the 12th largest contributor of samples to GEO.

2.7. Access

The database and associated query and analysis tools can be accessed at http://www.stembase.ca. Access requires free registration, which can be accomplished online.

3. Methods and Notes

This section describes some properties of the data in StemBase and aims to describe methods to study the expression of particular genes in stem cells. We are developing web-based tutorials on the use of StemBase and general gene expression data analysis. These will be made available at http://www.ogic.ca in mid-November 2006, under the auspices of the Stem Cell Network.

3.1. Query Tools Within StemBase

The simplest method for exploring the data available in StemBase is through the "Browse" feature available on the home page (*see* **Fig. 2**; left side). This lists the name and description of all experiments in the database, giving an overview of the data available. Details of any experiment can be viewed, providing access to a full description of the experiment, a description of the experimental samples, and quality control data about each individual array hybridization. Links on the page allow downloading of expression values for the whole experiment in a tab-delimited file suitable for analysis in Excel or downloading of the individual. CEL files containing the raw data. These can

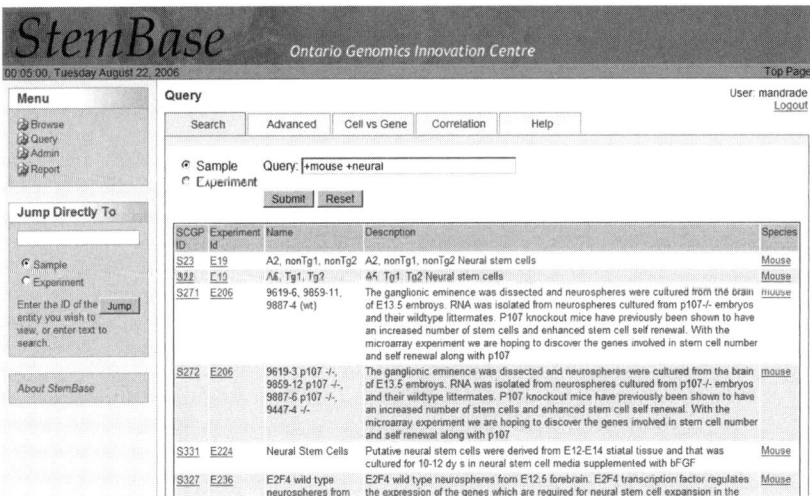

Fig. 2. StemBase snapshot. View of the result of a query in StemBase for mouse samples containing the word "neural" in the description. The table contains the results of the query with links to pages that provide a detailed description of the samples and access to the original data for each sample. Other query options are available at the top of the page, and the left side contains options for navigation of the database.

be analyzed using packages such as DChip *(11)*, BioConductor *(12)*, and other tools described later in this chapter.

A simple query tool allows experiment and sample descriptions to be searched by keyword, and an advanced query tool extends the search to specific fields or to search for experiments from a particular species (*see* an example in **Fig. 2**).

The *Cell vs Gene* tool allows for sections or "slices" of expression data to be extracted and downloaded from StemBase, allowing for further analysis. The "Cell" portion of this tool specifies a range of samples of interest. Samples can be selected based on any combination of species, chip name, origin tissue, cell type/line, stem cell ontology, and SCGP Id. Sample specifications can be simple, such as searching for "Mouse" samples or more focused such as searching for "Blastocyst" in "Mouse" samples. Samples are cross-referenced to the "Gene" part of the tool, which specifies Affymetrix expression probe sets (e.g., 1415673_at), SAGE tags (e.g., TCCATCAAGAAGCTATG), or gene symbols (e.g., Psmc5) of interest. The results returned can include SAGE tags, Affymetrix expression values, or both.

The results of the search are displayed in a series of tables, with the total number of matches found displayed at the top of each table. (i) The "Query Map" cross-references SAGE tags to probe sets and gene names across all the chip types selected. A search using a probe set will return only that probe set and associated SAGE tags, plus the symbol for the associated gene. Searching with a SAGE tag or gene name may find several matching probe sets. (ii) The "Affymetrix Results" table shows the selected probe sets and gene symbols tabulated against the specified samples. The expression values can be shown as either a call (present, absent, or marginal) or as the signal value, averaged among the replicates for each sample. Links to relevant MEDLINE bibliographic references are also provided and marked with an "M" or with an "S" if the corresponding article is relevant to the field of stem cell research [computed as described in Suomela et al. *(13)*]. (iii) The "SAGE Results" table shows the number of tags per million (normalized to allow comparison between libraries) for each relevant SAGE tag in each library that matches the cell parameters.

If the table of Affymetrix results contains multiple probe sets, and contains only probe sets from one chip type, there is an option to generate a heat map of the results. A heat map provides a graphical representation of the expression values of the specified samples and genes with color representing the level of expression. Heat maps can also be generated from SAGE data when multiple tags are returned. After data have been downloaded from StemBase, software such as Cluster and TreeView *(14)* can be used to generate and visualize heatmaps.

3.2. Integration of SAGE Data

The ability to return results from both Affymetrix and SAGE experiments from the *Cell vs Gene* query tool requires a mapping to be made between SAGE tags, Affymetrix probe sets, and gene names. Although both techniques provide measures of gene expression, the issue of comparing their results is not trivial and deserves some explanation.

To generate the Affymetrix–SAGE mapping, we make use of the NetAffx *(15)* annotation of the Affymetrix probe sets. NetAffx is a database of probe set annotations provided by Affymetrix, which identifies the sequence from which the probe set was derived and information such as the gene to which this sequence corresponds, DNA and protein sequences, and functional information. We use the associated sequences in NetAffx to identify expected tags for a given probe set. For each probe set on each array, we obtain the representative ID (the accession for the sequence from which the probe set was designed), the RefSeq ID (the accession for the reference sequence to which the probe set has been assigned), and the UniGene ID (the accession for the UniGene cluster to which the probe set has been assigned). From each sequence we identify expected SAGE tags (the 17 bp distal to the most 3′ NlaIII site in the sequence) and store them in a database, which is then used to associate observed SAGE tags with probe sets and their associated genes.

3.3. Expression Correlation Measures in StemBase

The large number and variety of samples in StemBase, combined with uniformly high level of data quality, offers the opportunity to observe variations in expression levels of a particular gene across many different stem cell types. This can provide insights into the specificity of gene expression in a particular stem cell type. Moreover, it is possible to detect pairs, or by extension groups, of genes with coordinated expression values across the different samples; this may indicate that they participate in the same biological process. These groups will comprise genes that are either up or down regulated together (positive correlation) or genes that have complementary patterns of expression (negative correlation). Assuming that the relationships are linear, we can compute Pearson correlation coefficients between the hybridization values of any pair of probe sets on a chip (e.g., MOE430A), by normalizing all the samples for the given chip after averaging the expression values of the available replicates.

This calculation can provide clues to the function of a non-annotated gene with expression strongly correlated to an annotated gene. The StemBase query page provides an option to search for probe sets with expression positively and/or negatively correlated to a user-specified probe set across all samples in StemBase. After the user specifies the probe set of interest, the array type, and the parameters for list length, the database will return a list of the most positively and/or negatively correlated probe sets.

3.4. Analysis of Significant Differential Expression

StemBase does not currently support the online comparison of microarray data from different samples to find genes with significant differential expression. This type of analysis is, however, possible after downloading expression data from StemBase; one of many methods for performing this analysis is described below. This method can also be used to compare StemBase data with microarray data obtained from elsewhere.

SAM, or Statistical Analysis of Microarrays *(16)*, is a modified version of the t-test used for identifying microarray probe sets showing significant changes of expression between conditions. SAM assigns a score to each probe set on the basis of the change in gene expression relative to the standard deviation of repeated measurements within the replicates for that probe set. SAM modifies the gene-specific t-test by adding a small positive constant to the denominator, reducing the likelihood that genes showing small changes will be identified as changing significantly between conditions. The value of the constant added is calculated so as to minimize the total coefficient of variation for all probe sets. SAM also generates an estimate of the false discovery rate, based on the number of probe sets exceeding a specified cut-off in permutations of the data being examined.

We use the implementation of the SAM algorithm provided in the Bioconductor *(12)* package siggenes. To compare samples by SAM, the .CEL files (from StemBase or elsewhere; all must be from the same array type) are loaded into Bioconductor and normalized using RMA *(17)* or GCRMA (http://www.bepress.com/jhubiostat/paper4). SAM is then run, assigning the replicates for the two samples to test groups for scoring. The software will output lists of probe sets ranked by the likelihood of the observed difference being significant; a cutoff value, Δ, is used to limit the size of this list. Reported values include the observed fold change and its associated p-value.

We have used SAM to compare StemBase Affymetrix expression values obtained in triplicate from mouse R1 Embryonic Stem Cells prior to, and 12 h after, initiation of undirected differentiation following removal of Leukemia inhibitory factor (LIF) from the culture medium. Using a cut-off value of 1.7 returns a list of 103 probe sets, with a maximum p-value of 0.00094. The most significantly different probe sets include a number of genes known to be involved in early embryonic differentiation, including Socs3, Klf4, Klf5, and Bcl3.

3.5. Gene Ontology Annotations of Affymetrix Probe Sets

Study of the expression data in StemBase is facilitated by information about the function and cellular localization of transcripts. For example, a user might be interested in examining the level of expression of transcription factors. Or examination of a set of genes differentially expressed between two cellular types could reveal that most genes involved in a particular signaling cascade are down regulated. Functional and cellular localization data of this type is provided by the Gene Ontology *(18)* (GO, http://www.geneontology.org/), which is used to annotate genes with molecular function, cellular localization, and the biological processes in which they are involved. GO annotation can be obtained from several sources, some of which are used to generate a basic level of GO annotation of probe sets in the NetAffx database mentioned above. NetAffx GO annotations are updated regularly, as new versions of NetAffx are released. However, the annotations appear in many cases to be incomplete and in need of improvement.

In order to extend and improve the GO annotation available for probe sets, we identified two strategies [described in **ref.** *19*]. (i) Obtaining GO annotations from databases to which NetAffx provides direct links: Entrez Gene *(20)*, InterPro *(21)*, SwissProt *(22)*, and Gene Ontology Annotation [GOA *(23)*] via SwissProt links to UniProt. (ii) Inferring GO annotations from associations between keywords and GO terms in other databases extracted by data mining. Using this approach results in a large increase in the number of annotations: 33,821 new GO terms associated with probe sets on the MOE430 array and 35,608 on HG-U133. This resulted in annotation of 2759 probe sets on MOE430 and 2987 probe sets on HG-U133 that previously had no associated GO terms. These new annotations can be accessed through a web server (http://www.ogic.ca/p2g), where the predicted annotations can be retrieved, along with their supporting evidence. These extended annotations have been recently implemented in the query tool of StemBase.

Beyond StemBase, GO annotations can be also used to interpret microarray results. Given a selection of probe sets or genes obtained by analysis of microarray data (e.g., genes correlated to a particular probe set of interest in StemBase), data useful for interpretation can be obtained by an examination of the associated GO annotations. Several methods have been derived to assign significance to the enrichment of GO terms for sets of genes using variants of a t-test analysis. One example, GOstat *(24)* provides a web server where the analysis can be performed on a list of probe sets from any of the Affymetrix microarrays currently used in StemBase.

3.6. Biomarker Discovery

Interest in identifying molecular markers has surged in recent years, with the availability of high-throughput technologies such as expression arrays. These are ideal technologies for biomarker discovery due to their ability to assess the levels of tens of thousands of RNA species simultaneously with low cost per assay. Large volumes of data can be generated and used to identify candidate transcripts for future development into biomarkers.

Once identified, biomarkers can be extremely useful as indicators of cellular identity, for example, in cell sorting experiments *(25)* or for the prognosis of disease phenotypes *(26,27)*. In the latter case, a biomarker could be used to predict the probability of recurrence or the aggressiveness of certain cancers.

The most straightforward method to identify biomarkers for samples of interest is to identify sample-specific up regulated genes with a technique such as significance analysis of microarrays (*see* **Subheading 3.4.**). More complex methods can include the discovery of multi-gene signatures which are associated with the phenotype of interest. In either case, the essence of a good biomarker is that it can distinguish between two types of sample based on a threshold of some quantifiable property, here, the level of expression on a microarray.

It is simple to contrast two groups of samples if one is only interested in a specific comparison, for example, between two types of hematopoietic cells at different stages of differentiation. However, contrasting a group of interest (e.g., undifferentiated hematopoietic cells) against a larger, heterogeneous data set (such as StemBase) can generate a more robust set of biomarkers because of the increased specificity of the putative molecular markers for the selected group (e.g., What genes distinguish hematopoietic stem cells from 40 other cell types?).

We have generated a method and a web server to facilitate Biomarker discovery in StemBase *(28)*. Larger datasets like StemBase allow for a larger number of comparisons but create a computational problem as this number of comparisons rises exponentially with the size of the database. This problem can be mitigated by estimating possible comparisons with the data itself. By first identifying individual genes that have biomarker-like properties (testing for a demarcation between the expression levels of two arbitrary groups of samples), we generate a number of possible two-state arrangements of samples across the dataset.

For each unique arrangement of samples, we can then use a non-parametric test to compute sets of genes which are able to demarcate samples in a similar way. The results are predicted associations between groups of samples based on supporting expression data, along with sets of genes predicted to be co-markers for each group of samples. We have used this methodology to analyze the distributions of known stem cell related markers in StemBase mouse samples. StemBase data can be studied using this method at http://www.ogic.ca/projects/markerserver/enter.php. The web site includes detailed instructions and examples.

References

1. Ivanova NB, Dimos JT, Schaniel C, Hackney JA, Moore KA, Lemischka IR. (2002) A stem cell molecular signature. *Science* **298**: 601–604.
2. Ramalho-Santos M, Yoon S, Matsuzaki Y, Mulligan RC, Melton DA. (2002) "Stemness": transcriptional profiling of embryonic and adult stem cells. *Science* **298**: 597–600.
3. Vogel G. (2003) Stem cells. 'Stemness' genes still elusive. *Science* **302**: 371.
4. Perez-Iratxeta C, Palidwor G, Porter CJ, et al. (2005) Study of stem cell function using microarray experiments. *FEBS Lett* **579**: 1795–1801.
5. Lockhart DJ, Dong H, Byrne MC, et al. (1996) Expression monitoring by hybridization to high-density oligonucleotide arrays. *Nat Biotechnol* **14**: 1675–1680.
6. Velculescu VE, Zhang L, Vogelstein B, Kinzler KW. (1995) Serial analysis of gene expression. *Science* **270**: 484–487.
7. Saha S, Sparks AB, Rago C, et al. (2002) Using the transcriptome to annotate the genome. *Nat Biotechnol* **20**: 508–512.
8. Liu WM, Mei R, Di X, et al. (2002) Analysis of high density expression microarrays with signed-rank call algorithms. *Bioinformatics* **18**: 1593–1599.
9. Hubbell E, Liu WM, Mei R. (2002) Robust estimators for expression analysis. *Bioinformatics* **18**: 1585–1592.
10. Wheeler DL, Barrett T, Benson DA, et al. (2006) Database resources of the National Center for Biotechnology Information. *Nucleic Acids Res* **34**: D173–180.
11. Li C, Wong WH. (2001) Model-based analysis of oligonucleotide arrays: expression index computation and outlier detection. *Proc Natl Acad Sci USA* **98**: 31–36.
12. Gentleman RC, Carey VJ, Bates DM, et al. (2004) Bioconductor: open software development for computational biology and bioinformatics. *Genome Biol* **5**: R80.
13. Suomela BP, Andrade MA. (2005) Ranking the whole MEDLINE database according to a large training set using text indexing. *BMC Bioinformatics* **6**: 75.

14. Eisen MB, Spellman PT, Brown PO, Botstein D. (1998) Cluster analysis and display of genome-wide expression patterns. *Proc Natl Acad Sci USA* **95**: 14863–14868.

15. Liu G, Loraine AE, Shigeta R, et al. (2003) NetAffx: Affymetrix probesets and annotations. *Nucleic Acids Res* **31**: 82–86.

16. Tusher VG, Tibshirani R, Chu G. (2001) Significance analysis of microarrays applied to the ionizing radiation response. *Proc Natl Acad Sci USA* **98**: 5116–5121.

17. Irizarry RA, Hobbs B, Collin F, et al. (2003) Exploration, normalization, and summaries of high density oligonucleotide array probe level data. *Biostatistics* **4**: 249–264.

18. Harris MA, Clark J, Ireland A, et al. (2004) The Gene Ontology (GO) database and informatics resource. *Nucleic Acids Res* **32**: D258–261.

19. Muro EM, Perez-Iratxeta C, Andrade-Navarro MA. (2006) Amplification of the Gene Ontology annotation of Affymetrix probe sets. *BMC Bioinformatics* **7**: 159.

20. Maglott D, Ostell J, Pruitt KD, Tatusova T. (2005) Entrez Gene: gene-centered information at NCBI. *Nucleic Acids Res* **33**: D54–58.

21. Mulder NJ, Apweiler R, Attwood TK, et al. (2005) InterPro, progress and status in 2005. *Nucleic Acids Res* **33 Database Issue**: D201–205.

22. Bairoch A, Apweiler R, Wu CH, et al. (2005) The Universal Protein Resource (UniProt). *Nucleic Acids Res* **33 Database Issue**: D154–159.

23. Camon E, Magrane M, Barrell D, et al. (2004) The Gene Ontology Annotation (GOA) Database: sharing knowledge in Uniprot with Gene Ontology. *Nucleic Acids Res* **32 Database issue**: D262–266.

24. Beissbarth T, Speed TP. (2004) GOstat: find statistically overrepresented Gene Ontologies within a group of genes. *Bioinformatics* **20**: 1464–1465.

25. Singh SK, Clarke ID, Hide T, Dirks PB. (2004) Cancer stem cells in nervous system tumors. *Oncogene* **23**: 7267–7273.

26. Sieuwerts AM, Look MP, Meijer-van Gelder ME, et al. (2006) Which cyclin E prevails as prognostic marker for breast cancer? Results from a retrospective study involving 635 lymph node-negative breast cancer patients. *Clin Cancer Res* **12**: 3319–3328.

27. Wang Y, Jatkoe T, Zhang Y, et al. (2004) Gene expression profiles and molecular markers to predict recurrence of Dukes' B colon cancer. *J Clin Oncol* **22**: 1564–1571.

28. Krzyzanowski PM, Andrade-Navarro MA. (2007) Computational detection of molecular markers in sets of gene expression data. (submitted)

12

Isolation of Stem Cells from Human Umbilical Cord Blood

Nishanth P. Reddy, Mohan C. Vemuri, and Reddanna Pallu

Summary

Umbilical cord blood (UCB) is gaining more prominence in recent times as a source of non-embryonic multipotent stem cells. Global annual human birth rate (100 million) presents UCB as the largest non-controversial stem cell source, with an added advantage of naive immune status. Cord blood stem cells are routinely utilized in stem cell transplantation in leukemia patients and carry huge potential to treat other human diseases with less concern of rejection. Because UCB contains low number of stem cells, their use is associated with significant delays in engraftment of neutrophils and platelets. Development of reliable methods for isolation and expansion of cord blood stem cells is critical for consequent clinical application. The focus of this chapter is to review the methods currently used by different research groups and to recommend an isolation protocol that yields optimal number of UCB stem cells.

Key Words: KD Umbilical cord blood; $CD34^+$; Stem cells; Ex vivo expansion; Growth factors.

1. Introduction

Development of methods to isolate stem cells from umbilical cord blood (UCB) and enrich them through ex vivo expansion can potentially benefit clinical transplantation and gene therapy. Multipotent stem cells derived from UCB are gaining more attention as a means of ethically acceptable source of stem cells (1). Clinical application of UCB stem cells is at present recognized as a valid approach for treating malignant and non-malignant hematopoietic disorders (see **Table 1**) (2). UCB was originally used as an alternative source of hematopoietic cells for transplantation in a child with Fanconi's anaemia (3).

From: *Methods in Molecular Biology, vol. 407: Stem Cell Assays*
Edited by: M. C. Vemuri © Humana Press, Totowa, NJ

Table 1
Diseases Treated by Cord Blood Transplantation

Malignant diseases	Non-malignant diseases
Acute lymphocytic leukemia	Fanconi's anemia
Acute myelocytic leukemia	Adrenoleukodystrophy
Chronic myelogeneous leukemia	Hunter syndrome
	Blackfan–Diamond syndrome
Myelodysplastic syndrome	Dyskeratosis congenital
Juvenile chronic myelogeneous leukemia	A megakaryocytic thrombocytopenia
	Globoid cell leukodystrophy
Neuroblastoma	Gunther disease
	Severe combined immune deficiency
	Hurler syndrome
	Idiopathic aplastic anaemia
	Kostmann syndrome
	Osteoporosis
	Lesch–Nyhan syndrome
	Thalassaemia
	X-linked lymphoproliferative syndrome

Preclinical in vitro studies demonstrated the proliferative advantage of primitive cord blood stem cells, compared with the bone marrow *(4)*. Several research groups have evaluated a number of formulations for their ability to support survival and expansion of hematopoietic stem cells (HSCs) from UCB. New approaches to enhance the expansion of human cord blood stem cells include co-culture with mesenchymal stem cells *(5)*, human umbilical vein endothelial cells *(6)*, angiopoietin-like proteins (Angpt12 or Angpt13) *(7)* and optimization of cultures in serum-free defined medium *(8)*. The focus of this chapter is to evaluate existing approaches and recommend an optimal simplified isolation protocol that yields optimal number of UCB stem cells.

1.1. Transplanted Stem Cells from UCB Restore Normal Hematopoiesis in Pediatric Leukemia/Lymphomas

The high proliferative capacity of UCB stem cells to repopulate bone marrow has been confirmed in clinical transplants *(9)*. The two large patient series that have been analyzed in detail were collected by the *New York*

Placental/Umbilical Cord Blood Program (562 patients) *(10)* and by *Eurocord* (331 patients) *(11)*. In both groups, more than 80% of transplants were performed in pediatric recipients suffering from leukemia or lymphoma. The clinical data from the *New York Placental/Umbilical Cord Blood Program* show that platelet and myeloid engraftment was significantly related to transplant cell dose *(10)*. In particular, platelet engraftment was prolonged compared to transplant recipients of hematopoietic progenitor cells (HPC) from bone marrow or mobilized peripheral blood *(9)*. The frequency of transplant-related events such as the occurrence of death, autologous reconstitution or second graft was significantly related to cell dose and the recipient age and diagnosis.

Graft versus host disease (GvHD) is a frequent complication in clinical transplants, and the incidence of grade III–IV acute GvHD and chronic GvHD were 23 and 25%, respectively. In addition, the frequency of relapse, which was 26% by 1 year in acute leukemia recipients, was significantly related to the disease stage at transplantation. The Kaplan–Meier survival estimate at 100 days was 49%. The outcome of unrelated UCB transplantation in children with acute leukemia was investigated by Locatelli et al. *(12)*, who examined 40 patients with lymphoblastic leukemia and 20 patients with myeloid leukemia, reported to the *Eurocord Registry* during April 1990–December 1997. This patient group included 42 and 18 patients transplanted in good risk and poor risk conditions. Children receiving transplantation during first and second complete remission were considered as belonging to good risk group. By contrast, patients in third or subsequent remission, relapse or partial remission were grouped under poor risk. Kaplan–Meier estimates of 2-year event-free survivals in the good risk and poor risk groups were 40 and 7%, respectively *(12)*.

1.2. Immune Phenotype of UCB Stem Cells

UCB is a rich source of hemopoietic stem and progenitor cells but contains fewer T cells than bone marrow, which may permit greater degree of mismatch without increased GvHD *(13)*. Many cancer centers developed techniques for the removal of T lymphocytes, in order to reduce the severity of GvHD. T-cell depletion did decrease the incidence of GvHD in patients undergoing allogeneic stem cell transplantation. Unfortunately, removal of T lymphocytes also caused an increased risk of graft failure, as selective T lymphocytes are necessary for engraftment. UCB shows the lower proportion of $CD4^+/CD45^+$ T cells, and their response to the stimulation is decreased when compared with human adult peripheral blood (APB) *(14)*. The notable difference between the T lymphocytes in UCB and those derived from adult marrow or mobilized peripheral blood stem cells is the maturational status. Because the fetus is exposed to few foreign antigens, the T lymphocytes in UCB are almost exclusively naive. Naive T lymphocytes express a phenotype that is identifiable as CD45RA$^+$, CD45R0$^-$ and CD62L$^+$, naïve T-cell marker. As individuals age, there is an increase in the frequency of T cells that have differentiated into a memory phenotype, CD45RA$^-$ CD45R0$^+$ and CD62L$^{-/low}$ as a result of antigenic exposure. The predominant naive phenotype of T lymphocytes in UCB may contribute to the reduced alloreactivity observed in UCB transplant recipients *(15)*. According to the phenotypic studies, the total B-cell numbers of UCB are comparable to APB but are primarily of the immature phenotype (CD5$^+$/CD19$^+$) *(16)*. The total number of natural killer (NK) cells is also comparable to those found in APB. But the activity of cord blood (CB) NK cells has been reported to be low, compared with APB *(17)*.

Cairo et al. *(18)* demonstrated a significant reduction in mRNA expression and decreased protein production of granulocyte macrophage-colony stimulating factor, granulocyte colony stimulating factor (G-CSF), interleukin (IL)-3 *(19)*, macrophage colony-stimulating factor *(20)*, transforming growth factor-B1 and macrophage inflammatory protein-1a *(21)* in activated UCB mononuclear cells (MNCs) compared with adult MNC. Similarly, the same group demonstrated a significant reduction in mRNA and protein expression of IL-12, IL-15, and IL-18 in activated CB MNC compared with activated APB MNC *(22,23)*. Expression and protein production of IL-11 *(24)*, stem cell factor (SCF) *(25)* and thrombopoietin *(26)* are significantly increased in UCB compared with adult fibroblasts and endothelial cells.

1.3. Ex vivo Expansion of UCB Stem Cells

Major limitations in the use of stem cells from UCB for bone marrow transplantation are often the availability of a suitable donor, and when a donor is available, the number of stem cells that can be isolated from donor. Hence, the development of reliable methods for the expansion of stem cells from UCB is mandatory to ensure effective treatment to a wider group of patients. Each umbilical cord has cells enough only to transplant a small child. In order to transplant an adult or more fetal transplants, the stem cells have to be expanded ex vivo, retaining their stem cell phenotype, self-renewal and lineage-specific differentiation abilities. The differentiated cells allow short-term engraftment that reduces the effect of neutropenia and thrombocytopenia, potentially preventing the risk of early mortality and graft failure in transplant recipients *(27)*. Studies have shown that the rate of neutrophil and platelet engraftment correlates with the number of CD34$^+$cells and total nucleated cells infused.

The undifferentiated primitive stem cells will allow long-term engraftment and reconstitution of hematopoietic system for the patient, an observation that is supported by a higher prevalence of long-term repopulating hematopoietic stem cells in UCB relative to short-term repopulating hematopoietic stem cells.

Several groups have worked on UCB stem cells and evaluated a number of formulations for their ability to support survival and expansion of HSCs (*see* **Table 2**). Key cytokines have been identified as extracellular regulators of hematopoiesis and have been routinely used in the isolation and expansion of human UCB stem cell population. Some cytokine combinations are G-CSF, SCF, Flt-3 ligand (FL), thrombopoietin (TPO), megakaryote growth and development factor, several ILs (IL-1β, IL-2, IL-3, IL-6 and IL-7) and interferon-γ. Some cytokines or cytokine combinations also exhibited a negative effect on the self-renewal ability of HSCs in vitro *(34,35)*. The combinations of IL-6 in concert with FL, SCF, IL-3 and G-CSF have supported the expansion of stem cells from UCB in vitro *(36,37)*. In particular, SCF, TPO and G-CSF were found to have beneficial effect during ex vivo expansion of human UCB stem cells grown in a serum/animal protein-free medium *(29)*. The expanded cells were assessed for surface antigen analysis, colony-forming cell (CFC) assay, long-term culture-initiating cell (LTC-IC) assay and competitive repopulating unit assay. Many studies rely on in vitro and in vivo assays for the assessment of HSC activity. Among the assays currently employed, the CFC assay, also referred to as the methylcellulose assay, and LTC-IC assay are the commonly employed in vitro assays for the quantification of committed hematopoietic progenitors. The non-obese diabetic/severe combined immunodeficient (NOD/SCID) mouse is widely accepted as an in vivo model to assess the stem cell potential and optimization before proceeding to human clinical transplantation/trials.

2. Materials
2.1. Collection and Processing of Cord Blood

1. Standard blood collection bags containing citrate phosphate dextrose adenine (CPDA) with 20-gauge syringe
2. Ficoll-paque PLUS (cat. no. 17-1440-02, Amersham Biosciences, Uppsala, Sweden).
3. Dulbecco's phosphate-buffered Saline (PBS) without Ca^{+2} and Mg^{+2} (cat. no. TS1006, Himedia, Mumbai, India).
4. Ammonium chloride solution (cat. no. A0171, Sigma, St. Louis, MO, USA).
5. Trypan blue solution, 0.4% (cat. no. 93595, Sigma).

Table 2
Methods Used by Different Research Groups for Ex-vivo Expansion

Method Employed	Key Components of Media	Fold Expansion (or) Results	Phenotype Profile	References
Expansion and differentiation of UCB products using two-step expansion culture	DM, SCF,G-CSF, MGDF (100 ng/mL)	TNC: 438 $CD34^+$: 29	CD33, CD19, CD14, CD61, CD3 and Glycoporin A	Niece et al. (28)
IL-3 improves ex vivo expansion	X-vivo 10, rhIL-3, TPO, SCF, FL 1% Glutamine	$CD34^+$: 20 LTC-IC: 16	$CD34^+ CXCR_4$	Robmanith et al. (26)
Expansion of UCB $CD34^+$ hematopoietic progenitor cells in co-culture with autologus umbilical vein endothelial cells	X-vivo medium, DMEM, SCF(50ng/mL), FL-3 (50ng/mL), IL-6 (10ng/mL), IL-1 (10ng/mL), vascular endothelial growth factor	5 growth factors $CD34^+$: 22.7 ± 1.9 3-Growth factors $CD34^+$: 22.6 ± 1.6	CD45 CD34 CDG1A GLyA CD14 $CD15^+$	Yildirim et al. (29)
Serum-free expansion of $CD34^+$ UCB using Stemline™ HSC expansion medium	Stem line™ SCF TPO and G-CSF (100 ng/mL)	Greater overall expansion of TNC and $CD34^+$ is seen. $CD34^+$: 30.5 ± 2.02	CD34, CD41, CD15	Allison et al. (30)

Production of human UCB stem cells with embryonic characteristics	IMDM+10% FCS, TPO (10 ng/mL), FL (50 ng/ml), c-Kit ligard (20 ng/mL)	Second generation CBE population significantly expanded to 168-fold	TRA-1-60, TRA-1-81, SSEA-4, SSEA-3, OCT-4	McGuckin et al. (5)
Ex vivo expansion of UCB–MNC on mesenchymal stem cells	DM, rh-SCF, rhG-CSF, rh-MGDF (100 ng/ml)	TNC: 10–20 GM-CFC: 7–18 HPP-CFC: 2–5 CD34$^+$: 16–37	CD34, CD19, CD13	McNiece et al. (8)
A systematic strategy using factorial design and steep ascent method to optimize ex vivo expansion medium for HSCs derived from UCB mononuclear cells	IMDM, 4 g/L BSA, 0.71 g/L insulin, 27.8 µg/mL transferrir, 53 ng/mL TPO, 2.05 ng/mL IL-3, 2.36 ng/mL IL-6, 0.69 ng/mL IL-11, 16 ng/mL SCF, 4.43 ng/mL FL, 1.56 ng/mL GM-CSF, 2.64 ng/rrL SCGF	CD34$^+$: 30.4 CD34$^+$/CD38$^-$: 63.9 CFC: 10.7 LTC − IC : 2.8	CD34$^+$ CD34$^+$/CD38$^-$ CFC	Yao et al. (31)

(Continued)

Table 2
(Continued)

Method Employed	Key Components of Media	Fold Expansion (or) Results	Phenotype Profile	References
Expansion of SCID repopulating cells and increased engraftment capacity in UCB following X-vivo culture with human brain endothelial cells	M199, IMDM, FBS 10%, 50 µg/mL Heparin, L-glutamine, 100 µg/mL, SCF 120 ng/mL, Flt-3 50 ng/mL, TPO 20 ng/mL, endothelial cell growth supplement	TNC: 29 CD34[+]: 19 CD34[+]/CD38[−]: 156	CD13, CD19, CD45	Chute et al. (**32**)
Ex vivo expansion of UCB CD34[+] cells using the DIDECO Pluricell system	X-vivo medium FLT3, SCF, TPO	TNC: 230.4 ± 91.5 CD34[+]: 21.0 ± 11.9	CD34, CD33, & CD61	Astori et al. (**37**)
Angiopoietin-like proteins stimulate ex vivo expansion of hematopoietic stem cells	Angptl2 100 ng/mL, SCF 10 ng/mL, TPO 20 ng/mL, IGF-2 20 ng/mL	LT-HSCs: 24–30	CD45[+], CD34[+]	Zhang et al. (**33**)

BSA, bovine serum albumin; DM, defined media; DMEM, Dulbecco's Modified Eagle's Medium; FCS, Fetal Calf Serum; FL, Flt-3 ligand; G-CSF, granulocyte colony stimulating factor; HSC, hematopoietic stem cell; IL, interleukin; MGDF; megakaryote growth and development factor; MNC, mononuclear cell; SCF, stem cell factor; TPO, thrombopoietin; UCB, umbilical cord blood.

2.2. Flow Cytometry

1. Iscove's modified Dulbecco's medium (IMDM)/2%/FBS: IMDM with 2% fetal bovine serum (FBS) (cat. no. 14-502F Cambrix Bioscience, Verviers, Belgium). Store at 4 °C.
2. HF buffer (HBSS+FBS): Phenol red-free Hank's Balanced Salt Solution (HBSS) (cat. no. H6648, Sigma) containing 2% FBS. Filter Sterilized.
3. HF/PI: HF with $2 \mu g/mL$ propidium iodide (PI) (cat. no. P4170, Sigma).
4. Anti-human Phycoerythrin (PE)-conjugated CD34 antibody (cat. no. 550761, BD Biosciences).
5. PE-tagged IgG_1k as isotype control (cat. no. 555749, BD Biosciences Pharmingen, San Diego, CA, USA).

3. Method

3.1. Collection of Cord Blood

1. An ideal sample is fresh and anti-coagulated.
2. Human UCB was obtained with the informed consent of the donors from the department of gynecology, Vijaya Hospital, Hyderabad, India.
3. Carry out infectious disease screening for syphilis, hepatitis B & C, HIV I & II, for the mother.
4. Collect the blood following institutional guidelines. CB can be collected in vivo, after deliver of the baby but before the placenta is delivered, or in vitro after the placenta is delivered. (*see* **Fig. 1**) *(38)* (*see* **Note 1**).
5. Collect UCB in clean aseptic conditions using standard blood collection bags, containing CPDA as anticoagulant (*see* **Note 3**).
6. Add approximately 25 mL CPDA for every 100–120 mL of UCB collected.
7. Gently shake the bag during collection of blood, so that the anticoagulant freely mix with UCB.
8. Transport the sample immediately from maternity units and store at 4 °C, till its use in further steps (*see* **Note 2**).

3.2. Separation of Mononuclear Cells

Separation of MNCs from CB is performed by density gradient centrifugation using Ficoll. NH_4Cl can also be used for red blood cell lyses, as CB possess high platelet count.

3.2.1. Density Gradient Separation Procedure (Ficoll)

1. Keep the sample at room temperature for 1 h before processing.
2. Dilute the CB sample 1:1 in PBS with out Mg^{++} and Ca^{++}
3. Pour 20 mL Ficoll into 50-mL tube (cat. no. 546040, Tarsons Kolkata, WB, India) and slowly layer 25 mL of diluted UCB on top (*see* **Note 4**).

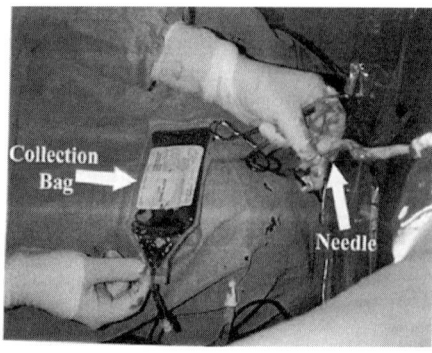

In vivo collection of cord blood *(Moise 2005)*

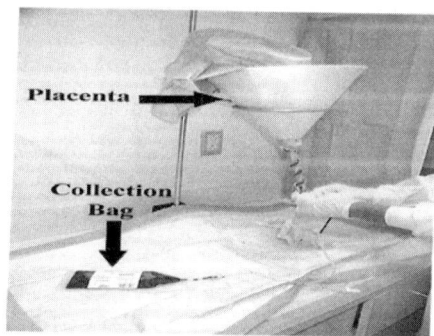

In vitro collection of cord blood. *(Moise 2005)*

Fig. 1. In vivo and in vitro collection of cord blood *(38)*.

4. Centrifuge at room temperature, 360g for 20 min.
5. Carefully pipette off cloudy interface layer (buffy coat) and transfer into a clean 50-mL centrifuge tube (*see* **Note 5**). Wash these cells with 50-mL PBS without Mg^{++} and Ca^{++}
6. Wash the MNCs twice with the 20 mL of PBS (*see* **Note 6**).
7. MNC count and viability enumerated by hemocytometer using Trypan blue.
8. Resuspend cells in media with serum or HBSS with 2–5% FBS.

3.2.2. Red Cell Lysis Procedure (NH_4Cl)

1. Centrifuge cells and wash twice in PBS without Mg^{++} and Ca^{++}.
2. Resuspend in cold NH_4Cl solution at 3–4 times the original sample volume.
3. Incubate on ice for 10 min.
4. Centrifuge cells and wash twice in PBS without Mg^{++} and Ca^{++}. Resuspend cells in media with serum or HBSS with 2–5% FBS.

3.3. Flow Cytometric Analysis

1. After isolating, the light density fraction of UCB cells was calculated ($< 1.077\,g/mL$) by centrifugation on Ficoll.
2. Acquisition of cells was performed using a FACS Scan flowcytometer (FACS Caliber, Becton Dickinson, San Jose, CA, USA).
3. Add 10^5 cells to each of three labeled tubes. Add PE-conjugated isotype control antibody to one tube, PE-conjugated CD34 antibody to a second tube and HF only to the third tube.
4. Place the remaining cells in the fourth tube and add anti-CD34–PE. Incubate all the tubes for 30 min on ice and protect from light (*see* **Note 7**).
5. Add HF to all tubes and centrifuge the cells at 300–350 g for 10 min. Discard the supernatants and repeat adding HF/PI to the second wash. Finally, resuspend the cells in HF for flow cytometric analysis (*see* **Note 8**).
6. Gate the low-density cells on the forward scatter (FSC) versus side scatter (SSC) dot plot (*see* **Fig. 2A**).
7. Evaluate non-specific antibody binding by plotting SSC versus IgG_1_PE, cells stained by control antibody (*see* **Fig. 2B**).
8. To estimate the percentage of CD34$^+$ cells, draw a third plot (SSC versus CD34$^+$ fluorescence) in order to gate the PE-positive cells. Cells, 0.44%, were gated as CD34$^+$–PE (*see* **Fig. 2C**).

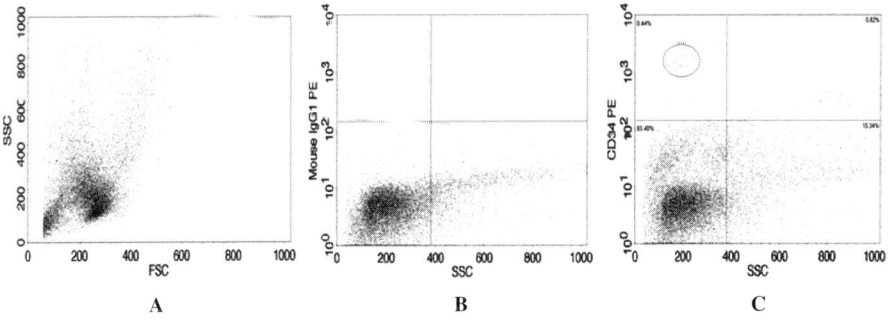

Fig. 2. Flow cytometric analysis of umbilical cord blood-derived CD34$^+$ cells. (**A**) Gating of cells on a forward scatter (FSC)/side scatter (SSC) dot plot. (**B**) Evaluation of the frequency of non-specific antibody binding on a SSC versus IgG_1–phycoerythrin (PE) fluorescence dot plot of cells stained with control antibody. (**C**) Evaluation of the frequency of CD34$^+$ cells on a SSC versus CD34–PE dot plot of cells stained with anti-CD34 monoclonal antibody.

4. Notes

1. CB should be collected and processed following the local ethical guidelines and with the approval of Institutional Review Board.
2. After the collection of CB, it is carefully shifted from maternity house to lab in a box maintaining at 4 °C and processed within 24 h.
3. The use of heparin as anticoagulant tends to cause clumping and lower separation efficiency. If the cells are collected in heparin, wash twice with PBS (without Mg^{++} and Ca^{++}) and continue the procedure, using medium without heparin
4. It is important to layer the diluted CB on Ficoll gently and slowly. Plastic disposable droppers are used to layer CB suspension Ficoll. The maximum ratio of diluted CB and Ficoll should be 1:1. Care should be taken while dropping Ficoll; it should not touch the walls of centrifuge tube.
5. It is critical to collect the MNC interface with a minimum amount of Ficoll and plasma supernatant by using disposable droppers.
6. To optimize cell viability during processing, one should minimize exposure to conditions which will cause clumping and cell death (e.g., medium without serum/protein or prolonged contact with Ficoll).
7. Higher temperature and longer incubation time for fluorescent labeling would result in non-specific binding.
8. Maintaining a clean flow cytometer significantly reduces the chances for bacterial contamination of the sample and also helps in minimizing the pathogen-associated risk to instrument operator.

Acknowledgments

This research work was supported by a grant from the Pacific Stem Cell Bank & Biomedical Research Center, unit of Pacific Hospitals Pvt. Ltd. Hyderabad, India. We gratefully acknowledge their financial support.

References

1. Broxmeyer HE. (1996) Primitive hematopoietic stem and progenitor cells in human umbilical cord blood: an alternative source of transplantable cells. Cancer Treat Res **84**, 139–148.
2. Ballen K, Broxmeyer HE, McCullough J, et al. (2001) Current status of cord blood banking and transplantation in the United States and Europe. Biol Blood Marrow Transplant **7**, 635–645.
3. Gluckman, E, et al. (1989) Hematopoietic reconstitution in a patient with Fanconi's anemia by means of umbilical cord blood from an HLAidentical sibling. N Engl J Med **321**, 1174–1178.
4. Broxmeyer HE, et al. (1989) Human umbilical cord blood as a potential source of transplantation hematopoietic stem/progenitor cells. Proc Natl Acad Sci USA **86**, 3828–3832.

5. Guckin CP, et al. (2005) Production of stem cells with embryonic characteristics from human umbilical cord blood. Cell Prolif **38**, 245–255.

6. Robmanith T, et al. (2001) Interleukin 3 improves ex vivo expansion of primitive human cord blood progenitor cells and maintain the engraftment. Stem Cells **19**, 313–320.

7. Astori G, et al. (2005) Evaluation of ex vivo expansion and engraftment in NOD-SCID mice of umbilical cord blood CD34Þ cells using the DIDECO 'Pluricell System' Bone Marrow Transplant **35**, 1101–1106.

8. McNiece IK, et al. (2004) Ex vivo expansion of cord blood mononuclear cells on mesenchymal stem cells. Cytotherapy **6**, 311–317.

9. Gluckman E, et al. (1997) Outcome of cord-blood transplantation from related and unrelated donors. N Engl J Med **337**, 373–381.

10. Rubinstein P, Carrier C, Scaradavou A, et al. (1998) Outcomes among 562 recipients of placental-blood transplants from unrelated donors. N Engl J Med **22**, 1565–1577.

11. Gluckman E, Rocha V, Chastang C, on behalf of Eurocord- Cord Blood Transplant Group. (1998) Cord blood hematopoietic stem cells: biology and transplantation. Hematology, American Society of Hematology, Washington, DC. **1998**, 1–14.

12. Locatelli F, et al. (1999) Factors associated with outcome after cord blood transplantation in children with acute leukemia. Blood **93**, 3662–3671.

13. Kutzberg J, et al. (1996) Placenta blood as a source of hematopoietic stem cells for transplantation in to un related recipients. N Engl J Med **335**, 157–166.

14. Hassan J, Reen DJ. (1997) Cord blood CD4_ CD45RA_ T cells achieve a lower magnitude of activation when compared with their adult counterparts. Immunology **90**, 397–401.

15. Chao NJ, Emerson SG, Weinberg KI (2004) Stem cell transplantation (cord blood transplants). Hematology, American Society of Hematology Education Program Book **1**, 354–371.

16. Harris DT, et al (1992) Phenotypic and functional immaturity of human umbilical cord blood T lymphocytes. Proc Natl Acad Sci USA **89**,10006–10010.

17. Tanaka H, Kai S, Yamaguchi M, Misawa M, Fujimori Y, Yamamoto M, Hara H. (2003) Analysis of natural killer (NK) cell activity and adhesion molecules on NK cells from umbilical cord blood. Eur J Haematol **71**, 29–38.

18. Cairo MS, Suen Y, Knoppel E, van de Ven C, Nguyen A, Sender L. (1991) Decreased stimulated GM-CSF production and GM-CSF gene expression but normal numbers of GMCSF receptors in human term new borns compared with adults. Pediatr Res **30**, 362–367.

19. Cairo MS, Suen Y, Knoppel E, Dana R, Park L, Clark S, van de Ven C, Sender L. (1992) Decreased G-CSF and IL-3 production and gene expression from mononuclear cells of newborn infants. Pediatr Res **31**, 574–578.

20. Suen Y, Lee SM, Schreurs J, Knoppel E, Cairo MS. (1994) Decreased macrophage colony-stimulating factor mRNA expression from activated cord

versus adult mononuclear cells: altered post transcriptional stability. Blood **84**, 4269–4277.

21. Chang M, Suen Y, Lee SM, Baly D, Buzby JS, Knoppel E, Wolpe S, Cairo MS. (1994) Transforming growth factor beta1, macrophage inflammatory protein-1 alpha, and interleukin-8 gene expression is lower in stimulated human neonatal compared with adult mononuclear cells. Blood **84**, 118–124.

22. Lee SM, Suen Y, Chang L, Bruner V, Qian J, Indes J, Knoppel E, van de Ven C, Cairo MS. (1996) Decreased interleukin-(IL-12) from activated cord versus adult peripheral blood mononuclear cells and up regulation of interferon gamma, natural killer, and lymphokine-activated killer activity by IL-12 in cord blood mononuclear cells. Blood **88**, 945–954.

23. Cairo MS, Ayello J, Suri M, Wu J, Lillianthal S, van de Ven C. (2002) IL-12 _ IL-18 significantly enhances CB NK and LAK cytotoxicity but is significantly inferior compared to IL-12 _ IL-18 induced APB NK_LAK cytotoxicity: implications for differential immune reconstitution and/or graft-vs.-tumor effect post UCBT [abstract]. Blood **100**, 2427.

24. Suen Y, Chang M, Lee SM, Buzby JS, Cairo MS. (1994) Regulation of interleukin-11 protein and mRNA expression in neonatal and adult fibroblasts and endothelial cells. Blood **84**, 4125–4134.

25. Buzby JS, Knoppel EM, Cairo MS. (1994) Coordinate regulation of Steel factor, its receptor (Kit), and cytoadhesion molecule (ICAM-1 and ELAM-1) mRNA expression on in human vascular endothelial cells of differing origins. Exp Hematol **22**,122–129.

26. Chang M, Suen Y, Buzby J, Cairo M. (1995) Thrombopoietic (TPO) mRNA expression by RT-PCR in neonatal endothelial cells and fibroblasts is increased compared to adults: implications in the regulation of neonatal thrombopoiesis [abstract]. Pediatr Res **37**, 280a.

27. McNiece IK, Almedia-Porada G, Shpall EJ, Zanjani E. (2002) *Ex vivo* expanded cord blood cells provide rapid engraftment in fetal sheep but lack long-term engrafting potential. Exp Hematol **30**, 612–616.

28. McNiece IK, et al. (2000) Increased expansion and differentiation of cord blood products using a two step expansion culture. Exp Hematol **28**, 1181–1186.

29. Yildirim S, et al. (2005) Expansion of cord blood CD34 þ hematopoietic progenitor cells in coculture with autologous umbilical vein endothelial cells (HUVEC) is superior to cytokine-supplemented liquid Culture. Bone Marrow Transplant. **36**, 71–79.

30. Allison D, et al. *(2004)* Serum-free Expansion of CD34+ umbilical cord blood using Stemline™ hematopoietic stem cell expansion medium. BMT Program, University of Colorado Health Sciences Center, Denver, USA. Cell Transmissions *20*.

31. Yao C-L, et al. (2004) A systematic strategy to optimize ex vivo expansion medium for human hematopoietic stem cells derived from umbilical cord blood mononuclear cells. Ex. Hematol **32**, 720–727.

32. Chute JP et al. (2004) Quantitative analysis demonstrates expansion of SCID repopulating cells and increased engraftment capacity in human cord blood following. Stem Cells **22**, 202–215.

33. Zhang CC, et al. (2006) Angiopoietin-like protein stimulate ex vivo expansion of hematopoetic stem cells. Nat. Med. **12**, 240–245.

34. Yonemura Y, Ku H, Hirayama F, Souza L, Ogawa M. (1996) Interleukin 3 or interleukin 1 abrogates the reconstituting ability of hematopoietic stem cells. Proc Natl Acad Sci USA **93**, 4040–4044.

35. Yonemura Y, Ku H, Lyman S, Ogawa M. (1997) *In vitro* expansion of hematopoietic progenitors and maintenance of stem cells: comparison between Flt3/Flk-2 ligand and Kit ligand. Blood **89**, 1915–1921.

36. Bhatia M, Bonnet D, Kappu U, Wang J, Murdoch B, Dick J. (1997) Quantitative analysis reveals expansion of human hematopoietic repopulating cells after short term ex vivo culture. J Exp Med **186**, 619–624.

37. Conneally E, Cashman J, Petzer A, Eaves C. (1997) Expansion in vitro of transplantable human cord blood stem cells demonstrated using a quantitative assay of their lympho-myeloid repopulating activity in nonobese diabetic -scid/scid mice. Proc Natl Acad Sci USA **94**, 9836–9841.

38. Moise KJ, Jr. (2005) Umbilical cord blood stem cells. Obstet Gynecol. **106**, 1393–1407.

13

Ex Vivo Expansion of Hematopoietic Stem Cells from Human Cord Blood in Serum-Free Conditions

Chao-Ling Yao and Shiaw-Min Hwang

Summary

Human cord blood (CB), collected from the postpartum placenta and umbilical cord, has been identified as a rich source of hematopoietic stem cells (HSCs) and provides an alternative to bone marrow or mobilized peripheral blood transplantation. However, the major restriction of CB transplantation is the low number of HSCs in each CB unit and limits its use in clinical transplantation. The development of ex vivo culture systems that facilitate the expansion of HSCs is crucial to stem cell research and clinical application. In this chapter, we describe the protocols to isolate HSCs from CB, expand HSCs in serum-free condition, and analyze HSCs. This information is benefical for successful use of CB stem cells in therapeutic studies.

Key Words: Hematopoietic stem cell; Cord blood; Serum-free; Ex vivo expansion; Immuno-phenotyping.

1. Introduction

Hematopoiesis is the process of generating mature blood cells, which are produced at an average rate of 400 billion per day in humans *(1)*. All mature blood cells originate from a small population of hematopoietic stem cells (HSCs), which are characterized by their capacity to self-renew with an ability to differentiate into multiple hematopoietic cell lineages *(2,3)*. Empirically, HSCs are functionally assayed by their potential to produce hematopoietic colonies in vitro, to expand in long-term culture, and to engraft nonobese diabetic/severe combined immunodeficiency (NOD/SCID) mice in vivo *(4–7)*. Several studies have proposed many sets of cell-surface antigens to identify HSCs, such as CD34, CD133, CD38, and CXCR4. CD34+ cell

From: *Methods in Molecular Biology, vol. 407: Stem Cell Assays*
Edited by: M. C. Vemuri © Humana Press, Totowa, NJ

dose is an important prognostic factor of clinical HSC transplantation *(8)*. Additionally, some studies have demonstrated that the CD34$^+$CD38$^-$ or CD34$^+$CD133$^+$ subpopulation cells contain more clonogenic cells that can repopulate NOD/SCID mice than CD34$^+$ cells *(9,10)*. The expression of CXCR4, stromal cell-derived factor-1 receptor, is critical for HSC homing and repopulation *(11)*. Even when HSCs can be maintained and expanded in culture, engraftment defects still occur if the HSCs cannot home to bone marrow (BM) efficiently *(12)*.

Human cord blood (CB), collected from the postpartum placenta and umbilical cord, has been identified as a rich source of HSCs and provides an alternative to BM or mobilized peripheral blood (MPB) transplantation *(13,14)*. CB transplantation has been used for treatment of hematologic disorders, congenital immunodeficiencies, metabolic disorders, and autoimmune diseases *(15)*. However, the transplantation of CB cells has two major disadvantages: (i) the low number of HSCs in each CB unit limits its application to children only (usually less than 20 kg body weight) and (ii) neutrophil and platelets in CB-transplanted patients need a longer recovery time than in BM or MPB following transplantation *(16)*. Ex vivo expansion of the CB HSCs may solve the above shortages and is an important issue in clinical transplantation *(17–19)*.

This chapter contains three aspects. Each aspect in turn describes a method for isolation, serum-free expansion, and analysis of HSCs from human CB, respectively. Isolated and expanded HSCs provide an important cell source for hematopoietic studies in vitro, as well as transplantation studies to explore their potential for therapeutic purposes.

2. Materials

2.1. Isolation of Hematopoietic Stem Cells

2.1.1. Preparation of Mononuclear Cells

1. Blood bags, 250 ml (cat. no. KBS-250CA7, Kawasumi Laboratories Co. Ltd, Pratumtanee, Thailand).
2. CB.
3. Tubes, 50 ml (cat. no. 352070, Becton Dickinson, San Jose, CA, USA).
4. Wash buffer: Dulbecco's phosphate-buffered saline (D-PBS) (D5652, Sigma, cat. no. St. Louis, MO, USA) supplemented with 2 mM ethylenediaminetetraacetic acid (EDTA) (Sigma, cat. no. E6635).
5. Ficoll-Paque solution (cat. no. 17-1440-03, Amersham Biosciences, Uppsala, Sweden).
6. Magnetic activated cell separation (MACS) buffer: D-PBS supplemented with 0.5% bovine serum albumin (BSA) (Sigma, cat. no. A9418) and 2 mM EDTA.

2.1.2. Enrichment of CD34⁺ Cells

1. MACS buffer.
2. Human CD34 cell isolation kit (cat. no. 130-046-702, Miltenyi Biotech, Bergish Gladach, Germany), including human CD34 Microbeads and FcR Blocking Reagent.
3. Magnetic columns (cat. no. 130-042-401, Miltenyi Biotech).
4. Magnetic cell separator: Miltenyi VarioMACS device (cat. no. 130-043-102, Miltenyi Biotech).

2.2. Serum-Free Expansion of Hematopoietic Stem Cells

1. Serum-free (SF)-HSC medium. The following ingredients should be added to Iscove's modified Dulbecco's medium (cat. no. SH30005.02, Hyclone, Logan, UT, USA) to give the final indicated concentrations: 8.46 ng/ml thrombopoietin (cat. no. 300-18, PeproTech EC Ltd, London, UK), 4.09 ng/ml interleukin-3 (IL-3), (cat. no. 200-03, PeproTech EC Ltd), 15 ng/ml stem cell factor (cat. no. 300-07, PeproTech EC Ltd), 6.73 ng/ml flk-2/flt3 ligand (FL) (cat. no. 300-19, PeproTech EC Ltd), 0.78 ng/ml IL-6 (cat. no. 200-06, PeproTech EC Ltd), 3.17 ng/ml granulocyte colony-stimulating factor (cat. no. 300-23, PeproTech EC Ltd), 1.30 ng/ml granulocyte-macrophage colony-stimulating factor (cat. no. 300-03, PeproTech EC Ltd), 1.5 g/l BSA, 4.39 μg/ml human insulin (cat. no. I9278, Sigma), 60 μg/ml human transferrin (cat. no. T4132, Sigma), and 25.94 μM 2-mercaptoethanol (cat. no. 21985-023, GIBCO/BRL, Carlsbad, CA, USA). SF-HSC medium was filtered by 0.22-μm filter (cat. no. MSLGS025, Millipore, Billerica, MA, USA).
2. Six-well plates (cat. no. 353046, Becton Dickinson).
3. Tissue culture flasks, 75 cm² (cat. no. 353110, Becton Dickinson).

2.3. Analysis of Hematopoietic Stem Cells

2.3.1. Immunophenotyping Analysis by Flow Cytometry

1. Fluorescent activated cell sorting (FACS) buffer: D-PBS supplemented with 1% fetal bovine serum (FBS) (cat. no. SH30070.03, Hyclone).
2. Fluorescein isothiocyanate (FITC)-conjugated anti-human CD34 (cat. no. 120-000-427, Miltenyi Biotech), CD38 (cat. no. 555459, Becton Dickinson, San Joes, CA, USA), CXCR4 (cat. no. FAB170F, R&D System, Minneapolis, MN, USA), and FITC isotype control antibody.
3. Phycoerythrin (PE)-conjugated anti-human CD34 (cat. no. 120-000-428, Miltenyi Biotech), CD133 (cat. no. 120-001-243, Miltenyi Biotech), and PE isotype control antibody.
4. Flow cytometry: Two-color FACSCalibur flow cytometry (Becton Dickinson).
5. CellQuest software (Becton Dickinson).

2.3.2. In Vivo Repopulation Assay in NOD/SCID Mice

1. NOD/SCID mice.
2. X-ray or ^{137}Cesium source.
3. D-PBS.
4. ammonium chloride solution, 6% (cat. no. A0171, Sigma).

3. Methods

3.1. Isolation of Hematopoietic Stem Cells

3.1.1. Preparation of Mononuclear Cells

1. Collect donor CB in a standard 250-ml blood bag with mother's informed consent and process within 24 h (*see* **Note 1**).
2. Draw CB from blood bag and place into 50-ml tubes.
3. Centrifuge at 200 g for 20 min with the brake off (*see* **Note 2**).
4. Identify the buffy coat at the interface and carefully aspirate into a new 50-ml tube.
5. Dilute with twice volume of wash buffer.
6. Separately, add 20 ml of Ficoll in a 50-ml tube. Carefully layer the buffy coat suspension on Ficoll. Do not disturb or mix the layer (*see* **Note 3**).
7. Centrifuge at 400 g for 40 min with the brake off at room temperature.
8. Identify the mononuclear cells (MNCs) at the interface. Draw off the upper plasma layer and leave the interface undisturbed.
9. Use a clean pipette to transfer the MNC layer into a new 50-ml tube (*see* **Note 4**).
10. Wash the MNCs twice with 20 ml of wash buffer.
11. Resuspend the MNC pellet with MACS buffer for CD34$^+$ cell enrichment.

3.1.2. Enrichment of CD34$^+$ Cells

1. Adjust cell density in the 300 µl of MACS buffer per 10^8 MNCs. If cell number is less than 10^8 cells, use the same volume of MACS buffer. Then proceed to magnetic labeling (*see* **Note 5**).
2. Add 100 µl of FcR Blocking Reagent to 10^8 MNCs/300 µl of MACS buffer.
3. Label cells by adding 100 µl of CD34 MicroBeads (final volume is 500 µl per 10^8 MNCs).
4. Mix well and incubate for 30 min at 4 °C (*see* **Note 6**).
5. Wash cells with 20 ml of MACS buffer. Centrifuge at 200 g for 5 minutes.
6. Aspirate off the supernatant. Resuspend the cell pellet in 1 ml degas MACS buffer per 10^8 MNCs for magnetic separation process (*see* **Note 7**).
7. Place a magnetic column in the magnetic field of MACS cell separator. Rinse column with 3 ml of MACS buffer.
8. Load cell suspension through the column. Wash column with 5 ml of MACS buffer. Allow the unlabeled cells to elute (*see* **Note 8**).

9. Remove column from MACS cell separator. Wash column with 5 ml of MACS buffer. Allow the labeled CD34+ cells to elute.

10. Proceed to **steps 7–9** again. Repeated steps of magnetic separation can get higher purity of CD34+ cells (*see* **Fig. 1**) (*see* **Note 9**).

3.2. Serum-Free Expansion of Hematopoietic Stem Cells

1. Centrifuge freshly purified CD34+ cells at 200 g for 5 min. Resuspend the CD34+ cell pellet with SF-HSC medium for expansion.

2. Seed CD34+ cells at 5×10^4 cells/ml (3 ml/well in six-well plate or 20 ml in 75-cm^2 culture flask) and incubate at 37 °C in an atmosphere of 5% CO_2 and humidified incubator.

3. After 7 days of culture, harvest the cells to analyze or split the cells with fresh SF-HSC medium to maintain a cell density of 5×10^5 cells/ml (*see* **Note 10**).

3.3. Analysis of Hematopoietic Stem Cells

3.3.1. Immunophenotyping Analysis by Flow Cytometry

1. Centrifuge approximately 5×10^5 cells (freshly purified CD34+ cells or expanded cells) at 200 g for 5 min. Resuspend the cell pellets with the following antibody combinations to the aliquots for immunophenotyping analysis (*see* **Note 11**).

   a. Aliquot A: No antibody in 100 µl of FACS buffer.

   b. Aliquot B: 10 µl of FITC-conjugated anti-human CD34, CD38, or CXCR4 in 90 µl of FACS buffer.

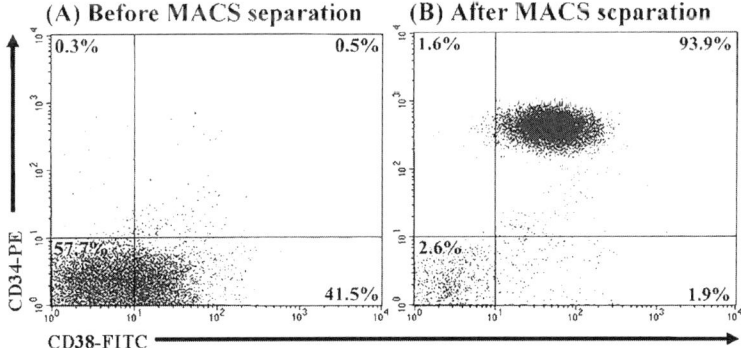

Fig. 1. Flow cytometry analysis of surface antigen expression of CD34+CD38− cells before and after CD34 MACS separation. The numbers within dot plots indicate the percentage of total cells that fell within the particular quadrant. (**A**) Before CD34 MACS separation. (**B**) After CD34 MACS separation.

    c. Aliquot C: 10 μl of PE-conjugated anti-human CD34 or CD133 in 90 μl of FACS buffer.

    d. Aliquot D. 10 μl of FITC isotype control antibody and 10 μl of PE isotype control antibody in 80 μl of FACS buffer.

    e. Aliquot E. 10 μl of FITC-conjugated anti-human CD34, CD38, or CXCR4, 10 μl of PE-conjugated anti-human CD34 or CD133 in 80 μl of FACS buffer.

2. Mix well and incubate at 4 °C for 30 min in the dark (*see* **Note 12**).

3. Wash each aliquot twice with 5 ml of FACS buffer by using a 200 g centrifugation for 5 min.

4. Remove supernatant carefully and resuspend the cell pellet with 1 ml of FACS buffer. Proceed to analyze by flow cytometry immediately (*see* **Note 13**).

5. Identify the lymphocyte population on the flow cytometry using the forward scatter (FSC)/side scatter (SSC) dot plot. Use aliquot A to adjust the gain in the FSC and SSC channels so that the cell population locates at 400–600 of FSC value and 50–200 of SSC value. Draw a gate around the cells that excludes debris, erythrocytes, and platelets. Analyses of flow cytometry should be performed on this region of cells.

6. Use aliquot D to adjust the gain in each channel on the flow cytometry so that the mean fluorescence intensity of the isotype control is less than 10 in each channel.

7. Use aliquot B and aliquot C in turn, compensate the flow cytometry appropriately.

8. Use aliquot E to identify the HSCs as the $CD38^-$, $CD133^+$, or $CXCR4^+$ population of $CD34^+$ cells on the dot plot. If cells are readily identified, acquire more than 30,000 cells on the cytometry from each aliquot and store data for analyses.

9. Determine the percentage of $CD38^-$, $CD133^+$, or $CXCR4^+$ subsets of $CD34^+$ HSCs in the freshly purified $CD34^+$ cells or expanded cells by CellQuest software (*see* **Fig. 2**).

3.3.2. In Vivo Repopulation Assay in NOD/SCID mice

1. Handle all NOD/SCID mice under sterile conditions and maintain under microisolators.

2. Irradiate 8-week-old NOD/SCID mice with 400 cGy using X-ray or ^{137}Cesium source (*see* **Note 14**).

3. Let NOD/SCID mice rest for 4 h following irradiation.

4. Wash freshly purified $CD34^+$ cells or the expanded cells with D-PBS. Resuspend appropriate cell number in 200 μl of D-PBS (*see* **Note 15**).

5. Inject cell suspension through mouse tail vein. Control group of mice was injected with 200 μl of D-PBS only (*see* **Note 16**).

6. Sacrifice transplanted mice at week 6 for engraftment analysis.

7. Collect BM cells by flushing femurs and tibias with D-PBS.

8. Centrifuge cells at 200 g for 5 min. Resuspend cells with 6% ammonium chloride solution to lyse red blood cells.

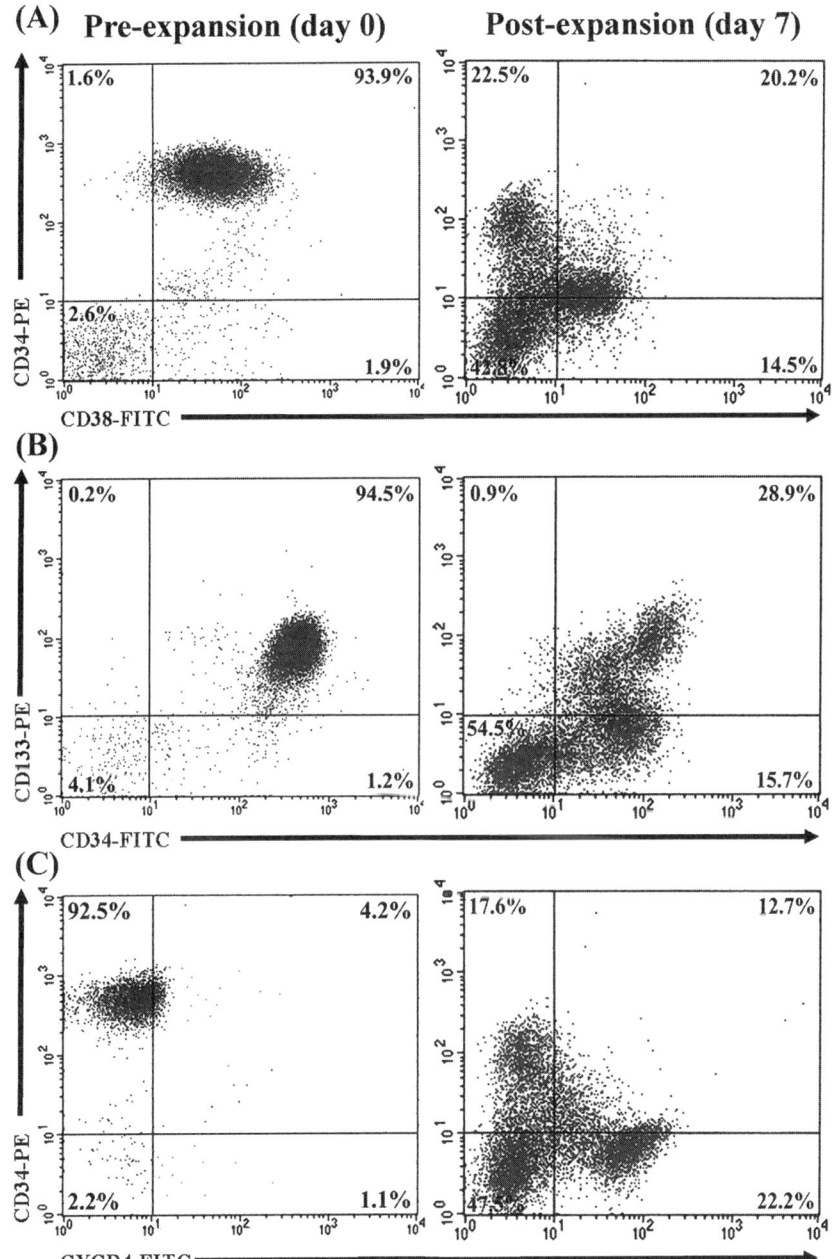

9. Wash cells twice with FACS buffer and then proceed immunophenotyping analysis by flow cytometry to determine the percentage of engrafted human cells (*See* **Subheading 3.3.1.**).

4. Notes

1. CB should be collected according to the institutional guidelines, appropriate local ethical committee approval, and mother's consent. After delivery of newborn, CB is collected from the umbilical cord attached to the placenta by gravity, yielding approximately 60–150 ml.

2. This step can concentrate the volume of the leukocytes and reduce the volume of the Ficoll used in the step 6 (*see* **Subheading 3.1.1.**). It is necessary to centrifuge with the brake off, or the interface of the buffy-coat cells and MNCs will be hard to identify in the **steps 3 and 7** (*see* **subheading 3.1.1.**), respectively.

3. It is important to create a clear interface when layering buffy coat suspension on Ficoll. The maximum ratio of buffy coat versus Ficoll volume is 1:1. We usually use plastic disposable droppers to layer buffy coat on Ficoll to prevent interface disturbance.

4. It is critical to take as much as the MNC interface (including those that adhere to the side of the tube) by using plastic disposable aspirators.

5. According to the conditions of the experiments, BSA can be replaced by HSA, human serum, or FBS. EDTA can be replaced by other supplements such as anticoagulant citrate phosphate dextrose or citrate dextrose formula-A. Buffer containing Ca^{2+} or Mg^{2+} ion is not recommended for use.

6. Higher temperature and longer incubation time for magnetic labeling would result in unspecific cell labeling.

7. MACS buffer can be degassed by using ultrasonic cleaning device or applying vacuum. It is important to use degas MACS buffer for magnetic separation process. Bubble in the MACS buffer may clog the magnetic column and result in the loss of viable cells.

8. Cell clumps and aggregates can be removed by using 30-μm nylon mesh filters (cat no. 130-041-407, Miltenyi Biotech). Cell aggregates or other large particles

---

Fig. 2. Flow cytometry analysis of surface antigen expression of CD34$^+$ cells before and after ex vivo expansion in the Serum-free haematopoietic stem cell (SF)-HSC medium. Fresh CD34$^+$ cells isolated from mononuclear cells and their expanded cells for 7–day culture in the SF-HSC medium are shown at the left and right of figure, respectively. The numbers within dot plots indicate the percentage of total cells that fell within the particular quadrant. (**A**) CD38-FITC and CD34-PE, (**B**) CD34-FITC and CD133-PE, (**C**) CXCR4-FITC and CD34-PE.

in cell suspension may clog the magnetic column and lose the quality of cell separation.

9. The first elution of MACS separation can only reach 70–85% purity of CD34$^+$ cells. Repeat separation on a new, freshly prepared column can reach 95% purity of CD34$^+$ cells.

10. SF-HSC medium is a serum-free, stroma-free, and cytokine-limited defined medium for CD34$^+$ cells and colony forming cell (CFC) expansion *(20,21)*. The number of CD34$^+$ cells reached the maximum on the 7th day in the SF-HSC medium. All the cells in this culture system are in suspension. The cell viability will drop abruptly after 1-week in case of no replenishment of fresh medium. However, CD34$^+$ cells can be expanded continuously for at least 2 months by replenishing fresh medium twice a week. After ex vivo expansion, cells can be functionally analyzed in vitro by CFC assay (cat. no. 04434, StemCell Technologies, Vancouver, BC, Canada) and long-term culture-initiating cell (LTC-IC) assay (cat no. 05100, StemCell Technologies).

11. It is important to reduce the final volume during labeling cells. This facilitates a reduction in the amount of each antibody and saturation of all antibody-binding sites of cells. The antibodies need to be titrated before experiments.

12. Higher temperature and longer incubation time for fluorescent labeling would result in unspecific cell labeling.

13. If the cells cannot proceed to analyze by flow cytometry immediately, resuspend and fix each aliquots of cells in 1 ml of FACS buffer supplemented with 3% paraformaldehyde at 4 °C in the dark. Fixed cells can be stored for up to 3 days prior to analyses without affecting the results.

14. The dose of 400 cGy radiation is sub-lethal to NOD/SCID mice. If NOD/SCID mice are irradiated over 400 cGy (lethal radiation), mice will die rapidly after irradiating (in a few days). This would cause a high variable result of engraftment, because the transplanted cells might not reconstitute BM of mice in the short time.

15. It is not suggested to inject larger volume (more than 500 μl) of cell solution or D-PBS through mouse tail vein, which may lead to increase in death lates due to injection process.

16. Irradiated NOD/SCID mice that are injected with D-PBS only will die within 2 weeks due to the hematopoiesis, and microenvironment in BM is destroyed. If NOD/SCID mice are engrafted successfully, they can survive at least 2 months. We usually scarifice mice at week 6 to analyze the donor cells.

Acknowledgments

The authors thank the financial support from the Ministry of Economic Affairs, Taiwan (95-EC-17-A-17-R7-0525) and Drs Tin-Yin Liu and Chii-Cherng Liao for their constant encouragement to this project.

References

1. McAdams, T.A., Miller, W.M., and Papoutsakis, E.T. (1996) Hematopoietic cell culture therapies (part I): cell culture considerations. *Trends Biotechnol.* **14**, 341–349.
2. Moore, K.A., Ema, H., and Lemischka, I.R. (1997) In vitro maintenance of highly purified, transplantable hematopoietic stem cells. *Blood* **89**, 4337–4347.
3. Danet, G.H., Lee, H.W., Luongo, J.L., Simon, M.C., and Bonnet, D.A. (2001) Dissociation between stem cell phenotype and NOD/SCID repopulating activity in human peripheral blood CD34$^+$ cells after ex vivo expansion. *Exp. Hematol.* **29**, 1465–1473.
4. Hows, J.M., Bradley, B.A., Marsh, J.C., Luft, T., Coutinho, L, Testa, N.G., et al. (1992) Growth of human umbilical cord blood in long term haemopoietic cultures. *Lancet* **340**, 73–76.
5. Sutherland, H.J., Lansdrop, P.M., Henkelman, D.H., Eaves, S.C., and Eaves, C.J. (1990) Functional characterization of individual human hematopoietic stem cells at limiting dilution on supportive marrow stromal layers. *Proc. Natl. Acad. Sci. U. S. A.* **87**, 3584–3588.
6. Prosper, F., Vanoverbeke, K., Stroncek, D., and Verfaillie, C.M. (1997) Primitive long-term culture initialing cells (LTC-ICs) in granulocyte colony-stimulating factor mobilized peripheral blood progenitor cells have similar potential for ex vivo expansion as primitive LTC-ICs in steady state bone marrow. *Blood* **89**, 3991–3997.
7. Larochelle, A., Vormoor, J., Hanenberg, H., Wang, J.C., Bhatia, M., Lapidot, T., et al. (1996) Identification of primitive human hematopoietic cells capable of repopulating NOD/SCID mouse bone marrow: implication for gene therapy. *Nat. Med.* **2**, 1329–1337.
8. Burt, R.K. (1999) Clinical utility in maximizing CD34$^+$ cell count in stem cell grafts. *Stem Cells* **17**, 373–376.
9. Zandstra, P.W., Conneally, E., Petzer, A.L., Piret, J.M., and Eaves, C.J. (1997) Cytokines manipulation of primitive human hematopoietic cell self-renewal. *Proc. Natl. Acad. Sci. U. S. A.* **94**, 4698–4703.
10. Bhatia, M., Wang, J.C., Kapp, U., Bonnet, D., and Dick, J.E. (1997) Purification of primitive human hematopoietic cells capable of repopulating immune-deficient mice. *Proc. Natl. Acad. Sci. U. S. A.* **94**, 5320–5325.
11. Denning-Kendall, P., Singha, S., Bradley, B., and Hows, J. (2003) Cytokine expansion culture of cord blood CD34$^+$ cells induce marked and sustained changes in adhesion receptor and CXCR4 expressions. *Stem Cells* **21**, 61–70.
12. Hart, C., Drewel, D., Mueller, G., Grassinger, J., Zaiss, M., Kunz-Schughart, L.A., et al. (2004) Expression and function of homing-essential molecules and enhanced in vivo homing ability of human peripheral blood-derived hematopoietic progenitor cells after stimulation with stem cell factor. *Stem Cells* **22**, 580–589.

13. Gluckman, E., Rocha, V., and Chastang, C. (1998) Cord blood banking and transplant in Europe. *Bone Marrow Transplant.* **22**, 68–74.

14. Gluckman, E. (2000) Current status of umbilical cord blood hematopoietic stem cell transplantation. *Exp. Hematol.* **28**, 1197–1205.

15. Rubinstein, P., Carrier, C., Scaradavou, A., Kurtzberg, J., Adamson, J., Migliaccio, A.R., et al. (1998) Outcomes among 562 recipients of placental-blood transplants from unrelated donors. *N. Engl. J. Med.* **339**, 1565–1577.

16. Laughlin, M.J., Eapen, M., Rubinstein, P., Wagner, J.E., Zhang, M.J., Champlin, R.E., et al. (2004) Outcomes after transplantation of cord blood or bone marrow from unrelated donors in adults with leukemia. *N. Engl. J. Med.* **351**, 2265–2275.

17. Williams, S.F., Lee, W.J., Bender, J.G., Zimmerman, T., Swinney, P., Blake, M., et al. (1996) Selection and expansion of peripheral blood CD34[+] cells in autologous stem cell transplantation for breast cancer. *Blood* **87**, 1687–1691.

18. Yao, C.L., Chu, I.M., Hsieh, T.B., and Hwang, S.M. (2004) A systematic strategy to optimize ex vivo expansion medium for human hematopoietic stem cells derived from umbilical cord blood mononuclear cells. *Exp. Hematol.* **32**, 720–727.

19. Ali, M.Y., Oyama, Y., Monreal, J., Winter, J.N., Tallman, M.S., Williams, S.F., et al. (2003) Ideal or actual body weight to calculate CD34[+] cell doses for autologous hematopoietic stem cell transplantation? *Bone Marrow Transplant.* **31**, 861–864.

20. Yao, C.L., Liu, C.H., Chu, I.M., Hsieh, T.B., and Hwang, S.M. (2003) Factorial designs combined with the steepest ascent method to optimize serum-free media for ex vivo expansion of human hematopoietic progenitor cells. *Enzyme Microb. Technol.* **33**, 343–352.

21. Yao, C.L., Feng, Y.H., Lin, X.Z., Chu, I.M., Hsieh, T.B., and Hwang, S.M. (2006) Characterization of serum-free ex vivo-expanded hematopoietic stem cells derived from human umbilical cord blood CD133[+] cells. *Stem Cells Dev.* **15**, 70–78.

14

Hematopoietic Colony-Forming Cell Assays

Carla Pereira, Emer Clarke, and Jackie Damen

Summary

Hematopoiesis is the process by which stem cells divide and differentiate to produce the multiple types of mature cells found in blood. The process begins in early embryonic development and continues throughout adult life, primarily in the bone marrow. Various in vivo and in vitro assays have been developed to detect and assess stem cells and early multi-potential progenitors. While highly informative about primitive hematopoietic cells these assays are long and labour intensive. Alternatively, colony-forming cell (CFC) assays may be used to quantify more lineage-restricted progenitors in a simple in vitro assay. When cultured in a semi-solid medium containing the appropriate cytokines, CFCs are able to divide and differentiate into a colony of more mature cells that can be detected by light microscopy. This allows for the quantification of erythroid, myeloid, lymphoid, megakaryocytic, and multi-potential cell lineages from various cell sources. This chapter outlines the materials and methods used for the culture and assessment of CFC from humans, mice, and other species.

Key Words: Hematopoiesis; Colony-forming cell assays; Methylcellulose; Collagen; Erythroid; Myeloid; Megakaryocytic; Lymphoid; Cytokines; CFU-GEMM; BFU-E; CFU-E; CFU-GM; CFU-Pre-B; CFU-Mk.

1. Introduction

Since their introduction in 1961, hematopoietic colony-forming cell (CFC) assays have been used extensively for research and clinical applications in humans and animal models as a way of quantifying and assessing the hematopoietic progenitor content of a cell sample. In humans, the colony assay has been used to identify stimulatory and inhibitory growth factors, as

From: *Methods in Molecular Biology, vol. 407: Stem Cell Assays*
Edited by: M. C. Vemuri © Humana Press, Totowa, NJ

supportive diagnostic assays of myeloproliferative disorders and leukemias, to evaluate the hematopoietic proliferative potential of bone marrow (BM), cord blood (CB), and mobilized peripheral blood (MPB) samples for clinical transplantation *(1)* and more recently as a method to determine the maximum tolerated dose (MTD) of new lead compounds before phase 1 clinical trials *(2)*.

Owing to the diversity of CFC content and frequency in cell samples from different sources, the selection of the appropriate cell population for specific applications must be considered. Often the cell source selected requires processing to obtain the cell population of interest. For those interested in identifying the CFC frequency of a specific cellular phenotype, pre-enrichment of mononuclear cells by ficoll gradient density centrifugation followed by cell purification can be used. Although this process is excellent at removing unwanted cells, this is not achieved without the loss of progenitors. For others, the sample may be moderately processed [by red blood cell (RBC) lysis] to remove specific cell populations that make the detection of the colonies within the matrix difficult. Where cell numbers are limiting (CB transplantation for example), any loss of progenitors may compromise its use as a clinical product and minimally invasive processing methods are used albeit making the detection and enumeration of the colonies within the red blood cell background a challenge.

Standard procedures to quantify CFCs involve culturing a suspension of BM, CB, or MPB in a semi-solid matrix supplemented with the appropriate cytokines for 7–14 days. The increased knowledge on recombinant growth factors and their specific use in the last 10 years has resulted in the optimization of the yield of CFC such that hematopoietic progenitors of the myeloid, erythroid, and platelet lineages can be quantified *(3)*. With an ever increasing understanding of the cytokines that regulate the proliferation and differentiation of specific CFC, cytokine cocktails can be specifically chosen to address various scientific questions. Media formulations can be prepared in the absence of cytokines to detect activity of a new potential cytokine or mimetic, with minimal cytokine concentrations to assess synergistic activity or with saturating concentrations of cytokines to assess potential toxicity of a compound for any specific progenitor type.

Microscopic enumeration of CFC-derived colonies is still the method to identify these progenitors despite being inherently subjective. Because of this, variable measurements of CFC between centers may explain in part why some groups have seen correlations between the progenitor content and the engraftment *(4)*, and others have found no such associations *(5)*. The colony assay however has remained the benchmark functional assay to assess the ability of various hematopoietic cell sources to divide and differentiate, especially following ex vivo manipulations such as T-cell depletion, CD34[+] cell enrichment, gene therapy protocols, and cryopreservation.

The study of hematopoietic stem and progenitor cells is aided by the use of animal models, especially mice. For example, the development of the competitive repopulating unit (CRU) assay (also known as the SCID repopulating cell assay) allows the frequency of the true human- or mouse-derived hematopoietic stem cell to be assessed by testing the ability of that cell to repopulate both myeloid and lymphoid compartments of a compromised host *(6–8)*. There are also models of myelosuppression, where after the administration of 5-fluoruracil, one can assess the decline and recovery of hematopoietic cell populations and can thereby compare chemotherapeutic agents or assess the effects of compounds on the recovery of hematopoietic lineages in a stressed environment *(9,10)*. In addition, one can assess the effect of knocking out a specific gene on the number and function of immature and mature hematopoietic cell populations by comparing the cell populations between genetically altered mice and their wild-type controls *(11)*.

Using a mouse model, determination of the hematopoietic cell content and functionality in a number of organs from the same mouse can assess if the changes are organ specific or more global. The femur within a mouse provides a very discrete entity, where cell and progenitor frequencies remain relatively constant for the life of the mouse in unperturbed situations. Assessment of changes of the hematological cell populations within the femur provides both quantitative and qualitative assessment of a number of parameters that may be of interest in basic as well as clinical research. Mouse CFC assays offer a great advantage in that many mouse-specific cytokines are well characterized and available commercially, allowing for the assessment of erythroid, myeloid, lymphoid, and megakaryocytic progenitor populations *(12–15)*.

The use of canine models was pioneered in Seattle where Storb's group *(16)* gained an extensive understanding of the effects of leukocyte antigens on transplantation outcome. Dogs were also used to provide insight on the effects of various forms of irradiation on the hematopoietic cell populations. Although there are limited numbers of canine-specific cytokines, the development of the CFU-GM assay using dog-derived cells provided quantitative measurements of progenitor functionality. More recently, the dog is commonly one of the larger animal models used in the toxicity assessment of lead compounds before phase 1 clinical trial with the CFU-GM assay used to detect BM toxicity *(17,18)*.

Another larger animal model, which has provided significant insight into various clinically designated protocols, is the non-human primate. As these animals respond to cytokines in a manner similar to humans, and as their cells may be cultured in media formulations containing human cytokines, cells derived from non-human primates can be assessed for various immature and mature hematopoietic cell populations. The resulting data have been informative in evaluating mobilization regimens *(19)*, in the evaluation of cell expansion protocols *(20)* or in determining the ability of compounds to enhance engraftment following irradiation-induced aplasia *(21)*. These large animal models have also been used to assess hematological parameters following the administration of monoclonal antibodies for the assessment of potential toxicity *(22)*.

Overall, the CFC assay has been and remains a very useful research tool to allow assessment of specific primitive hematopoietic cell populations within multiple organs of many animals. It has provided us with insight into the functionality of various hematopoietic cell populations and an understanding of the regulatory molecules that control hematopoietic cell proliferation and differentiation. In vitro assessment of the frequency of hematopoietic CFC can determine the feasibility of using certain chemotherapeutics for clinical trials and has been shown to correlate with clinical outcome in the identification of hematotoxic compounds *(23)*. The methods to assess the various hematopoietic CFCs are described in detail below.

2. Materials

Listed are suggested supplies and reagents for the growth and assessment of hematopoietic CFC, including erythroid (CFU-E, BFU-E), myeloid (CFU-GM), lymphoid (CFU-Pre-B), and megakaryocytic (CFU-Mk) CFC (*see* **Note 1**).

2.1. Cell Processing

2.1.1. Isolation of Mononuclear Cells using Ficoll-Paque® PLUS

1. Iscove's Modified Dulbecco's Medium (IMDM) containing 2% fetal bovine serum (FBS) [cat. no. 07700, StemCell Technologies (STI, Vancouver BC Cananda)].
2. Ficoll-Paque® PLUS (cat. no. 07907, 07957, STI).

2.1.2. RBC Lysis Using Ammonium Chloride (ACK) Lysis Buffer

1. IMDM 2% FBS (cat. no. 07700, STI).
2. Ammonium chloride solution, 0.8% NH_4Cl with 0.1 mM ethylenediaminetetraacetic acid (EDTA) in water (cat. no. 07800, 07850, STI).

2.2. Erythroid, Myeloid, and Lymphoid CFC Assays

2.2.1. Culture Setup

1. Methylcellulose-based media containing cytokines (MethoCult™, *see* **Note 2**).

    a. Store at −20 °C, stable for 2 years. Stable for 1 month at 4 °C. Repeat freeze thawing is not recommended.

    b. Thaw bottles of media at room temperature or overnight at 4 °C. Shake vigorously for 30–60 s and let stand for at least 5 min to allow bubbles to dissipate.

    c. Dispense into tubes using a 16-gauge blunt end needle and 6- or 12-mL luer lock syringe at 1.6, 3.2, 3.6, or 4.0 mL per 14 mL (17 × 100 mm) tube depending on media formulation, and store at −20 °C until use (*see* **Note 3**). Blunt-end needles should be used for safety reasons and accurate dispensing of the viscous methylcellulose-based media.

2. IMDM 2% FBS (cat. no. 07700, STI).
3. 3-mL syringes with luer-lock fitting (cat. no. 90311, STI).
4. 16-gauge blunt-end needles (cat. no. 28110, 28120, STI).
5. 35-mm pre-tested culture dishes (cat. no. 27100, 27150 STI).
6. 150 × 25 mm tissue culture dishes (cat. no. 353025, BD Biosciences Falcon, Franklin Lakes NJ USA) or 245-mm square bioassay dishes (cat. no. 7200600, Fisher, Pittsburgh PA USA).

2.2.2. Colony Enumeration

60-mm gridded scoring dish (cat. no 27500, STI).

2.3. Megakaryocytic CFC Assays

2.3.1. Culture Setup

1. MegaCult®-C Serum-Free Medium (*see* **Note 4**).

    a. Store medium at −20 °C, stable for 1 year. Stable for 2 weeks at 4 °C.

    b. Thaw bottles of media at room temperature or overnight at 4 °C.

    c. Mix well and dispense into tubes at 2.55 or 3.0 mL per 14-mL tube depending on media formulation, and store at −20 °C until use (*see* **Note 3**).

2. Collagen solution.
   Store at 4 °C. Denaturation may occur at temperatures greater than 26 °C and below freezing.

3. Double chamber slides.
   Store at room temperature.

4. Recombinant cytokines.

Recombinant human (rh) Tpo (cat. no. 02522, 02822, STI), recombinant mouse (rm) interleukin-3 (IL-3) (cat. no. 02733, 02903), rh IL-6 (cat. no. 02506, 02606), rh IL-11 (cat. no. 02511, 02611).

2.3.2. Dehydration

1. Polypropylene Spacers, pre-cut to the size of the chamber slide (cat. no. 04911, STI).
2. Filter cards, pre-cut to the size of the chamber slide (cat. no. 04911, STI).

2.3.3. Fixation

1. Methanol–Methanol ACS (cat. no. BDH1135-4LG, VWR, West Chester PA USA 4 L/bottle).
2. Acetone–Acetone optima (cat. no. A929-4, Fisher, 4 L/bottle).
3. 2.5 L plastic container with tight fitting lid.

2.3.4. Staining

2.3.4.1. HUMAN MEGAKARYOCYTE STAINING

1. 0.05 M Tris/NaCl buffer, pH 7.6 is prepared as follows:

   a. 0.15 M isotonic saline: 8.766 g NaCl dissolved in 1 L deionized, distilled water. Stable for 1 month at room temperature.
   b. 0.5 M Tris/HCl, pH 7.6: 78.8 g Tris/HCl (cat. no.T-3253 Sigma, St. Louis MO USA) dissolved in 1 L deionized, distilled water. Adjust pH to 7.6 using 5 M NaOH. Stable for 1 month at room temperature.

   On the day of use, mix nine parts solution A plus one part solution B. One liter of this buffer is enough to stain 48 slides. Discard remaining buffer at the end of the day. Phosphate buffers should not be used as the staining reaction may be inhibited.

2. 5% Human serum (cat. no. 04807, STI).

   a. Dilute human serum to a concentration of 5% in 0.05 M Tris/NaCl buffer. Human serum provides a source of human immunoglobulin to block non-specific binding of the anti-GPIIb/IIIa antibody.

3. Primary antibody (cat. no. 04803, STI).

   a. Mouse anti-human GPIIb/IIIa antibody (IgG_{2a}), supplied at 1.0 mg/mL (0.36 mL/tube).
   b. Dilute 1/100 in 5% human serum for use at 10 μg/mL.
   c. Store at 4 °C. Do not freeze.

4. Negative Control antibody (cat. no. 04804, STI).

   a. Mouse anti-TNP isotype control antibody (IgG_{2a}), supplied at 0.5 mg/mL (0.1 mL/tube).

   b. Dilute 1/100 in 5% human serum for use at 5 μg/mL.

   c. Store at 4 °C. Do not freeze.

5. 1% bovine serum albumin (BSA) buffer (cat. no. 04915, STI).

   a. Dilute 10% BSA 1/10 in 0.05 M Tris/NaCl buffer to a concentration of 1%.

6. Secondary antibody (cat. no. 04906, STI).

   a. Biotin-conjugated goat anti-mouse IgG, supplied at 3.0 mg/mL (0.125 mL/tube).

   b. Dilute 1/300 in 0.05 M Tris/NaCl buffer with 1% BSA for use at 10 μg/mL.

   c. Contains 15 mM sodium azide as a preservative (see material safety data sheet, MSDS for complete safety instructions). Store at −20 °C. To aliquot the stock solution, thaw on ice and dispense 25 μL into 0.5 or 1.5 mL tubes. Close tightly and store at −20 °C. To make a working solution, thaw aliquots and dilute 1/10 with sterile 0.05 M Tris/NaCl buffer with 1% BSA. Diluted solutions are stable for one month at 4 °C. Further dilute 1/30 to achieve final concentration. Do not refreeze.

7. Avidin-Alkaline Phosphatase conjugate (cat. no. 04905, STI).

   a. Supplied at 2.8 mg/mL (0.2 mL per tube).

   b. Dilute 1/150 in 0.05 M Tris/NaCl buffer with 1% BSA for use at 18 μg/mL.

   c. Store at 4 °C. Do not freeze. Contains 15 mM sodium azide as a preservative (see MSDS). Working solutions can be prepared by diluting avidin-alkaline phosphatase conjugate 1/10 with sterile 0.05 M Tris/NaCl with 1% BSA. Diluted aliquots are stable for 1 month at 4 °C. Further dilute 1/15 to achieve final concentration.

8. Alkaline phosphatase substrate tablets (cat. no. 04809, STI).

   a. The alkaline phosphatase substrate solution is prepared in several steps:

      i. Determine the number of tablets required to make the appropriate volume of substrate (0 5 mL per slide). Each tablet set makes 1 mL of substrate solution. Allow tablets to come to room temperature (1 or 2 h).

      ii. Add the tablet(s) marked "tris buffer" (in gold foil) to the required volume of water. Vortex until fully dissolved.

      iii. Next add the tablet(s) marked "naphthol" (in silver foil) and vortex until dissolved. The solution should be pale pink in color.

      iv. Use the prepared substrate within 1 h.

   b. The substrate tablets are toxic. Do not touch them with your bare hands. See MSDS for complete safety instructions.

   c. Store below 0 °C.

9. Evans Blue 1% w/v (cat. no. 04913, STI).

   a. Dilute stock 1/6 in methanol (3–4 drops Evans Blue/mL methanol)

   b. Store at room temperature.

2.3.4.2. MOUSE MEGAKARYOCYTE STAINING.

1. Acetylthiocholiniodide (cat. no.A5751, Sigma).
   Store in −20 °C desiccator.
2. 0.1 M sodium citrate solution ($C_6H_5Na_3O_7$, FW 294.1; cat. no. S4641, Sigma)
   Dissolve 2.94 g in approximately 95 mL water. Adjust to pH 6.0 with 0.1 M citric acid and q.s. to 100 mL.
3. 0.1 M sodium phosphate buffer, pH 6.0 (Na_2HPO_4, FW 268.07; cat. no. S373, Fisher)
   Dissolve 13.40 g in approximately 95 mL water and q.s. to 100 mL.
4. 30 mM copper sulfate solution ($CuSO_4$, FW 249.68; cat. no. C493, Fisher)
   Dissolve 749 mg in approximately 95 mL water and q.s. to 100 mL.
5. 5 mM potassium ferricyanide solution [$K_3Fe(CH)_6$, FW 329.2; ACS Reagent cat. no. P3667, Sigma].

   a. Dissolve 165 mg in 95 mL water and q.s. to 100 mL.
   b. Solution decays slowly on standing, protect from light.

6. Harris' hematoxylin solution (cat. no. HHS16, Sigma).

3. Methods

3.1. Cell Processing

When obtaining blood or bone marrow, anti-coagulants are used to prevent clotting of the cell samples. Heparin is routinely used to collect BM, CB, and PB samples for research use, and anti-coagulants including heparin and acid citrate dextrose (ACD) are used to collect hematopoietic cells for clinical applications. Be sure to follow guidelines set out by your institution regarding the safe handling of blood products.

Mouse hematopoietic cells can be isolated from BM, spleen, PB, or fetal liver, with attention paid to maintaining sterility of the tissue sample (*see* **Note 5**). Laboratory mouse strains are routinely used between 6 and 12 weeks of age. For younger or older animals, transgenic mouse strains, or those treated with various compounds, it is important to use strain and age-matched controls. Sacrifice mice using procedures recommended by your institution (*see* **Note 6**).

Cell processing is performed to both reduce the number of contaminating background cells going into culture (mostly red blood cells) and increase the frequency of progenitors in the assay. The appropriate method of cell processing is dependant on the cell source and the population of interest. Cells can be prepared by light-density separation to enrich for mononuclear cells (MNC), ammonium chloride treatment (ACK) to lyse red blood cells or CD34$^+$ selection or lineage depletion to enrich for CD34 expressing cells. Light-density separation using Ficoll-Paque® PLUS is recommended for human cell sources

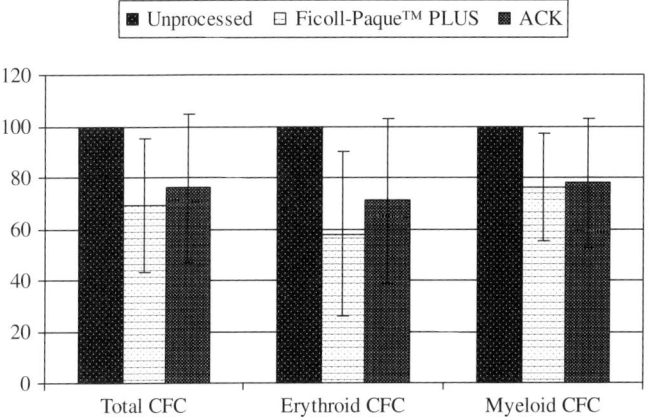

Fig. 1. The percentage recovery of erythroid and myeloid human progenitor cells following light-density separation using Ficoll-Paque® PLUS and red blood cell lysis using ACK lysis buffer ($n = 6$).

and when sample volumes are large. Where sample volumes are limiting and for mouse cell sources, red blood cell lysis using ACK lysis buffer is recommended. Unfortunately, both methods result in the loss of some progenitor cells, as illustrated in **Fig. 1**.

3.1.1. Isolation of MNCs Using Ficoll-Paque® PLUS

1. After mixing the sample well, measure the total volume. Perform a nucleated cell count to determine the number and concentration of nucleated cells in the original sample.
2. Dilute the sample at least 1:1 with IMDM 2% FBS, for example, add 2 mL PB to 2 mL IMDM 2% FBS for a total volume of 4 mL. Mix by inversion or by drawing the sample in and out of a Pasteur pipette. BM has a significantly higher cellular content than peripheral blood; therefore, dilute the sample at least 1:3 with IMDM 2% FBS, for example, add 2 mL BM to 6 mL IMDM 2% FBS for a total volume of 8 mL.
3. Pre-aliquot room temperature Ficoll-Paque® PLUS into an appropriately sized tube, for example, add 3 mL Ficoll-Paque® PLUS to a 14 mL tube for a 4 mL sample. The diameter of the centrifuge tube may vary, however, the approximate height of Ficoll-Paque® PLUS (2.4 cm) and sample (3.0 cm) should be maintained *(24)*.
4. Slowly layer the diluted sample over Ficoll-Paque® PLUS. It is best to hold the tube containing Ficoll-Paque® PLUS at an angle, then slowly and continuously

pipette the blood so it runs down the side of the tube allowing it to rest on top of the Ficoll-Paque® PLUS layer.

5. Handle the tube carefully as to not disturb the Ficoll-Paque® PLUS/blood interface. Centrifuge the sample by spinning at 400 g for 30 min at room temperature with the brake off.

6. Using a sterile pasteur pipette, remove and discard the top plasma layer, taking care not to disturb the whitish layer of mononuclear cells present at the interface of the Ficoll-Paque® PLUS layer. Remove the mononuclear cell layer and transfer it to a sterile 14-mL tube. It is critical to remove all the cells at the interface but a minimum amount of Ficoll-Paque® PLUS and supernatant. Removing excess Ficoll-Paque® PLUS causes granulocyte contamination, and removing excess supernatant results in unnecessary contamination by platelets and plasma proteins (*see* **Note 7**).

7. Wash the transferred cells by adding at least three volumes of IMDM 2% FBS and spinning at 400 g for 10 min at room temperature with the brake on.

8. Remove the supernatant by quickly but carefully decanting the tube in one smooth motion. Resuspend the pellet by vortexing gently and make up the tube volume to approximately 10 mL for a second spin. If multiple tubes were set up for the same sample, they can be pooled together at this stage.

9. Remove the supernatant from the second wash and resuspend cells in 1–2 mL IMDM 2% FBS and record the final volume (a larger volume may be used if the initial cell number was high). If cells start to clump after being resuspended, DNAse may be added to the cell suspension (100 μL DNase to 1 mL of cell suspension).

10. The final nucleated cell count can now be performed. The percent recovery of mononuclear cells from the original sample can be calculated as follows:

Percent cell recovery = total cell number following ficoll

÷ total cell number in original sample × 100

3.1.2. Red Blood Cell Lysis Using Ammonium Chloride (ACK) Lysis Buffer

1. After mixing the sample well, measure the total volume. Perform a nucleated cell count to determine the number and concentration of nucleated cells in the original sample.

2. Transfer the sample to 14-mL tubes, adding no more than 1–2 mL of sample per tube (use 50-mL tubes for larger volumes).

3. Add ACK lysis buffer (NH_4Cl) to the sample to give a minimum 4:1 volume: volume ratio of NH_4Cl: sample for human BM (i.e., 4 mL NH_4Cl to 1 mL of sample) or a 9 : 1 ratio for human or mouse PB and mouse BM (i.e., 9 mL NH_4Cl to 1 mL of sample). Mix by inversion or by drawing the sample in and out of a Pasteur pipette.

4. Incubate on ice for 10 min, mixing once or twice during the incubation period. All the red blood cells should be lysed within 10 min. The mixture should appear translucent in color when the lysis is complete. If lysis is not complete, mix and incubate on ice for a further 5–10 min for human cells. Mouse cells should not be left in lysis buffer for longer than 10 min.
5. Make up the volume in each tube with IMDM 2% FBS and centrifuge at 400 g for 10 min at room temperature with brake on.
6. Remove the supernatant by quickly but carefully decanting the tube in one smooth motion. Resuspend the cell pellet by vortexing gently and wash the sample by adding approximately 10 mL IMDM 2% FBS.
7. Centrifuge sample at 400 g for 10 min at room temperature with brake on. Decant the supernatant and wash cells once more with 10 mL IMDM 2% FBS. Cells from the same samples may be pooled before this final wash.
8. Decant the supernatant from this final wash and resuspend cells in 1–2 mL IMDM 2% FBS for final nucleated cell counts. A larger volume may be used if the initial cell number was high.

$$\text{Percent cell recovery} = \text{total cell number following ACK lysis}$$

$$\div \text{total cell number in original sample} \times 100$$

3.2. CFC Assays

CFC assays may be used to quantify the number of specific progenitors from various cell sources in both human- and animal-derived hematopoietic tissues. The media formulations that support the growth of the various progenitors for diverse species are summarized in **Table 1** (*see* **Note 8**). As the relative frequencies of the progenitors vary between tissues and from one species to another, **Tables 2** and **3** summarize an appropriate plating concentration to achieve suitable plating efficiency to acquire quantitative data. The general method for the assessment of progenitors is described below.

3.2.1. Erythroid, Myeloid, and Lymphoid CFC Assays

3.2.1.1. CULTURE SETUP

1. Thaw appropriate methylcellulose-based media formulation, previously aliquotted into tubes at the appropriate volume.
2. Prepare culture dishes by placing three 35-mm culture dishes with lids inside a 150 × 25 mm tissue culture dish with a lid. Add an additional 35-mm culture dish without a lid to serve as a water dish. This set of dishes is sufficient for one triplicate assay. If plating multiple assays, the 35-mm dishes may be placed in 245-mm bioassay trays with two extra culture dishes without lids for water dishes.

Table 1
Cytokine Content of Methylcellulose-Based Medium Formulations for the Assessment of Various CFC from Human, Mouse, Rat, Monkey, and Canine Cell Sources

Progenitors	Cytokines	MethoCult® medium
Human cell sources		
CFU-GM	rh GM-CSF, rh IL-3, rh IL-6, rh G-CSF, rh SCF	H4536
BFU-E	rh GM-CSF, rh IL-3, rh SCF, rh Epo	H4434
CFU-E		
CFU-GM		
CFU-GEMM		
CFU-Mk	rh Tpo, rh IL-6, rh IL-3	MegaCult®-C
Mouse cell sources		
CFU-E	rh Epo	M3334
Mature BFU-E		
BFU-E	rm SCF, rh IL-6, rm IL-3, rh Epo	M3434
CFU-GM		
CFU-GEMM		
CFU-Pre-B	rh IL-7	M3630
CFU-Mk	rh Tpo, rm IL-3, rh IL-6, rh IL-11	MegaCult®-C
Rat cell sources		
CFU-GM	GM-CSF, SCF, IL-3	GF R3774
Monkey cell sources		
CFU-E	rh GM-CSF, rh IL-3, rh SCF, rh Epo	H4434
BFU-E		
CFU-GM		
CFU-GEMM		
Canine cell sources		
CFU-GM	rc GM-CSF, rc SCF	Not commercially available

GM-CSF, granulocyte monocyte-colony stimulating factor; G-CSF, granulocyte stimulating factor IL, interleukin; SCF, stem cell factor; Epo, erythropoietin; Tpo, thrombopoietin.

Table 2
Recommended Cell Plating Concentrations for Erythroid and Myeloid Progenitor Assessment in MethoCult®, Including the Expected Number of Colonies Obtained Per Dish

	Final plating concentration (cells per 35-mm dish)	Expected colony number (per 35-mm dish)			
		CFU-E	BFU-E	CFU-GM	GEMM
Human cell sources					
Bone marrow (NH$_4$Cl treated)	2×10^4	8 ± 2	21 ± 15	48 ± 39	1 ± 1
Bone marrow (Light density)	1×10^4	9 ± 4	38 ± 8	47 ± 10	1 ± 1
Bone marrow (CD34+ cell enriched)	500	3 ± 3	28 ± 7	73 ± 17	2 ± 1
Cord blood (fresh unprocessed)	3×10^4	ND	21 ± 8	18 ± 9	7 ± 7
Cord blood (frozen unprocessed)	5×10^4	ND	23 ± 14	20 ± 16	3 ± 3
Cord blood (CD34+ cell enriched)	500	ND	59 ± 5	122 ± 47	5 ± 3
Peripheral blood (Light density)	1×10^5	ND	31 ± 19	11 ± 7	ND
Mobilized PB	2×10^4	4 ± 2	$72 \perp 30$	69 ± 36	4 ± 3
Mouse cell sources (no.)					
Bone marrow (untreated)	2×10^4		8 ± 3	$64 \perp 16$	$3 + 1$
Bone marrow (untreated)	2×10^5	340 ± 80			
Spleen (NH$_4$Cl treated)	2×10^5		15 ± 10	23 ± 10	1 ± 1
Spleen (NH$_4$Cl treated)	2×10^5	18 ± 8			
Peripheral blood (NH$_4$Cl treated)	3×10^5		4 ± 4	11 ± 5	1 ± 1
Day 14 fetal liver	2×10^4		9 ± 3	55 ± 10	3 ± 2

(*Continued*)

Table 2
(*Continued*)

	Final plating concentration (cells per 35-mm dish)	Expected colony number (per 35-mm dish)			
		CFU-E	BFU-E	CFU-GM	GEMM
Rat cell sources					
Bone marrow	2×10^4	ND	ND	53 ± 4	ND
Monkey cell sources[a]					
Bone marrow (NH$_4$Cl treated)	1×10^4				
Peripheral blood	2×10^5				
Mobilized peripheral blood	3×10^4				
Dog cell sources[a]					
Bone Marrow (NH$_4$Cl treated)	$1–2 \times 10^5$				

ND, none detected.

no.—the numbers quoted for mouse may vary depending on the species used.

[a] The plating concentrations listed have been acquired from references *(19–21)*; colony numbers are not listed for individual progenitors.

It is good practice to wipe the trays down with 70% isopropanol before each use to maintain sterility.

3. Label the edge of each 35-mm culture dish lid with experiment and assay number using a permanent (water resistant) fine tip felt marker.

4. Dilute the cells with IMDM 2% FBS to 10× the final concentration(s) required for plating. Refer to **Table 2** for the recommended plating concentrations. When it is difficult to anticipate the correct plating cell concentration, the use of two or more 2–3-fold serially diluted cell concentrations is advised.

5. Add 0.4 mL of diluted cells to the 4 mL media tube (*see* **Note 9**)

6. Vortex the tube vigorously for approximately 4 s and let it stand for at least 5 min to allow bubbles to dissipate.

7. Using a 3 cc syringe with and a 16 gauge blunt end needle, draw up approximately 1 mL of media and dispense it completely back into the tube to remove air bubbles from the syringe (repeat if necessary). Methylcellulose is a viscous solution and cannot be accurately dispensed using pipettes because of adherence of the medium to the inside of the pipette. Blunt end needles should be used to prevent injury due to needle pricks and to facilitate the accurate dispensing of methylcellulose-based medium.

Table 3
Recommended Cell Plating Concentrations for Lymphoid and Megakaryocytic Progenitor Assessment in MethoCult® and MegaCult®, Including the Expected Number of Colonies Obtained Per Dish

	Final plating concentration (cells per dish or slide)	Expected colony number (per dish or slide)		
		CFU-Pre-B	CFU-Mk	Non-CFU-Mk
Human cell sources				
Bone Marrow (light density)	1×10^5		68 (10–150)	3 (1–10)
Cord blood (light density)	5×10^4		27 (0–55)	3 (0–8)
BM, CB, MPB (CD34+ cell enriched)	1500–5000		65 (0–140)	3 (0–8)
Mouse cell sources (no.)				
Bone marrow (untreated)	1×10^5		35 ± 5	49 ± 9
Bone marrow (untreated)	1×10^5	37 ± 40		

no.—the numbers quoted for mouse may vary depending on the species used.

8. Draw up media containing cells to the 2.6 mL mark of the syringe. Use the opposite hand to remove the culture dish lid and dispense 1.1 mL into the center of the first 35-mm dish (plunger now at the 1.5 mL mark). Replace lid. Try to not touch the end of the needle to the bottom of the dish when plating.

9. Similarly, dispense another 1.1 mL into the second 35-mm dish (plunger now at 0.4 mL mark). Draw up more media to the 1.5 mL mark and dispense 1.1 mL into the third dish. For each tube plated, use a new sterile disposable 3-mL syringe with a new 16 gauge blunt end needle.

10. Tilt and rotate the dishes to spread the medium evenly over the surface of each dish. Allow the meniscus to attach to the dish wall evenly on all sides, but avoid getting media too far up the side wall.

11. Place the dishes in the large tissue culture dish with lid or tray with lid, and add approximately 3 mL of sterile water to the uncovered 35-mm dishes to maintain humidity.

12. Put the tray of cultures in an incubator maintained at 37°C, 5% CO_2 in air, and > 95% humidity for 2–16 days, depending on the duration of the specific

Table 4
Optimal Incubation Times for the Enumeration and Morphological Assessment of Erythroid, Myeloid, Lymphoid, and Megakaryocytic Progenitors

Species	Progenitor type	Incubation time (days)
Human	CFU-E, BFU-E, CFU-GM, CFU-GEMM	14–16
	CFU-Mk	10–12
Mouse	CFU-E	2–3
	BFU-E, CFU-GM, CFU-GEMM	10–12
	CFU-Pre-B	7
	CFU-Mk	6–8
Rat	CFU-GM	9–11
Monkey	CFU-E, BFU-E, CFU-GM, CFU-GEMM	14–16
Canine	CFU-GM	7–10

assay (*see* **Table 4**). Culture conditions are very important to ensure optimal hematopoietic colony growth. A water jacketed incubator with an open pan of water placed in the incubator chamber is recommended. A suitable additive (i.e., copper sulphate crystals) can be added to the water pan to inhibit microbial growth. Incubator temperature should be confirmed using a thermometer placed in the incubator chamber, and CO_2 levels should be routinely monitored using a fyrite CO_2 device.

13. If the cultures will not be enumerated by the appropriate day they may be placed in a 33 °C incubator at this time to slow colony growth and temporarily maintain optimal morphology. Refill water dishes if necessary and count as soon as possible (3–4 days). See **Subheadings 3.2.1.2** and **3.2.1.3** for colony enumeration guidelines and criteria.

3.2.1.2. Colony Enumeration

It is important to use a high-quality inverted microscope equipped with low power ($\times 2.5$) and higher power ($\times 4$–5, $\times 10$) objectives, $\times 10$–12.5 ocular eyepieces, and a blue filter to enhance the red color of hemoglobinzed erythroblasts.

When plated at the correct concentration, there should be sufficient spacing between colonics to be able to identify individual clusters of cells as being derived from a common precursor cell. Fewer than 30 colonies on a dish indicate that the cultures have been under plated and therefore may not be an accurate assessment of the progenitor content of the sample. Greater than 150 colonies on a dish reduce the counter's ability to distinguish individual colonies.

1. For counting 35-mm dishes, prepare a 60-mm gridded scoring dish by drawing two perpendicular lines across the center of the dish using a permanent fine felt marker on the bottom of the dish.
2. Take the cultures to be scored from the 37 °C incubator. Only take the number of dishes that can be scored within 1 h.
3. Center the 35-mm dish on the 60-mm gridded scoring dish. Adjust the focus under low power ($\times 2.5$ objective) until the colonies are in focus. Scan the entire dish and make note of the overall appearance of the culture to help with the scoring and evaluation. Considerations to keep in mind are the placement of colonies relative to one another, the approximate number of colonies on the dish, background debris, general morphology, health of the colonies, and so on.
4. Starting at one side of the dish, count all colonies on the dish including those on the edge. Counting up and down rather than side to side will minimize the sensation of motion sickness common to individuals new to scoring.
5. Count all the colonies of interest under high power (using a $\times 4$–5 objective for a total magnification of $\times 40$–50 with a $\times 10$ ocular eyepiece). It is necessary to continually focus up and down to identify all colonies present in the 3D culture, those at the edges, and to distinguish individual colonies that are close together but present on different planes.

Accurate enumeration of hematopoietic progenitors takes training and experience. Subjectivity of microscopic evaluation of colonies can cause variability in the reported progenitor content. **Table 5** summarizes the intra- and inter-assay variability of hematopoietic CFC enumeration using the same cell source. In the previous tables (*see* **Tables 2** and **3**), the frequency of progenitors may be dependent on the cell source, but, even within one cell source (i.e., human PB), there is a large variability in the hematopoietic CFC content

Table 5
Coefficient of Variation (CV) in the Enumeration of Human Erythroid and Myeloid CFC

Progenitor	Intra-Assay CV (%) ($n = 1$ participant, 20 replicate samples)	Inter-Assay CV (%) ($n = 54$ participants)
CFU-E	13.6	46.0
BFU-E	11.5	44.2
CFU-GM	9.1	36.9
CFU-GEMM	10.0	80.0
Total erythroid	8.1	25.2
Total CFC	4.9	23.2

Table 6
Variability in Human Erythroid and Myeloid CFC from the Same Cell Source from 30 Different Donors

	BFU-E derived colonies/mL PB	CFU-GM derived colonies/mL PB	Total CFC/mL PB
Mean ± SD[a]	483 ± 263	173 ± 122	680 ± 380
Range	127–1118	28–464	151–1511

[a] $n = 30$ normal individuals.

between different normal individuals. **Table 6** summarizes data of the total erythroid and myeloid progenitors per mL of PB from 30 normal individuals, age range 21–55 years.

3.2.1.3. COLONY IDENTIFICATION

Colony descriptions provide information on the typical colony morphologies that will be seen in optimal culture conditions. Red blood cell background, toxicity, or disease states could all potentially alter colony morphology and make colony enumeration more difficult *(25)*.

*3.2.1.3.1. Human Erythroid and Myeloid Colonies (see **Fig. 2**)*

1. Colony-forming unit-erythroid (CFU-E): Produces 1–2 clusters containing 8–200 erythroblasts. CFU-E are more mature erythroid progenitors with limited proliferative potential and require erythropoietin (EPO) for differentiation (*see* **Fig. 2, Panel 1**).
2. Burst-forming unit-erythroid (BFU-E): Produces a colony containing > 200 erythroblasts in a single or multiple clusters and can be sub-classified based on the number of cells or cell clusters per colony if desired. BFU-E are more immature progenitors than CFU-E and require EPO and cytokines with burst-promoting activity such as IL-3 and stem cell factor (SCF) for optimal colony growth (*see* **Fig. 2, Panel 2**).
3. Colony-forming unit-granulocyte, macrophage (CFU-GM): Produces a colony containing at least 20 granulocyte cells (CFU-G), macrophages (CFU-M), or cells of both lineages (CFU-GM). CFU-GM colonies arising from primitive progenitors may contain thousands of cells in single or multiple clusters. The monocytic lineage cells are large cells with an oval to round shape and appear to have a grainy or gray center. The granulocytic lineage cells are round, bright, and are much smaller and more uniform in size than macrophage cells (*see* **Fig. 2, Panel 3** and **4**).

Human CFC derived colonies

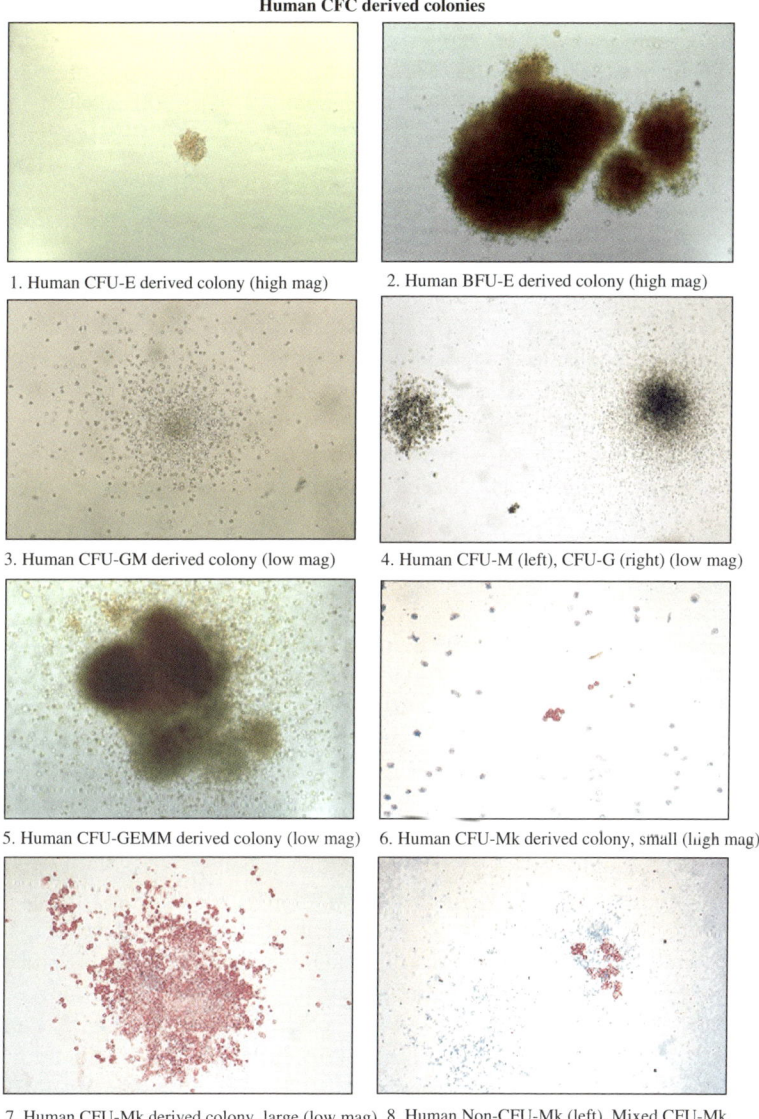

1. Human CFU-E derived colony (high mag)

2. Human BFU-E derived colony (high mag)

3. Human CFU-GM derived colony (low mag)

4. Human CFU-M (left), CFU-G (right) (low mag)

5. Human CFU-GEMM derived colony (low mag)

6. Human CFU-Mk derived colony, small (high mag)

7. Human CFU-Mk derived colony, large (low mag)

8. Human Non-CFU-Mk (left), Mixed CFU-Mk (right) (low mag)

Fig. 2. Human CFC-derived colonies. (**1**) Human CFU-E-derived colony (high mag); (**2**) human BFU-E-derived colony (high mag); (**3**) human CFU-GM-derived colony (low mag); (**4**) human CFU-M (left), CFU-G (right) (low mag); (**5**) human CFU-GEMM-derived colony (low mag); (**6**) human CFU-Mk-derived colony, small (high mag); (**7**) human CFU-Mk-derived colony, large (low mag); (**8**) human non-CFU-Mk (left), Mixed CFU-Mk (right) (low mag).

4. Colony-forming unit-granulocyte, erythroid, macrophage, megakaryocyte (CFU-GEMM): A multi-potential progenitor that produces a colony containing erythroblasts and cells of at least two other recognizable lineages. Owing to their primitive nature, CFU-GEMM tend to produce large colonies of > 500 cells (*see* **Fig. 2, Panel 5**).

3.2.1.3.2. Mouse Erythroid, Myeloid, and Lymphoid Colonies (see **Fig. 3**)

1. CFU-E: Produces 1–2 clusters containing 8–32 maturing erythroblast cells. CFU-E are more mature erythroid progenitors and require EPO for differentiation. These colonies are very tiny as seen under ×40–50 magnification. Erythroblast cells within the cluster are irregular in shape and appear fused together. CFU-E are optimally counted at day 2 of culture (*see* **Fig. 3, Panel 9**).

2. BFU-E: Produces a colony containing erythroid clusters and a minimum of 30 cells. Each individual cluster contains a group of cells that are tiny, irregular in shape, and difficult to distinguish. The cells appear fused together. BFU-E do not usually have a dense core, and the clusters are relatively scattered. BFU-E are more immature progenitors than CFU-E and require EPO and cytokines with burst-promoting activity such as IL-3 and SCF for optimal colony growth (*see* **Fig. 3, Panel 10**).

3. CFU-GM: Produces a colony containing at least 30 granulocyte cells (CFU-G), macrophages (CFU-M), or cells of both lineages (CFU-GM). CFU-GM colonies arising from primitive progenitors may contain thousands of cells in single or multiple clusters. The monocytic lineage cells are large cells with an oval to round shape and appear to have a grainy or grey centre. The granulocytic lineage cells are round, bright and are much smaller and more uniform in size than macrophage cells (*see* **Fig. 3, Panel 11**).

4. CFU-GEMM: A multi-potential progenitor that produces a colony containing a highly dense core with an indistinct border between the core and the peripheral cells. Erythroblast clusters should be visible along the periphery of the CFU-GEMM colony. Monocytic and granulocytic cells should be easily identifiable and clusters of large megakaryocytic cells are usually seen. Owing to their primitive nature, CFU-GEMM tend to produce large colonies of > 500 cells (*see* **Fig. 3, Panel 12**).

5. CFU-Pre-B: A subset of B-lymphoid progenitors that can be detected in the presence of IL-7. Produces a colony containing at least 30 cells. CFU-pre-B colonies vary in size and morphology, and individual cells appear tiny and irregular to oval in shape. Some CFU-pre-B colonies are dense with very few cells in the periphery, and some have a smaller core with more cells in the periphery (*see* **Fig. 3, Panel 13**).

Following enumeration of mouse CFC-derived colonies, the number of colonies per dish and the number of cells plated from a specific organ can be used to back calculate the total number of CFC per organ. Calculations of the frequency of

Mouse CFC derived colonies

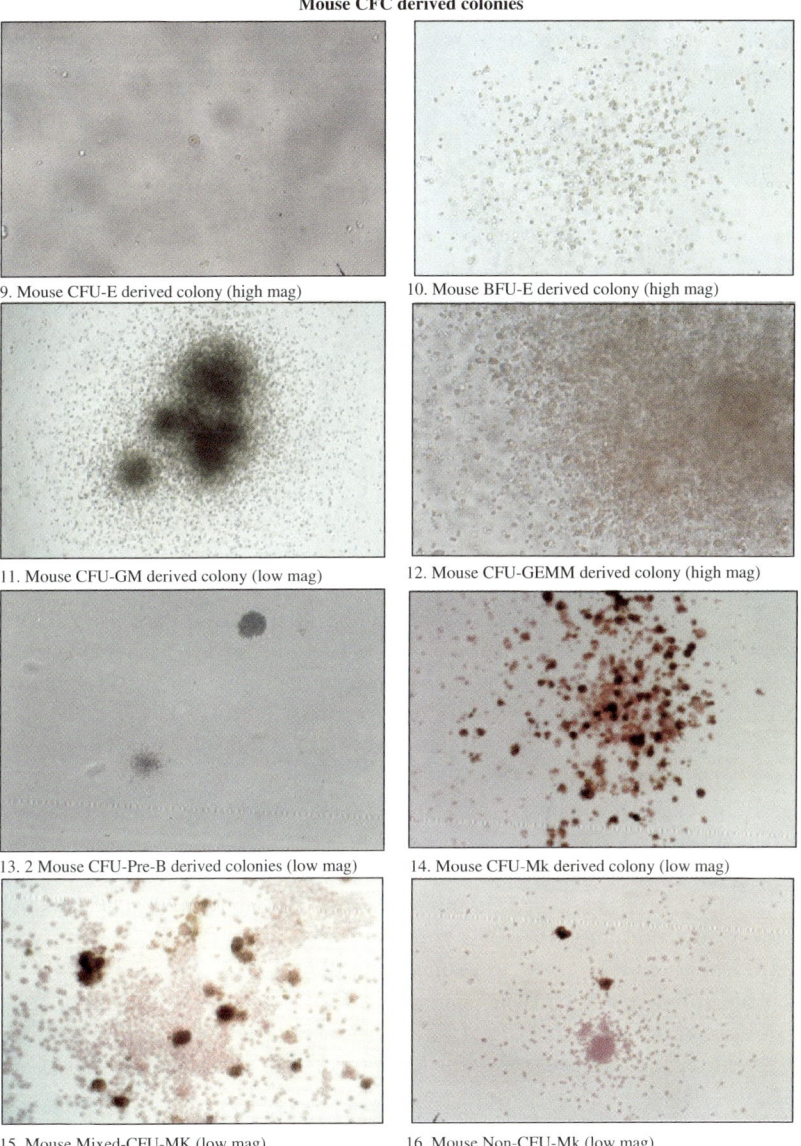

9. Mouse CFU-E derived colony (high mag)

10. Mouse BFU-E derived colony (high mag)

11. Mouse CFU-GM derived colony (low mag)

12. Mouse CFU-GEMM derived colony (high mag)

13. 2 Mouse CFU-Pre-B derived colonies (low mag)

14. Mouse CFU-Mk derived colony (low mag)

15. Mouse Mixed-CFU-MK (low mag)

16. Mouse Non-CFU-Mk (low mag)

Fig. 3. Mouse CFC-derived colonies. **(9)** mouse CFU-E-derived colony (high mag); **(10)** mouse BFU-E-derived colony (high mag); **(11)** mouse CFU-GM-derived colony (low mag); **(12)** mouse CFU-GEMM-derived colony (high mag); **(13)** mouse CFU-Pre-B derived colonies (low mag); **(14)** mouse CFU-Mk-derived colony (low mag); **(15)** mouse Mixed-CFU-MK (low mag); **(16)** mouse non-CFU-Mk (low mag).

progenitors per nucleated cell and the total number of CFC per femur or spleen can help identify disturbances to hematopoietic homeostasis in genetically altered or compound-treated mice.

3.2.2. Megakaryocytic CFC Assays

3.2.2.1. CULTURE SETUP

1. Thaw previously aliquoted tubes of medium at room temperature or overnight at 4 °C. Place media and collagen solution on ice. When using medium without cytokines (2.55 mL/tube), add the desired cytokines and IMDM (0.45 mL) to achieve a final volume of 3.0 mL.
2. Label slides using a pencil or diamond point pen. Ink labeling will become illegible during the fixation process.
3. Prepare a cell suspension in IMDM 2% FBS that is 22× the desired final plating concentration(s). Refer to **Table 3** for the recommended plating concentrations. When it is difficult to anticipate the correct plating cell concentration, the use of two or more 2–3-fold serially diluted cell concentrations is advised.
4. Add 0.15 mL of cell stock to each 3.0-mL tube of media and vortex (*see* **Note 9**).
5. Using a 2-mL pipet, transfer 1.8 mL of cold collagen into one tube and vortex. Using the same pipet, remove 1.5 mL of the mixture and dispense 0.75 mL into each well of the pre-labeled chamber slide. Avoid introducing bubbles into the well while dispensing. Repeat for the second and third chamber slides. The collagen will start to gel within minutes. If more than one tube is to be set up, collagen should be added one tube at a time and the contents of the tube dispensed into chamber slides before proceeding to the next tube.
6. Gently tip each slide using a circular motion to spread the media evenly over the surface of the slide.
7. Place each slide in a 150 × 25 mm tissue culture dish with lid or tray with lid, along with 1 or 2 uncovered 35-mm dishes containing approximately 3 mL sterile water to maintain humidity. It is good practice to wipe the trays down with 70% isopropanol before each use to maintain sterility.
8. Incubate cultures in a 37 °C incubator with 5% CO_2 and > 95% humidity for 6–12 days, depending on the duration of the specific assay (*see* **Table 4**). Gel formation will occur within approximately 1 h; it is important not to disturb the cultures during this time.
9. Cultures should be visually assessed for overall growth and morphology using an inverted light microscope before fixation and staining.

3.2.2.2. DEHYDRATION

1. Leave the cultures at 37 °C until just before dehydration as the collagen gel may become unstable at lower temperatures. Remove slides from the incubator singly or in small batches.

2. Remove plastic chamber from the slide. This can be done by slowly pulling up the chamber at the labeled end of the slide. Using forceps, pull up the rubber seal at each corner and hook it around the corners of the chamber. Gently pull the chamber and seal away from the rest of the slide, taking care not to disturb the gel.
3. If there are bubbles of the gel, they can be removed by lightly touching the bubble with a piece of filter paper card. If the gel has shifted, it can be straightened by gently adjusting it with forceps.
4. Place a pre-cut polypropylene spacer membrane onto each slide. Place a thick white filter card top to wick the liquid from the slide.
5. Once fully soaked, remove the filter card but leave the polypropylene spacer in place.

3.2.2.3. FIXATION

3.2.2.3.1. Human Megakaryocyte Fixation

1. Prepare fixative by combining 3 parts acetone (150 mL) with 1 part methanol (50 mL) and placing in a 2.5 L plastic container with lid.
2. Place the slides into the methanol-acetone fixative solution. Ensure slides are completely covered in fixative in a single layer to prevent slides from scratching each other. No more than approximately 12 slides should be placed in the container at one time. Fixative should be changed every 12 slides.
3. Leave slides in fixative for approximately 20 min at room temperature. Do not agitate fixative solution. Polypropylene spacers will float to the top of the fixative on their own or may be gently peeled off the slide with forceps.
4. Remove slides from fixative and allow to air dry for at least 15 min.
5. Cultures should be stained as soon as possible after fixation. It is recommended that slides be stained within three days.

3.2.2.3.2. Mouse Megakaryocyte Fixation

1. Prepare the fixative by adding 200 mL acetone in a 2.5 L plastic container and placing on ice for 15 minutes.
2. Place the slides into the cold acetone fixative solution. Ensure slides are completely covered in fixative in a single layer to prevent slides from scratching each other. No more than approximately 12 slides should be placed in the container at one time. The fixative solution should be changed every 12 slides.
3. Leave slides in fixative for approximately 5 min on ice. Polypropylene spacers will float to the top of the fixative on their own or may be gently peeled off the slide with forceps.
4. Remove slides from fixative and allow to air dry.
5. Slides can be stained immediately or stored at $-20\,°C$ or $4\,°C$ in the dark for up to 1 month.

3.2.2.4. Staining

3.2.2.4.1. Human Megakaryocyte Staining

1. Prepare all staining solutions except the alkaline phosphatase substrate.
2. Rehydrate the cultures on the slides by applying approximately 1.5 mL of 0.05 M Tris/NaCl buffer, pH 7.6 and incubate for 20 min at room temperature. The buffer can be gently applied using a wash bottle. Ensure that the cultures are completely covered with each solution. The use of covered containers is recommended to prevent cultures from drying out at any stage.
3. Remove buffer by pouring off into a waste container and blotting the short end of the slide on a piece of paper towel (or other absorbent material).
4. Apply 0.5 mL of 5% human serum in Tris/NaCl buffer to each slide for 20 min.
5. Remove blocking solution as in **step 3** and apply 0.5 mL of the primary antibody or negative control antibody and incubate for 30 min. One slide per batch should be stained with the negative control antibody rather than the primary antibody. All other steps of the staining procedure are the same.
6. Remove antibody as in **step 3** and rinse slide with 0.05 M Tris/NaCl buffer three times for 3 min each. Add buffer to one end of slide and let buffer run gently over slide to prevent the gel from dislodging.
7. Apply 0.5 mL of the biotin-conjugated goat anti-mouse IgG antibody and incubate for 30 min.
8. Remove antibody as in **step 3** and wash slide three times with 0.05 M Tris/NaCl buffer as outlined in **step 6**.
9. Apply 0.5 mL of the avidin alkaline phosphatase conjugate and incubate for 30 min.
10. Prepare the alkaline phosphatase substrate solution (*see* **Subheading 2.3.4.1.8.**).
11. Remove phosphatase conjugate as in **step 3** and rinse three times with 0.05 M Tris/NaCl as outlined in **step 6**.
12. Apply 0.5 mL of the alkaline phosphatase substrate solution and incubate for 15 min.
13. Remove substrate as in **step 3** and wash three times with 0.05 M Tris/NaCl as outlined in **step 6**.
14. Apply 0.5 mL of Evans Blue counterstain and incubate for a maximum of 10 min. Increasing the incubation time to greater than 10 min may result in a decrease in staining intensity.
15. While holding slide over waste container, use a wash bottle containing distilled water to gently rinse off excess Evan's blue.
16. Allow the slides to air dry. Slides can be stored in covered containers at room temperature or at 2–8 °C for prolonged storage. If the slides are to be cover-slipped, an aqueous mounting medium should be used. Most alkaline phosphatase complexes are soluble in organic solvents.

3.2.2.4.2. Mouse Megakaryocyte Staining

1. Allow slides to come to room temperature.
2. Dissolve 10 mg of acetylthiocholiniodide in 15 mL of 0.1 M sodium phosphate buffer.
3. Add in the following order with constant stirring:

   a. 1 mL of 0.1 M sodium citrate.
   b. 2 mL of 30 mM copper sulphate.
   c. 2 mL of 5 mM potassium ferricyanide solution.
      Total volume is 20 mL, sufficient for approximately 10–15 slides.

4. Flood slides with staining solution and incubate for 5 h in a humid chamber, ensuring the entire surface of the slide is covered. Use of covered containers is recommended to prevent cultures from drying out at any stage (*see* **Note 10**).
5. Remove staining solution by pouring off into a waste container and blotting the short end of the slide on a piece of paper towel (or other absorbent material).
6. Fix slides with 95% ethanol and let sit for 10 min, ensuring entire surface of slide is covered.
7. Rinse slides with lukewarm water and air dry.
8. Counterstain slides with Harris' hematoxylin solution for 30 s.
9. Rinse each slide with lukewarm water. Allow to air dry. Slides can be stored in covered containers at room temperature for prolonged storage.

3.2.2.5. ENUMERATION

1. First scan the entire slide using a ×5 objective lens, noting the distribution of the colonies on the slide. Scoring can then be performed with the same lens, and the ×10 objective lens can be used to examine colonies in greater detail.
2. Slides can be counted by "land marking" the edges of the field of view to prevent scanning overlapping areas (and counting colonies twice) or missing areas of the slide (and missing colonies altogether).

Similar to the variability in the human myeloid and erythroid progenitors, there is also variability in the enumeration of megakaryocytic progenitors. **Table 7** summarizes the inter- and intra-assay variability for CFU-Mk derived colony enumeration.

3.2.2.6. IDENTIFICATION

3.2.2.6.1. Human Megakaryocytic Colonies

1. Colony-forming unit-megakaryocyte (CFU-Mk): Produces a colony containing three or more megakaryocytic cells. Human megakaryocytes and platelets, which express Glycoprotein IIb/IIIa (CD41), will appear pink following fixation and staining. Counterstaining with Evans Blue causes the nuclei of all cells to appear pale blue,

Table 7
Percent Coefficient of Variability in the Enumeration of Human Megakaryocytic CFC

Progenitor	Intra-Assay CV (%) (n = 1 participant, 18 replicate samples)	Inter-Assay CV (%) (n = 41 participants)
CFU-Mk (3–20)	14.5	48.1
CFU-Mk (21–49)	21.1	41.6
CFU-Mk (> 50)	31.3	61.5
Total CFU-Mk	10.8	36.4

regardless of lineage. As a result, CFU-Mk-derived colonies appear as groups of cells, which have pink membrane staining with blue nuclei. Colonies will range in size from three to several hundred megakaryocytes per colony. It is therefore convenient to subdivide them by colony size as:

a. Small: 3–20 cells per colony, produced from a more mature Mk progenitor (*see* **Fig. 2, Panel 6**).
b. Medium: 21–49 cells per colony.
c. Large: ≥ 50 cells per colony, produced from a more primitive Mk progenitor (*see* **Fig. 2, Panel 7**).

2. Non-CFU-Mk: Non-megakaryocytic colonies (≥ 20 cells per colony) are usually of the granulocyte/monocyte lineage and have nuclear staining only. Numbers of these colonies can be recorded if desired, but media do not support optimal growth of CFU-GM (*see* **Fig. 2, Panel 8**).
3. Mixed-CFU-Mk: Mixed megakaryocytic colonies are distinguished by the presence of non-Mk cells and megakaryocytes within the same colony (*see* **Fig. 2, Panel 8**).

3.2.2.6.2. Mouse Megakaryocytic Colonies

1. CFU-Mk: Produces a colony containing 3 to approximately 50 megakaryocytic cells. Mouse megakaryocytes and early megakaryocyte progenitors express acetylcholinesterase and have brown granular deposits of copper ferrocyanide in the cytoplasm resulting from the enzymatic reaction. Granules may appear light redbrown in cells with low acetylcholinesterase content and may range from orangebrown to dark brown/black in cells with high cholinesterase content. Counterstaining with Harris' hematoxylin results in violet stained nuclei in all cells. However, this is less noticeable in megakaryocytes as the copper ferrocyanide precipitate from the acetylcholinesterase stain can mask the counterstain (*see* **Fig. 3, Panel 14**).

2. Mixed-CFU-Mk. Mixed megakaryocytic colonies are distinguished by the presence of non-Mk cells and megakaryocytes within the same colony. It is sometimes difficult to identify mixed Mk colonies, especially when the culture contains large numbers of CFU-Mk colonies and non-Mk colonies close together. Plating concentration should be adjusted accordingly (*see* **Fig. 3, Panel 15**).

3. Non-CFU-Mk. Non megakaryocytic colonies ($\geq$ 30 cells per colony) are usually of the granulocyte/monocyte lineage and have nuclear staining only, resulting in groups of cells with violet nuclei but no brown precipitate. Numbers of these colonies can be recorded if desired, but media do not support optimal growth of CFU-GM (*see* **Fig. 3, Panel 16**).

4. Notes

1. Methylcellulose, collagen, serum, albumin, and cytokines are all media components that are known to effect the proliferative potential of CFC and morphology of CFC-derived colonies. The media formulations described in this chapter have been developed with prescreened components to ensure optimal colony growth. If alternate media and reagents are to be used, it is suggested that they are pretested to ensure that acceptable growth and morphology are maintained.

2. Hematopoietic colony assays require both the use of culture medium that supports the growth and the differentiation of the progenitors of interest and a gelling agent that increases the viscosity of the medium without converting it to a solid. This allows a single progenitor to form clonal progeny, which stay together as a colony. Several types of semi-solid support media can be used for the culture of CFCs, including agar, collagen, and methylcellulose. Methylcellulose is now widely used as a gelling agent because it permits better growth of erythroid colonies than other agents (e.g., agar) and optimal myeloid colony formation. This allows erythroid, granulocyte, macrophage as well as multi-potential lineage progenitors to be assayed simultaneously in the same culture dish.

3. Media can be uniquely formulated for specific applications. MethoCult™ or MegaCult® "complete" formulations come as ready-to-use media containing recombinant cytokines required for the proliferation of specific progenitor lineages. Other MethoCult™ or MegaCult® formulations are provided without cytokines or other base components, allowing for the addition of exogenous cytokines or other components of interest for individual research needs. Depending on the formulation appropriate for individual applications, the volumes aliquoted into each tube must allow for the addition of supplementary components diluted in IMDM 2% FBS for a total volume of 4.0 mL per tube of MethoCult and 3.0 mL per tube of MegaCult® for triplicate cultures.

4. Unlike other hematopoietic lineages, megakaryocytes cannot easily be distinguished morphologically from other cell types and therefore require specific staining procedures for their identification. As methylcellulose-based media are not amenable to this, collagen gels have been used for the culture of megakaryocytic

progenitor cells. Collagen has been shown to support the growth of various cell types, and following incubation, the entire culture can be dehydrated and fixed for subsequent cytochemical or immunochemical staining.

5. Processing of mouse hematopoietic organs:

   a. BM—marrow cells can be removed from femurs and tibias by cutting off the ends of the bones and flushing the marrow out of the bone cavity using a 1 cc syringe containing IMDM 2% FBS and a 21 gauge needle. Cells can be drawn up into the syringe and expelled 3–4 times to dissociate the marrow into a single-cell suspension.

   b. Spleen—a single-cell suspension can be made by crushing the spleen using the rubber-ended plunger of a 3 cc syringe through a 70-μm cell strainer placed on a 50-mL falcon tube. The strainer is then washed with IMDM 2% FBS to collect additional cells from the mesh.

   c. PB—cells can be collected by cardiac puncture, tail bleeds, or orbital bleeds and transferred to microtainer tubes containing heparin or equivalent anti-coagulant (EDTA for example). Invert sample to mix with anti-coagulant.

   d. Fetal liver—mince fetal livers finely using scissors. Add 1–2 mL of IMDM 2% FBS. Cell aggregates can be disrupted by drawing cell suspension up and down 3–4 times using a 3 cc syringe and 21 gauge needle. Place cells in a 14-mL sterile tube and let stand for 3–5 min to allow tissue fragments to settle. Transfer cell suspension to a new sterile tube. Fill tube with IMDM 2% FBS and centrifuge for 7–10 min at 400 g. Resuspend cells in a small volume of media and mix.

6. At times it is not possible to process samples immediately. In the case of BM, studies indicate that processing the cells and maintaining them in IMDM plus 2% FBS at 4 °C minimizes the loss of CFC progenitors experienced when cells are maintained in the intact femurs overnight at 4 °C (*see* **Table 8**).

7. Alternately, directly remove the whitish layer of mononuclear cells sitting at the interface between the Ficoll-Paque® PLUS and plasma layers with a 5-mL pipette and transfer to a new 14-mL tube. As the MNCs are removed, the interface will become a clear sharp line, evidence that most cells have been collected.

8. **Table 1** summarizes media formulations containing various cytokines that have been designed for optimal colony morphology and number. Alternate cytokine formulations are available. In addition, serum-free medium formulations have been designed, which contain BSA, rh insulin, and human transferrin as a replacement for FBS.

9. If evaluation of a test compound is required, the compound may be added before the addition of cells directly to methylcellulose or collagen based media containing the appropriate cytokine formulation. The compound should be vortexed before the addition of the cells to ensure even distribution. The amount of compound added should not be more than 1% of the final volume. If compound has been

Table 8
Decrease in CFC Content in BM Cells Which Have Been Either (i) Flushed and Suspended in IMDM 2% FBS or (ii) Left in the Intact Femur.

Time (hours post sacrifice)	% viability (TB)		CFU-GM ($n = 4$)		BFU-E ($n = 4$)	
	Flush	Intact	Flush	Intact	Flush	Intact
1	92 ± 4	91 ± 3	100	100	100	100
2	90 ± 5	90 ± 4	101 ± 18	93 ± 12	123 ± 32	90 ± 40
4	91 ± 3	87 ± 12	105 ± 19	103 ± 30	106 ± 31	108 ± 16
8	91 ± 5	84 ± 14	115 ± 28	92 ± 27	110 ± 53	74 ± 29
24	88 ± 5	66 ± 13^a	119 ± 24	53 ± 40	126 ± 42	49 ± 34

Time (hours post sacrifice)	CFU-Mk ($n = 4$)		CFU-F ($n = 4$)		CFU-pre-B ($n = 3$)	
	Flush	Intact	Flush	Intact	Flush	Intact
1	100	100	100	100 ± 0	100	100
2	95 ± 19	97 ± 11	101 ± 33	122 ± 73	95 ± 44	77 ± 23
4	83 ± 29	94 ± 13	74 ± 10	93 ± 52	172 ± 42	163 ± 72
8	100 ± 9	93 ± 15	48 ± 16	77 ± 20	217 ± 118	207 ± 87
24	106 ± 8	68 ± 50	36 ± 10	34 ± 46^a	106 ± 107	13 ± 6^a

CFU-F, colony forming unit-fibroblast, a mesenchymal non-hematopoietic bone marrow progenitor. Data given as a percentage of 1 h growth (designated as control). Viability was assessed by trypan blue (TB) exclusion.
[a] Indicates a statistical difference less than $p = 0.01$.

solubilized in DMSO, the final volume of DMSO should not represent more than 0.1% of the final total volume of the culture as DMSO can inhibit hematopoietic colony growth.

10. It has been reported that 3.5–4 h is enough time to detect 100% of larger mature megakaryocytes, but only 30–60% of immature megakaryocytes. Slides can be stained for up to 6 h to detect all megakaryocytic progenitors (**26**). Granules appear light red-brown in cells with low acetylcholinesterase content and brown-black in the intensely stained megakaryocytes. The longer the staining solution is left on, the darker is the megakaryocytes stain.

References

1. Gordon MY. (1993) Human haemopoietic stem cell assays. *Blood Rev.* **7(3)**: 190–197.
2. Pessina A, Albella B, Bayo M, Bueren J, Brantom P, Casati S, Croera C, Gagliardi G, Foti P, Parchment R, Parent-Massin D, Schoeters G, Sibiril Y, Van Den Heuvel R, Gribaldo L. (2003) Application of the CFU-GM assay to

predict acute drug-induced neutropenia: an International Blind Trial to validate a prediction model for the maximum tolerated dose (MTD) of myelosuppressive xenobiotics. *Toxicol. Sci.* **75**: 355–367.

3. McNiece I, Andrews R, Stewart M, Clarke S, Boone T, Quesenberry P. (1989) Action of interleukin-3, G-CSF, and GM-CSF on highly enriched human hematopoietic progenitor cells: synergistic interaction of GM-CSF plus G-CSF. *Blood* **74(1)**: 110–114.

4. Bacigalupo A, Piaggio G, Podesta M, Figari O, Benvenuto F, Sogno G, Tedone E, Raffo MR, Grassia L, Ferrero R, et al. (1995) Influence of marrow CFU-GM content on engraftment and survival after allogeneic bone marrow transplantation. *Bone Marrow Transplant.* **15(2)**: 221–226.

5. Torres A, Alonso MC, Gomez-Villagran JL, Manzanares MR, Martinez F, Gomez P, Garcia JM, Andres P, Gomez C, Torre MA, et al. (1985) No influence of number of donor CFU-GM on granulocyte recovery in bone marrow transplantation for acute leukemia. *Blut* **50(2)**: 89–94.

6. Larochelle A, Vormoor J, Hanenberg H, Wang JC, Bhatia M, Lapidot T, Moritz T, Murdoch B, Xiao XL, Kato I, Williams DA, Dick JE. (1996) Identification of primitive human hematopoietic cells capable of repopulating NOD/SCID mouse bone marrow: implications for gene therapy. *Nat. Med.* **2(12)**: 1329–1337.

7. Eaves C, Miller C, Cashman J, Conneally E, Petzer A, Zandstra P, Eaves A. (1997) Hematopoietic stem cells: inferences from in vivo assays. *Stem Cells* **15 (Suppl 1)**: 1–5.

8. Bhatia M, Bonnet D, Murdoch B, Gan OI, Dick JE. (1998) A newly discovered class of human hematopoietic cells with SCID-repopulating activity. *Nat. Med.* **4(9)**: 1038–45.

9. Moore MA, Warren DJ. (1987) Synergy of interleukin 1 and granulocyte colony-stimulating factor: in vivo stimulation of stem-cell recovery and hematopoietic regeneration following 5-fluorouracil treatment of mice. *Proc. Natl. Acad. Sci. U.S.A.* **84(20)**: 7134–7138.

10. Clarke E, Rice GC, Weeks RS, Jenkins N, Nelson R, Bianco JA, Singer JW. (1996) Lisofylline inhibits transforming growth factor beta release and enhances trilineage hematopoietic recovery after 5-fluorouracil treatment in mice. *Cancer Res.* **56(1)**: 105–112.

11. Helgason CD, Damen JE, Rosten P, Grewal R, Sorensen P, Chappel SM, Borowski A, Jirik F, Krystal G, Humphries RK. (1998) Targeted disruption of SHIP leads to hemopoietic perturbations, lung pathology, and a shortened life span. *Genes Dev.* **12(11)**: 1610–1620.

12. Stephenson JR, Axelrad AA, McLeod DL, Shreeve MM. (1971) Induction of colonies of hemaglobin-synthesising cells by erythropoietin in vitro. *Proc. Natl. Acad. Sci. U.S.A.* **68**: 1542–1556.

13. Broxmeyer HE, Williams DE, Cooper S, Shadduch RK, Gillis S, Waheed A, Urdal DL, Bicknell DC. (1987) Comparative effects in vivo of recombinant murine

interleukin 3, natural murine colony-stimulating factor-1, and recombinant murine granulocyte-macrophage colony-stimulating factor on myelopoiesis in mice. *J Clin. Invest.* **79(3)**: 721–30.

14. Williams DE, Namen AE, Mochizuki DY, Overell RW. (1990) Clonal growth of murine pre-B colony-forming cells and their targeted infection by a retroviral vector: dependence on interleukin-7. *Blood* **75(5)**: 1132–1138.

15. Banu N, Wang JF, Deng B, Groopman JE, Avraham H. (1995) Modulation of megakaryocytopoiesis by thrombopoietin: the c-Mpl ligand. *Blood* **86(4)**: 1331–1338.

16. Sandmaier BM, Storb R, Santos EB, Krizanac-Bengez, Lian T, McSweeney PA, Yu C, Schuening FG, Deeg HJ, Graham T. (1996) Allogeneic transplant of canine peripheral blood stem cells mobilized by recombinant canine hematopoietic growth factors. *Blood* **87(8)**: 3508–3513.

17. Moneta D, Geroni C, Valota O, Grossi P, de Jonge MJ, Brughera M, Colajori E, Ghielmini M, Sessa C. (2003) Predicting the maximum-tolerated dose of PNU-159548 (4-demethoxy-3′-deamino-3′-aziridinyl-4′-methylsulphonyl-daunorubicin) in humans using CFU-GM clonogenic assays and prospective validation. *Eur. J. Cancer.* **39(5)**: 675–683.

18. Masubuchi N. (2006) Risk assessment of human myelotoxicity of anticancer drugs: a predictive model and the in vitro colony forming unit granulocyte/macrophage (CFU-GM) assay. *Pharmazie.* Feb; **61(2)**: 135–9.

19. MacVittie TJ, Farese AM, Davis TA, Lind LB, McKearn JP. (1999) Myelopoietin, a chimeric agonist of human Interleukin 3 and granulocyte colony-stimulating factor receptors, mobilizes CD34+ cells that rapidly engraft lethally x-irradiated nonhuman primates. *Exp. Hematol.* **27**: 1557–1568.

20. Tisdale JF, Hanazono Y, Sellers SE, Agricola BA, Metzger ME, Kato I, Donahue RE, Dunbar CE. (1998) Ex vivo expansion of genetically marked rhesus peripheral blood progenitor cells results in diminished long-term repopulating ability. *Blood* **92**: 1131–1141.

21. Farese AM, Herodin F, McKearn JP, Baurn C, Burton E, MacVittie TJ. (1996) Acceleration of hematopoietic reconstitution with a synthetic cytokine (SC-55494) after radiation-induced bone marrow aplasia. *Blood* **87**: 581–591.

22. Jonker M. (1990) The importance of non-human primates for preclinical testing of immunosuppressive monoclonal antibodies. *Semin. Immunol.* **2(6)**: 427–436.

23. Pessina A, Malerba I, Gribaldo L. (2005) Hematotoxicity testing by cell clonogenic assay in drug development and preclinical trials. *Curr. Pharm. Des.* **11(8)**: 1055–1065.

24. Ficoll-Paque® PLUS Instructions 71-7167-00AF, cell preparation, GE Healthcare, January 2006.

25. Nissen-Druey C, Tichelli A, Meyer-Monard S. (2005) Human hematopoietic colonies in health and disease. Reprint of *Acta Haematoplogica* (ISSN 0001-5792) **113(1)** (STI cat. no. 28760).
26. Testa NG, Molineaux G. (ed.) (1993) *Haematopoiesis: A Practical Approach.* Oxford University Press, New York.

15

Assays for Alloreactive Responses by PCR

Patrick Stordeur

Summary

In organ transplantation, major efforts are currently developed to minimize or even to withdraw immunosuppressive drugs. Different protocols have therefore been proposed to induce graft tolerance, that is, the survival of an allograft in the absence of continuing immunosuppression. They generally involve pre-transplant conditioning followed by hematopoietic stem cell infusion. Follow-up of immune status is mandatory in this kind of protocol, in order to investigate tolerance and consequently to reduce or withdraw immunosuppression without graft rejection. However, assessment of the alloreactive response using currently available techniques is labor-intensive and requires significant volume of patient's blood. The implementation of such tolerance protocols is thus difficult in routine practice. In this chapter, we describe a novel real-time polymerase chain reaction method based on interferon-γ and interleukin-2 mRNAs quantification that requires only 10 ml of blood. Moreover, results can be obtained within 48 h. A modified mixed lymphocyte reaction (MLR) using whole blood T lymphocytes as responsive cells and CD3 depleted peripheral blood mononuclear cells (PBMCs) as stimulators is described. An alternative, that is, the use of PBMC as responding cells instead of whole blood, is also depicted. We successfully applied the technique in the monitoring of anti-donor reactivity in living donor liver graft recipients. We suggest that this rapid MLR assay could be valuable for the monitoring of patients undergoing solid organ or hematopoietic stem cell transplantation.

Key Words: Blood; Alloreactivity; Transplantation; Tolerance; Immunosuppression; Immune monitoring; Methods; Messenger RNA; Real-time PCR; Cytokine; IL-2; IFN-gamma.

1. Introduction

In transplantation medicine, the finding of clinical protocols allowing to minimize or even to withdraw immunosuppressive drugs currently represents a major challenge (*1*). The success of these new approaches critically depends

From: *Methods in Molecular Biology, vol. 407: Stem Cell Assays*
Edited by: M. C. Vemuri © Humana Press, Totowa, NJ

on the availability of reliable assays to assess alloreactive responses. Several techniques were proposed for this purpose *(2–7)*. They include the mixed lymphocyte reaction (MLR) assessing T-cell proliferation, the measurement of precursor helper T cell or precursor cytotoxic T cell frequencies using limiting dilution analysis, the quantification of secreted cytokines, the ELISPOT assay and the tetrameric staining method. However, these techniques usually are labour-intensive and require significant volume of a patient's blood, which hampers their application in routine practice.

The technique described in this chapter only requires 10 ml of blood in a real-time polymerase chain reaction (PCR)-based method that is recently developed in our laboratory, which allows easy measurements of cytokine mRNA levels in whole blood *(8–9)*. More specifically, volumes as small as 100 to 500 µl of whole blood are cultured in the presence of the cells against which the alloreactive response is studied. Cytokine mRNAs are then quantified using real-time PCR. The cells used to induce the alloreactive response are donor-derived T-cell-depleted peripheral blood mononuclear cells (PBMCs), so that cytokine mRNAs can only be induced in recipient's whole blood T cells (the allogeneic response is a characteristic of T cells). The variation of mRNA levels mainly reflects the activation of the direct recognition pathway *(10,11)* and depends on the degree of human leucocyte antigens (HLA) disparity *(12)*. In other words, an increase of interleukin (IL)-2 and/or interferon (IFN)-γ mRNA levels reflects the ability of the recipient's T cells to recognize the donors cells as non-self. Such result indicates an inadequate immunosuppression leading to a rupture of tolerance, which is associated with a high risk of graft rejection. On the contrary, an absence of variation of mRNA level suggests adequate immunosuppression and graft acceptance.

The use of whole blood as "responding cells" simplifies the procedure and reduces the required blood volume. However, most of the techniques used to monitor immune responses are still performed on PBMC. Therefore, two alternatives are described here:the use of (i) whole blood and (ii) PBMC as responding cells. The stimulator cells are in both cases T-cell-depleted PBMC (*see* **Fig. 1**).

2. Materials

2.1. Blood Sample

Blood should be collected on sodium heparinate. Avoid lithium heparinate. We observed heterogeneous and sometimes incomprehensible results using lithium.

A

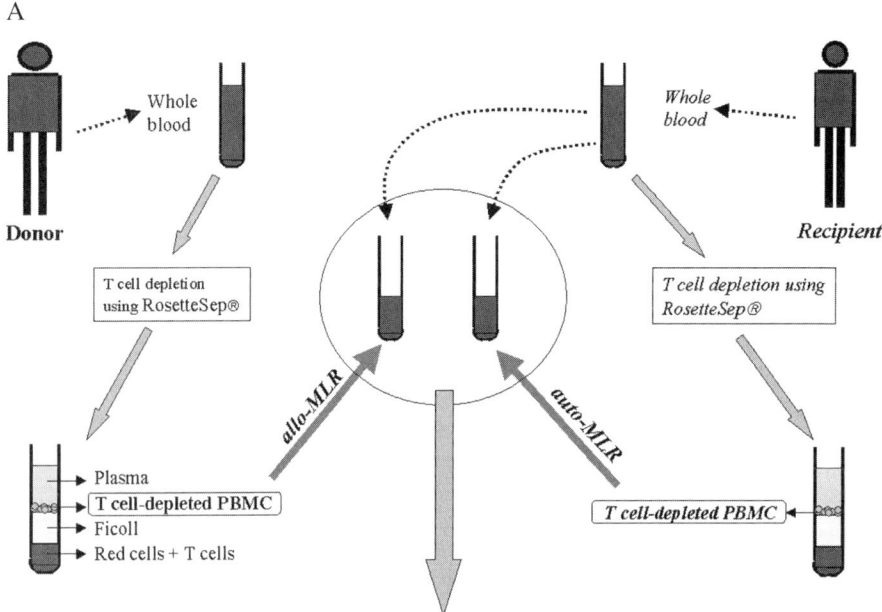

IL-2 and IFN-γ mRNAs quantification using real-time PCR

B

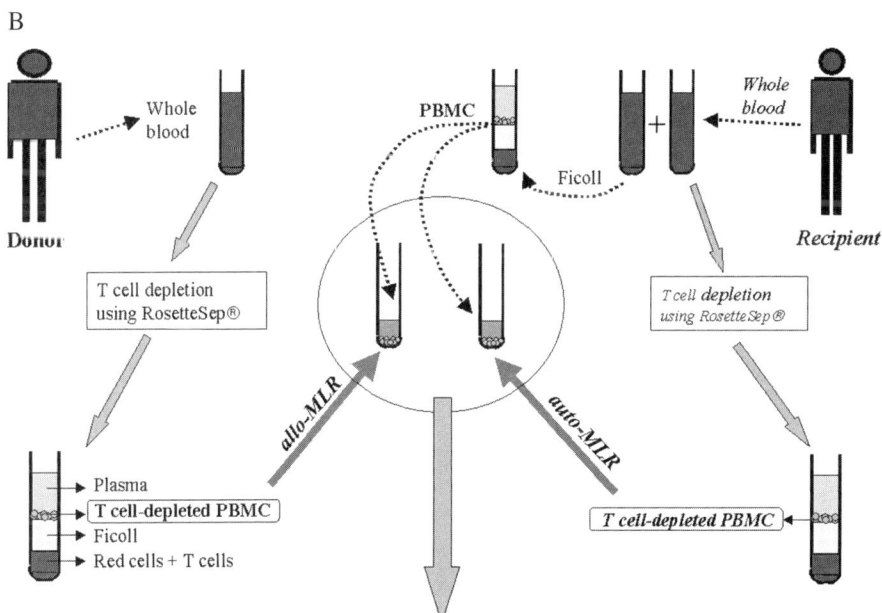

IL-2 and IFN-γ mRNAs quantification using real-time PCR

2.2. Cell Purification and Culture

1. Adoptive immunotherapy media-v (AIM-V®) medium (cat no. 12055-091, GIBCO, Invitrogen, Merelbeke, Belgium).
2. Hank's Balanced Salt Solution (HBSS), 10×, without calcium and magnesium (GIBCO, Invitrogen, cat no. 14180-046).
3. Phosphate-buffered Saline (PBS) without calcium and magnesium (BioWhittakker™, cat no. BE17-516F, Cambrex Bioscience, Verviers, Belgium).
4. Fetal bovine serum (FBS) (cat no. CH30160.03, HyClone, Perbio Science, Erembodegem, Belgium).
5. PBS supplemented with 2% FBS.
6. Lymphoprep™ (cat no. 1019819, Axis-Shield, Oslo, Norway).
7. Trypan blue 0.4% w/v (cat no. 00654, StemCell Technologies, Vancouver, Canada).
8. RosetteSep™ Human CD3$^+$ Cell Depletion kit (cat no. 15661, StemCell Technologies).
9. Phytohaemagglutinin (PHA) (cat no. R30852801, bioTRADING Benelux, Mijdrecht, The Netherlands).

2.3. mRNA Extraction

1. MagNA Pure™ instrument (cat no. 12 236 931 001, Roche Applied Science, Vilvorde, Belgium).
2. PAXgene tubes (cat no. 762165, Qiagen Benelux, Venlo, The Netherlands).
3. MagNA Pure™ mRNA extraction kit "MagNa Pure™ LC mRNA Kit I" (cat no. 03 004 015 001, Roche Applied Science).
4. TaqMan Pre-Developed Assay Reagent for human CD3 epsilon chain (cat no. 4329515F, Applied Biosystems, Foster City, CA, USA).
5. Facultative reagents: TriPure™ Isolation Reagent (cat no. 11 667 157 001, Roche Applied Science); TRIzol® Reagent (cat no. 15596-026, Invitrogen); PAXgene Blood RNA kit (cat no. 762174, Qiagen).

Fig. 1. **(A)** Monitoring of alloreactive immune responses using real-time polymerase chain reaction (PCR) for interleukin-2 and interferon-γ mRNAs quantification. T-cell-depleted peripheral blood mononuclear cells (PBMCs) are prepared from the donor [allo-mixed lymphocyte reaction (MLR)] and from the recipient (auto-MLR) using the RosetteSep™ Human CD3$^+$ Cell Depletion kit. These cell preparations are separately added to whole blood samples from the recipient. After incubation for 18 h, the cultures are stopped by adding the reagent contained in the PAXgene tube. Cytokine mRNA levels are then quantified using real-time PCR (*see* **Note 11**). **(B)** T-cell-depleted PBMCs from the donor and the recipient are added to PBMC of the recipient instead of its whole blood.

2.4. PCR

1. Lightcycler™2.0 instrument (cat no. 03 531 414 201, Roche Applied Science)
2. Lightcycler™ RNA Master Hybridization Probes Kit (cat no. 03 018 954 001, Roche Applied Science).
3. Oligonucleotides "OliGold®" (*see* **Table 1**) are from Eurogentec (Seraing, Belgium).
4. Wizard SV Gel Clean-up System (cat no. A9281, Promega, Leiden, The Netherlands).
5. DNA from fish sperm, MB-grade (cat no. 11 467 140 001, Roche Applied Science). Prepare a 10 mg/ml stock solution in Tris-HCL 10 mM, EDTA 1 mM (TE) buffer (pH 8.0). Keep at −20 °C.

3. Methods

3.1. Isolation of Peripheral Blood Mononuclear Cells (Responding Cells, See Note 1)

1. Transfer 12 ml of heparinized blood from the recipient in a 50-ml Falcon tube and make it up to 35-ml with HBSS 1× (*see* **Note 2**).
2. Layer the diluted sample on top of the Lymphoprep™ or layer the Lymphoprep™ underneath the diluted sample. Be careful to minimize mixing of Lymphoprep™ and sample. Use 15 ml of Lymphoprep™.
3. Centrifuge for 20 min at 800–1200 g (*see* **Note 3**) at room temperature, with the brake off.
4. Recover the PBMC fraction from the Lymphoprep™ : plasma interface and transfer it in a 50-ml Falcon tube. Suspend the fraction to 50 ml with HBSS 1×.
5. Centrifuge at 800 g for 10 min. Discard the supernatant and repeat this washing step.
6. Suspend cell pellet by adding 5 ml of AIM-V to the cell pellet (centrifugation of suspended cells in AIM-V must be done "without brake," to ensure complete sedimentation of all the cells).
7. Count the cells, evaluate their viability using Trypan blue (mix one volume of cell suspension with one volume of Trypan blue solution; dead cells appear light blue).
8. Adjust the volume with AIM-V to obtain one million cells per ml for fluorescent associated cell sorting analysis and for cell culture.
9. Analyze 100,000 cells by flow cytometry. Use an antibody cocktail that gives at least the percent of total, CD4 and CD8 T cells, B cells, natural killer (NK) cells.

3.2. Preparation of T-cell-depleted PBMC (Stimulator Cells, See Note 1)

1. The RosetteSep™ Kit uses bispecific tetrameric antibody complexes directed against CD3 on T cells and glycophorin A on red blood cells (RBCs). These antibody complexes crosslink unwanted T cells in human whole blood to multiple RBCs,

Table 1
Oligonucleotides[a].

mRNA target	Oligonucleotides (5′ → 3′)[b]	Product size (bp)	Final concentration (nM)[c]
Primers and probes for real-time PCR			
IL-2	F273: CTCACCAGGATGCTCACATTTA	95	F 900
	R367: TCCAGAGGTTTGAGTTCTTCTTCT		R 900
	P304: Fam-TGCCCAAGAAG GCCACAGAACTG-BHQ1		
IFN-γ	F464: CTAATTATTCGGTAA CTGACTTGA	75	F 600
	R538: ACAGTTCAGCCATCACTTGGA		R 900
	P491: Fam-TCCAACGCAAA GCAATACATGAAC-BHQ1		
β-actin	F976: GGATGCAGAAGGAGATCACTG	90[d]	F 300
	R1065: CGATCCACACGGAGTACTTG		R 300
	P997: Fam-CCCTGGCACCC AGCACAATG-BHQ1		
Primers for standard preparation by "classical" PCR[e]			
IL-2	F155: TGTCACAAACAGTGCACCTACT	518	
	R672: AGTTACAATAGGTA GCAAACCATACA		
IFN-γ	F154: TTGGGTTCTCTTGGCTGTTA	479	
	R632: AAATATTGCAGGCAGGACAA		
β-actin	F745: CCCTGGAGAAGAGCTACGA	509	
	R1253: TAAAGCCATGCCAATCTCAT		

IFN, interferon; IL, interleukin.

[a] For a full description, and other mRNA targets *see* **ref**. *(8,19)*.

[b] F, R and P indicate forward and reverse primers and probes, respectively; numbers indicate the sequence position from Genebank accession numbers X01586 for IL-2, X13274 for IFN-γ, and X000351 for β-actin. For Taqman probes, the quencher BHQ1 is preferable to TAMRA (it has no fluorescence of its own, which enhances the signal-to-noise ratio).

[c] Final concentration of forward (F) and reverse (R) primers.

[d] All primers were chosen to span intronic sequences so that genomic DNA amplification is not possible, except for β-actin for which a 112 bp longer band is obtained. This allows the detection of contaminating genomic DNA using this size difference on agarose gel.

[e] Standard curves were generated from serial dilutions of PCR products prepared by "classical" PCR, for which specific conditions were as follows: denaturation at 95 °C for 20 s, annealing at 58 °C for 20 s and elongation at 72 °C for 45 s, for a total of 35 cycles. MgCl$_2$ final concentration is 1.5 mM.

forming immunorosettes. This increases the density of the unwanted (rosetted) T cells, such that they pellet along with the free RBCs when centrifuged over the buoyant density medium Lymphoprep™. T-cell-depleted PBMCs are then easily collected as a highly enriched population at the interface between the plasma and the buoyant density medium.

2. Transfer 8 ml of heparinized blood from the donor into a 50-ml Falcon tube and add 400 μl of antibody complexes that are contained in the RosetteSep™ kit. Mix gently. Incubate for 20 min at room temperature.
3. Add an equal volume (i.e., 8 ml) of PBS supplemented with 2% FBS. Mix gently.
4. Layer the diluted sample on top of the Lymphoprep™ or layer the Lymphoprep™ underneath the diluted sample, avoiding mixing of Lymphoprep™ and sample. Use 15 ml of Lymphoprep™. Then proceed as described for PBMC purification (*see* **Subheading 3.1.**, **steps 3–5**).
5. Finally suspend cell pellet in 2 ml of AIM-V.
6. Count the cells, evaluate their viability using Trypan blue (*see* **Subheading 3.1.**, **step 7**).
7. Analyze 100,000 cells by flow cytometry. Adjust the volume with AIM-V to get one million cells per ml.
8. Use an antibody cocktail that gives the percent of total (CD3) T cells, B cells, NK cells, monocytes and HLA DR-positive cells. The cell preparation should not contain more than 1% of T cells.

3.3. Mixed Lymphocyte Reaction

1. Mix 200,000 recipient's PBMC as responding cells with 40,000 T-cell-depleted PBMC from the donor. Because one works with cell suspensions at one million per ml, the total culture volume should be 240 μl in AIM-V medium (*see* **Note 4**).
   Or:
   Add 40,000 T-cell-depleted PBMC from the donor to the recipient's blood volume (*see* **Note 1**) that contains 200,000 T cells (usually 200–300 μl in healthy subjects) (*see* **Note 4**).
2. Incubate for 18 h at 37 °C in a 5% CO_2 atmosphere (*see* **Note 5**).
3. Stop the reaction by adding, for PBMC, 600 μl (i.e., 2.5 volume) of the reagent contained in the PAXgene tubes (*see* **Note 6**).
   Or:
   For whole blood, add 2.5 volumes (i.e., 250 μl per 100 μl of blood) of the reagent contained in the PAXgene tubes.
4. Incubate at least 1 h at room temperature. The lysate is stable at room temperature up to 5 days and for months at −80 °C.

3.4. mRNA Extraction

Extracting mRNA on an automatic device such as the MagNA Pure™ (Roche Applied Science) provides reproducible amounts of high quality mRNA.

Moreover, the reverse transcription (RT)-PCR mixtures containing all reagents, oligonucleotides and samples are fully prepared directly in the PCR vessel (in the case of the Lightcycler™, a capillary), by the MagNA Pure™ instrument. The sampling of RT-PCR components is fully automated, avoids manual sampling errors and thereby enhances the reproducibility and accuracy of the qPCR. In this approach, the RNA sampled by the device will be mRNA instead of total RNA. Reverse transcription and the PCR are successively performed in one step RT-qPCR in the same reaction tube (or capillary). In the case of mRNA, the concentration is too low to be measurable at 260 nm. Hence, the same volume of mRNA (i.e., 5 μl) is always added to the PCR mixture, given the reproducibility ensured by the automation of the procedure. The technique described here is based on the use of the MagNa Pure™ LC mRNA Kit I (Roche Applied Science) on the MagNA Pure™ instrument. However, this expensive step in this method can be replaced by other conventional and less costly methods of RNA extraction. Different kits or reagents that provide high quality RNA exist. A good choice is the kit provided by the manufacturer of the PAXgene tubes, the PAXgene Blood RNA kit (Qiagen), but the use of reagents like TriPure™ (Roche Applied Science) and TRIzol® (Invitrogen) is also possible. If one of these alternatives is chosen, the concentration of total RNA should be measured on a spectrophotometer at 260 nm. Total RNA, 100 ng, are then engaged in the PCR.

1. Gently suspend the precipitate of the cell lysate (*see* **Subheading 3.3.**, **step 4**) and transfer 300 μl in a microtube. Centrifuge for 5 min at 16, 000 g. Discard the supernatant.
2. Dissolve thoroughly the nucleic acid pellet in 300 μl of the lysis buffer contained in the MagNa Pure™ LC mRNA Kit I. The pellet is difficult to dissolve. Vortexing for 45–60 s is often required. Some very small aggregates often remain undissolved (*see* **Note 7**).
3. Proceed with mRNA extraction following manufacturer's instructions. Elute in 100 μl and use 5 μl by PCR. If low number of PBMC (e.g., 50,000) or low blood volumes (e.g., 100 μl) are used, the elution volume can be reduced to 50 μl, and the volume added to the PCR kept at 5 μl.

3.5. RT-qPCR

Several assays are available to perform real-time PCR *(13)*. Their choice of use depends on the application of the technique (e.g., polymorphism detection and/or quantification). As far as mRNA quantification is concerned, three good choices are SYBR®Green, hybridization probes and hydrolysis (Taqman) probes. The first one is the cheapest, but also the less specific and sensitive (SYBR®Green binds all double-stranded DNAs, including unspecific PCR products and primer dimers). The use of probes eliminates this problem. Both formats can be used, but the design of the primers and probes is somewhat easier with the Taqman chemistry [only three oligonucleotides in all (two primers and one probe) instead of four with hybridization probes (two primers and two probes)]. The technique described below uses Taqman probes and the Lightcycler™ RNA Master Hybridization Probes Kit on the Lightcycler™ instrument (Roche Applied Science). For details on the design of oligonucleotides *see* **Note 8**.

3.5.1. Preparation and Execution of RT-qPCR Reaction

1. Prepare a RT-qPCR mixture as follows [each volume can be multiplied by the number of samples (+1 per 10 samples) to prepare the quantity needed for all samples of the experiment): (i) H_2O up to $20 \mu l$; (ii) $7.5 \mu l$ RNA Master Hybridisation Probes $2.7 \times$ concentration; (iii) $1.3 \mu l$ $50 mM$ $Mn(OAc)_2$; (iv) 1, 2 or $3 \mu l$ of 6 pmoles/μl forward and reverse primers (final concentration 300, 600 or 900 nM, depending on the mRNA target; *see* **Note 9** and **Table 1**); (v) $1 \mu l$ of 4 pmoles/μl TaqMan probe (final concentration 200 nM); (vi) $5 \mu l$ purified mRNA (or containing 100 ng of total RNA) or standard dilution.
2. Transfer the capillaries in the Lightcycler™ and start the RT-PCR.
3. The cycling conditions are as follows: after an incubation period of 20 min at 61 °C to allow mRNA reverse transcription, and then an initial denaturation step at 95 °C for 30 s, temperature cycling is initiated. Each cycle consists of 95 °C for 0 s (zero) and 60 °C for 20 s, the fluorescence being read at the end of this second step (F1/F2 channels when using FAM/TAMRA labelled probes, F1 channel when using FAM/BHQ1 labelled probes, no color compensation). Forty-five cycles are performed, overall.

3.5.2. Generation of External Standards

mRNA levels can be expressed either in absolute copy numbers or in relative copy numbers normalized against a house keeping gene (*see* **Note 10**). This is achieved by constructing, for each PCR run, a standard curve from serial dilutions of purified DNA. The latter consists in a PCR product that includes the quantified amplicon and that is prepared by "classical" PCR from cDNA positive for the concerned target mRNA. These PCR products used as standards are purified from agarose gel by the "Wizard SV Gel Clean-up System" following manufacturer's instructions. The copy number of the standards is calculated from the DNA concentration measured by spectrophotometry *(14)*. Dilute from 10:10 in TE buffer (pH 8.0) supplemented with $10 \mu g/ml$ fish DNA, to get dilutions from 10^8 to 10^2 copies. Detailed information concerning these standards is given in **Table 1**.

4. Notes

1. In the technique described here, the T cells from the recipient (referred as "responding cells") are contained in the PBMC isolated from the recipient as described in **Subheading 3.1.**, and the donor's cells (referred as "stimulator cells") are T-cell-depleted PBMC isolated from the donor as described in **Subheading 3.2.** These are T cell depleted to avoid a reaction of donor's T cells against recipient's PBMC. The proposed alternative (**Subheading 3.3., step 1**) is the use of whole blood as a source of recipient's T cells instead of PBMC. This alternative presents several advantages: (i) it avoids time-consuming purification of PBMC, this step being able in addition to either non-specifically activate the cells or modulate cytokine mRNA levels; (ii) it allows the use of smaller blood volumes; (iii) it is more representative of the in vivo situation than purified cells.

2. Blood volume depends on the number of PBMC needed for the experiment. Generally, at least one million PBMC can be obtained from 1 ml of blood from a subject with normal WBC count.

3. To convert g to rpm, use the following formula:

$$RPM = \text{square root of } [RCF/(1118 \times 10^{-5} \times Radius)],$$

where RCF = relative centrifugal force (g), RPM = centrifuge speed in revolutions per minute, and Radius = radius of rotor in cm.

4. The number of PBMC needed to perform RT-PCR mainly depends on the yield of the technique used to isolate RNA. The technique described in this chapter, that is, mRNA isolation on the MagNA Pure instrument, allows a high yield RNA recovery. The minimum number of PBMC required in this case is 100,000 or even 50,000. Manual RNA purification methods generally require larger samples. The numbers of PBMC and T-cell-depleted PBMC must be adapted while keeping the responding/stimulator cell ratio. The total culture volume is 100 µl per 100,000 cells. Although the best responding/stimulator cell ratio is 1 (one stimulator cell for one responding), most of the time, the limited number of available cells implies use of a ratio of 5. The MLR can also be optimized by mixing either the blood volume or the PBMC number that contains 200, 000 $CD3^+$ cells, with the number of T-cell-depleted PBMC that contains 40, 000 $HLA\ DR^+$ cells.

5. The incubation time cannot exceed 20 h with whole blood. After this time, spontaneous lysis of red and white cells occurs. When working with PBMC, longer incubation times (48–72 h) provide more consistent results, especially for IFN-γ (mRNA accumulates in the cells).

6. This reagent induces complete cell lysis and, at the same time, nucleic acid precipitation. The nucleic acids can then be dissolved in guanidium/thiocyanate solutions similar to that described in 1987 by Chomczynski and Sacchi *(15)* or to the guanidium/thiocyanate solution contained in the lysis buffer provided with the MagNA Pure LC kit for mRNA isolation (Roche Applied Science). The advantage

of adding this reagent directly to the cell culture is that there is no need to discard the supernatant, reducing the potential loss of cells that can occur when working with such small volumes.

7. If RNA is extracted using TriPure or TRIzol, simply replace the lysis buffer of the MagNA Pure LC mRNA Kit I by the TriPure or TRIzol reagent and proceed in the same way. Then follow the manufacturer's instructions as recommended for a "normal" cell lysate.

8. The primers and probes described here were designed with the Primer 3 software (http://www-genome.wi.mit.edu/cgi-bin/primer/primer3_www.cgi) *(16)*. The default parameters of the program were applied except for the following: product size 90–150 bp, primer size 20–27 bp, primer Tm 59 to 62 °C with a max Tm difference of 2.0 °C, Oligo Tm 69 to 72 °C, product Tm 0 to 85 °C, max self and 3' self complementarily = 6.00, max poly-X = 3, primer and Hyb Oligo penalty (penalty weights for primer pairs) = 2.0. In addition, the oligonucleotides were selected according to the following criteria (listed in order of importance): (i) intron spanning if possible, (ii) no G at the probe 5' end, (3) no more than two Gs or Cs within the five 3' nucleotides for primers and (4) more Cs then Gs in the probe. The expected size of the real-time PCR product must be checked by agarose gel electrophoresis for every new mRNA target.

9. The same protocol is used for all target mRNAs, the only adjustment being the primer concentration. Using the standard at 10^5 copies or a corresponding cDNA, primer titration is performed at 300, 600 or 900 nM, the three concentrations being checked for each primer (*see* **Fig. 2**). The conditions specific for each mRNA target are listed in **Table 1**. To use this protocol on other instruments that do not use capillaries and air for heating/cooling, the following adjustments are needed: increase denaturation time to 10–20 s and the combined hybridization/elongation step to 40–60 s. Keep the concentration of mRNA and oligonucleotides. Use the reagent mix provided by the manufacturer of the thermal cycler. Most of these reagents are formulated to be used at one "universal" concentration.

10. Because the aim of the method is the assessment of an allogencic T cell response, the most satisfactory house keeping gene can be the epsilon chain of the CD3 complex. In this case, use per PCR 1 μl of the TaqMan Pre-Developed Assay Reagent for human CD3 epsilon chain (a ready to use mix of probe and primers) from Applied Biosystem and express the results in copies of the quantified mRNA per million of CD3ε mRNA copies. An alternative to avoid external standard curve is to express the results in ΔΔCt *(8,17)*.

11. Whatever the manner of expressing the results, best is to compare those obtained with the six following conditions:

1. "Spontaneous": PBMC or whole blood from the recipient in culture medium alone.
2. "Auto": PBMC or whole blood from the recipient + T-cell-depleted PBMC from the recipient.

3. "Allo": PBMC or whole blood from the recipient + T-cell-depleted PBMC from the donor.
4. "Third party": PBMC or whole blood from the recipient + T-cell-depleted PBMC from a third party.
5. "Control of T-cell depletion of recipient's PBMC": T-cell-depleted PBMC from the recipient cultured in the presence of PHA at 1 µg/ml and in medium alone.
6. "Control of T-cell depletion of donor's PBMC": T-cell-depleted PBMC from the donor cultured in the presence of PHA at 1 µg/ml and in medium alone.

The condition (1) shows the spontaneous production of mRNAs, which is usually absent for IL-2 and basal for IFN-γ. The levels obtained with the condition (2) should be

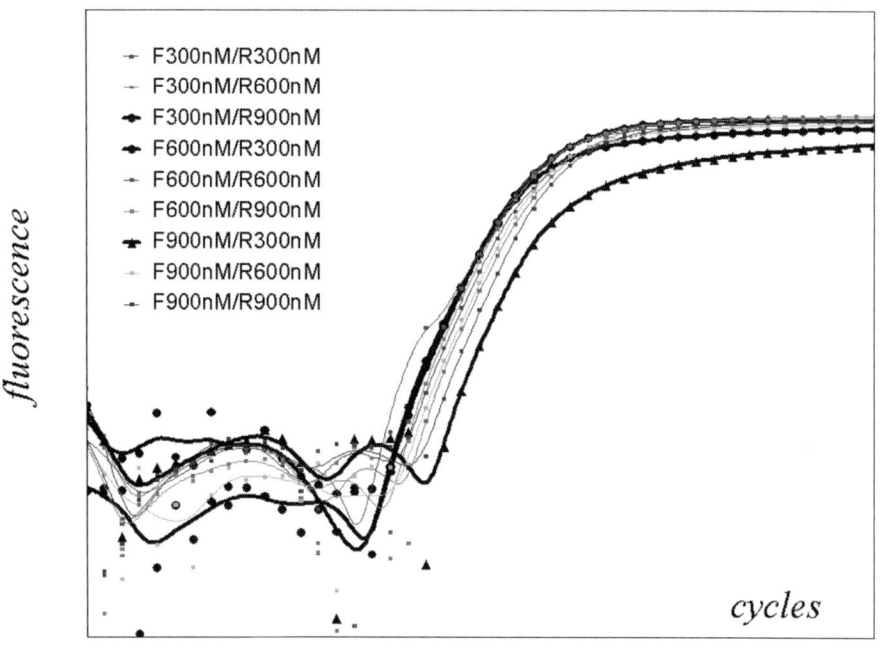

Fig. 2. Example of primer titration. Three different concentrations of each primer are tested: 300, 600 and 900 nM. Each concentration of forward primer is tested in combination with each concentration of reverse primer in a real-time polymerase chain reaction starting with 10^5 copies of purified standard. The preferred primer concentration gives the curve with the highest slope and the lowest CT value, in this example, the forward primer at 300 nM with the reverse at 900 nM, or the forward at 600 nM with the reverse at 300 nM (thick fluorescent curves with circles). The worst combination (thick fluorescent curve with triangles) being the forward primer at 900nM combined to the reverse at 300 nM.

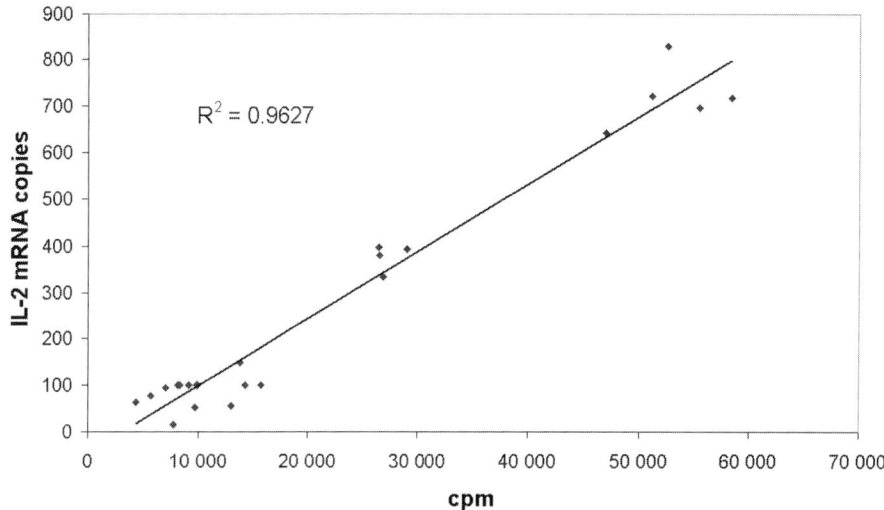

Fig. 3. Interleukin (IL)-2 mRNA quantification correlates with the thymidine incorporation assay. Thymidine incorporation assays using peripheral blood mononuclear cells (PBMCs) as responding cells were performed in parallel with mixed lymphocyte reaction (MLR) that use whole blood as responding cells and IL-2 mRNA quantification as read-out. In both assays, stimulator cells were T-cell-depleted PBMC and the responding/stimulator cell ratio adjusted to 5. Eleven auto-MLR and 11 allo-MLR were performed from 11 HLA-unrelated donors. The PCR results (*y*-axis) correlate with the proliferation assay results (*x*-axis) (Ling Zhou, personal communication).

similar. A positive result, that is, a reactivity of recipient's T cells against donor's cells, is considered as such if a significant difference is observed between conditions (2) and (3). Depending mainly on the HLA disparity and the responding/stimulator cell ratio, this difference varies from 5 to 200 times more mRNA for the condition "allo" compared to the condition "auto," for a 18-h incubation. The third party is used as positive control of allorecognition. It is chosen on the basis of its HLA phenotype, which must be different from that of the recipient. Therefore, a difference must be found between conditions (2) and (4) if the immune system of the recipient is effective, while no difference is found when the recipient is immunosuppressed. The best clinical situation is to make the recipient tolerant toward the graft while keeping a effective immune system. In this situation, no difference is found between conditions (2) and (3), while a significant difference is found between (2) and (4). Lastly, no difference can be observed with conditions (5) and (6) between PHA and medium alone. This to check that T cells were effectively depleted, PHA being a strong polyclonal activator of T cells.

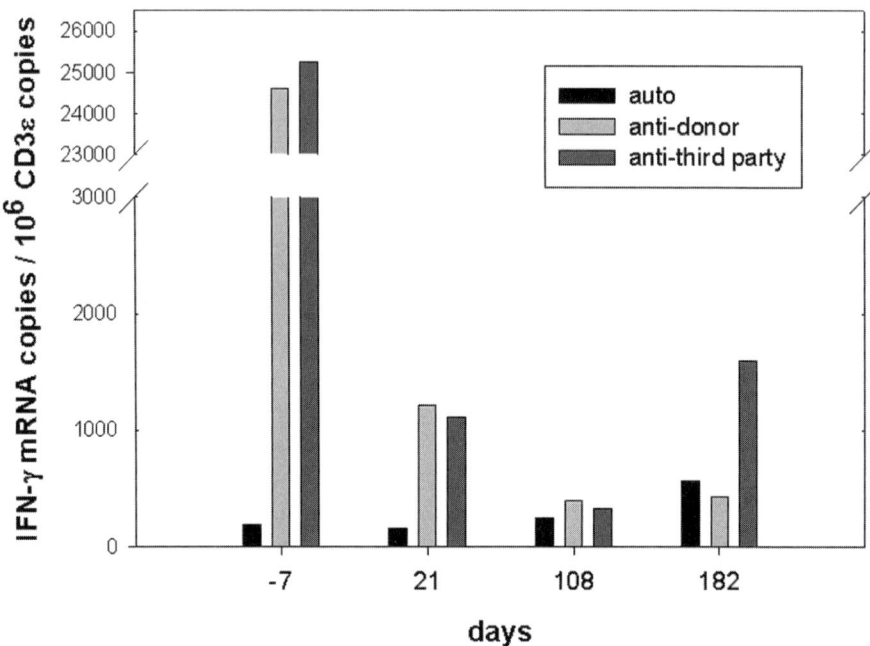

Fig. 4. Follow-up of adult living-related liver graft by real-time polymerase chain reaction. We applied the whole blood mixed lymphocyte reaction assay to monitor immune responses in cases of HLA mismatched adult living-related donor liver graft in the setting of a protocol for tolerance induction. This protocol consisted in a pre-transplant non-myeloablative conditioning followed by donor stem cell infusion, in patients with advanced liver cancers *(18)*. Some representative data from one case are shown here. In vitro anti-donor reactivity before and after living-related liver transplantation (LDLT) was evaluated by measuring interferon (IFN)-γ mRNA levels. Whole blood of the patient was incubated with autologous, donor and third party T-cell-depleted PBMC. Experiments were performed 7 days before and 21, 108 and 182 days after LDLT. IFN-γ mRNA levels were quantified and normalized to CD3ε mRNA. Similar mRNA amounts between "auto" and "anti-donor" conditions reflects a tolerance against the transplant. The specificity of this tolerance against the donor's cells is evidenced by the response observed against the third party. Indeed, 182 days after the liver graft, the patient begins to recover his reactivity against the third party, but not against the donor (Ligia Craciun and Vincent Donckier, personal communication).

References

1. Matthews J.B., Ramos E. and Bluestone J.A. (2003) Clinical trials of transplant tolerance: slow but steady progress. *Am J Transplant* **3**, 794–803.
2. Bouma G.J., Schanz U., Oudshoorn M., van der Meer-Prins P.M., Roelen D.L., van Rood J.J. and Claas F.H. (1996) A cell-saving non-radioactive limiting dilution analysis-assay for the combined determination of helper and cytotoxic T lymphocyte precursor frequencies. *Bone Marrow Transplant* **17**, 19–23.
3. van Besouw N.M., Zuijderwijk J.M., de Kuiper P., Ijzermans J.N., Weimar W. and van der Mast B.J. (2005) The granzyme B and interferon-gamma enzyme-linked immunospot assay as alternatives for cytotoxic T-lymphocyte precursor frequency after renal transplantation. *Transplantation* **79**, 1062–1066.
4. van der Mast B.J., van Besouw N.M., de Kuiper P., Vaessen L.M., Gregoor P.J., Ijzermans J.N., van Gelder T., Claas F.H. and Weimar W. (2001) Pretransplant donor-specific helper T cell reactivity as a tool for tailoring the individual need for immunosuppression. *Transplantation* **72**, 873–880.
5. Hornick P.I., Brookes P.A., Mason P.D., Taylor K.M., Yacoub M.H., Rose M.L., Batchelor R. and Lechler R.I. (1997) Optimizing a limiting dilution culture system for quantifying the frequency of interleukin-2-producing alloreactive T helper lymphocytes. *Transplantation* **64**, 472–479.
6. Hernandez-Fuentes M.P., Warrens A.N. and Lechler R.I. (2003) Immunologic monitoring. *Immunol Rev* **196**, 247–264.
7. Toungouz M., Denys C., de Groote D., Andrien M. and Dupont E. (1996) Optimal control of interferon-gamma and tumor necrosis factor-alpha by interleukin-10 produced in response to one HLA-DR mismatch during the primary mixed lymphocyte reaction. *Transplantation* **61**, 497–502.
8. Stordeur P., Poulin L.F., Craciun L., Zhou L., Schandene L., de Lavareille A., Goriely S. and Goldman M. (2002) Cytokine mRNA quantification by real-time PCR. *J Immunol Methods* **259**, 55–64. [Erratum in (2002) *J Immunol Methods*; **262**: 229].
9. Stordeur P., Zhou L., Byl B., Brohet F., Burny W., de Groote D., van der Poll T, and Goldman M. (2003) Immune monitoring in whole blood using real-time PCR. *J Immunol Methods* **276**, 69–77.
10. Whitelegg A. and Barber L.D. (2004) The structural basis of T-cell allorecognition. *Tissue Antigens* **63**, 101–108.
11. Cote I., Rogers N.J. and Lechler R.I. (2001) Allorecognition. *Transfus Clin Biol* **8**, 318–323.
12. Zhou L., Toungouz M., Donckier V., Andrien M., Troisi R., de Hemptinne B., Le Moine A., Dupont E., Goldman M. and Stordeur P. (2005) A rapid test to monitor alloreactive responses in whole blood using real-time polymerase chain reaction. *Transplantation* **80**, 410–413.
13. Bustin S.A. (2000) Absolute quantification of mRNA using real-time reverse transcription polymerase chain reaction assays. *J Mol Endocrinol* **25**, 169–193.

14. Overbergh L., Valckx D., Waer M. and Mathieu C. (1999) Quantification of murine cytokine mRNAs using real time quantitative reverse transcriptase PCR. *Cytokine* **11**, 305–312.
15. Chomczynski P. and Sacchi N. (1987) Single-step method of RNA isolation by acid guanidinium thiocyanate-phenol-chloroform extraction. *Anal Biochem* **162**, 156–159.
16. Rozen S. and Skaletsky H. (2000) WWW for general users and for biologist programmers, in *Bioinformatics Methods and Protocols: Methods in Molecular Biology* (Krawetz, S. and Misener, S. ed.), Humana, Totowa, NJ, pp. 365–386.
17. Wong M.L. and Medrano J.F. (2005) Real-time PCR for mRNA quantitation. *Biotechniques* **39**, 75–85.
18. Donckier V., Troisi R., Toungouz M., Colle I., Van Vlierberghe H., Jacquy C., Martiat P., Stordeur P., Zhou L., Boon N., Lambermont M., Schandene L., Van Laethem J.L., Noens L., Gelin M., de Hemptinne B. and Goldman M. (2004) Donor stem cell infusion after non-myeloablative conditioning for tolerance induction to HLA mismatched adult living-donor liver graft. *Transpl Immunol* **13**, 139–146.
19. Ocmant A., Michils A., Schandene L., Peignois Y., Goldman M. and Stordeur P. (2005) IL-4 and IL-13 mRNA real-time PCR quantification on whole blood to assess allergic response. *Cytokine* **31**, 375–381.

16

Immune Properties of Mesenchymal Stem Cells

Panagiota A. Sotiropoulou and Michael Papamichail

Summary

Mesenchymal stem cells (MSCs) are multipotent progenitor cells isolated by various relatively easily accessible tissues, such as bone marrow and cord blood. MSCs gained attention because of their ease for in vitro expansion together with their multilineage potential. More recently, in vitro and in vivo immunosuppressive properties have been ascribed to them, as they are able to modulate the function of all major immune cell populations, thus impeding immune responses. The underlying mechanisms of their differentiation and function are not thoroughly understood, but still they represent important candidates for tissue regeneration and manipulation of the immune response in graft rejection, graft versus host disease, and autoimmune disorders. Characteristics and immunogenic profile of MSCs, their interface with immune system and their potential use as immunosuppressive elements in cellular therapeutic protocols are reviewed in this chapter.

Key Words: Mesenchymal stem cells; Immune system; Immune suppression; GVHD; Autoimmunity; Cancer.

1. Introduction

Stem cell biology is currently at the center of scientific interest, providing a supply of cells possibly capable of tissue remodeling after trauma, disease, or aging. On the basis of the concept that tissue repair depends on cells circulating in blood in response to injury (*1*), stem cell biology has defined many types of stem cells and keeps evolving attributing new features to those already known and characterizing new ones.

From: *Methods in Molecular Biology, vol. 407: Stem Cell Assays*
Edited by: M. C. Vemuri © Humana Press, Totowa, NJ

Mesenchymal stem cells (MSCs) were the first non-hematopoietic stem cells to be characterized, by Friedenstein et al., as bone marrow (BM)-derived, clonal, plastic adherent cells, capable of differentiating into osteoblasts, adipocytes, and chondrocytes *(2)*. To date, MSCs are credited with more extended differentiation potential, including the ability to undertake myogenic, cardiomyogenic, hepatogenic, and neurogenic differentiation pathways as well *(3–6)*, whereas stem cells sharing features with MSCs have been retrieved from various tissues except BM, such as adipose tissue, umbilical cord blood, and peripheral blood (PB) *(7,8)*. MSCs represent a heterogeneous population of cells, lacking thus far a specific marker, thus making characterization and assessment of MSCs often a point of controversy. Their characterization depends on a combination of phenotypical and functional properties (for details, *see* Chapter 17). MSC features differ among species and depend on age, whereas their phenotype, developmental potential, and degree of homogeneity varies largely depending on the protocol used for their isolation and in vitro expansion.

Even though the stemness of MSCs has not yet been thoroghouly proven *(9)*, their availability, ease of in vitro expansion and genetic manipulation, as well as their multipotentiality and role as regulatory cells have designated them as one of the most promising cell types for cellular therapy.

2. MSC Characteristics

MSCs comprise only 0.0001%–0.01% of total BM nucleated cells. They can be identified using SH2, SH3, and SH4 antibodies (Abs), binding the first on CD105 and the two others on the CD73 molecule *(10,11)*. MSCs are characterized by the lack of hematopoietic markers, being negative for CD45, CD34, CD14, and CD31, while expressing CD44, CD71, CD29, CD90, CD106, CD166, MAB1470, and STRO-1. They express various adhesion-related antigens, such as integrins $\alpha v\beta3$ and $\alpha v\beta5$, integrin subunits $\alpha4$, $\alpha5$, and $\beta1$, intercellular adhesion molecule-1 (ICAM-1) and CD44H, enabling their adherence to extracellular matrix molecules *(12)*. They also express ICAM-2, vascular intercellular adhesion molecule-1 (VCAM-1), lymphocyte function-associated antigen-3 (LFA-3), whereas they are negative for L-, P- and E-selectins and ICAM-3 *(13)*. The CXCR4 receptor, essential for stem cell migration to the sites of injury in response to stromal cell-derived factor-1 (SDF-1), is highly expressed intracellularly but at a very low level on the cell surface *(14)*.

MSCs exhibit low expression of human leukocyte antigen (HLA)-class I molecules, whereas they are negative for HLA-class II. Incubation with interferon (IFN)-γ results in HLA-class I up-regulation and HLA-DR induction *(15)*.

Nonetheless, there is one study suggesting the constitutive expression of HLA-class II on human MSCs *(16)*, while Krampera et al. *(17)* developed mouse MSCs lacking the expression of both major histocompatibility complex (MHC) class I and II. Costimulatory molecules, such as B7-1, B7-2, CD80, CD86, CD40, and CD40L, are not expressed and cannot be induced on human MSCs *(15)*, whereas mouse MSCs express CD80 *(17)*. A recent report suggests that both human and murine MSCs can be induced by IFN-γ to vigorously up-regulate B7-H1 *(18)*.

MSCs, at initial plating and subsequent passages, exhibit a lag phase of growth, followed by a log phase, finally reaching a confluent growth-plateau stage *(12,19)*. During the lag phase, MSCs produce cytokines, chemokines, and growth factors, such as interleukin (IL)-6, IL-7, IL-8, IL-11, IL-12, IL-14, IL-15, leukemia inhibitory factor, granulocyte colony-stimulating factor, granulocyte macrophage colony-stimulating factor, stem cell factor, macrophage colony-stimulating factor (M-CSF) and fms-like tyrosine kinase-3 ligand, implicated mainly in hematopoiesis *(20–23)*. Upon co-culture with cells of the immune system, induction of the production of additional growth factors, such as hepatocyte growth factor (HGF), transforming growth factor (TGF)-β, IL-2, IL-4, and IL-10 *(24–26)*, or the endocellular enzyme indoleamine 2,3-dioxygenase (IDO) *(27,28)* has been reported.

3. MSC-Interference with Cells of the Immune System

Interactions of mesenchymal tissues with the cells of the immune system, occurring at multiple levels, are crucial for the development of the latter from the hematopoietic stem cells (HSC). Variations of the microenvironment provide different niches, resulting in diverse outcomes. MSCs interact with every type of cell of the immune system, either directly or through soluble factors, thus impeding all functions of the immune response.

3.1. Modulation of T-Cell Responses

MSCs share with the thymic epithelium the surface expression markers VCAM-1, ICAM-2, and LFA-3 *(12,15)*, which are essential for the interaction with T cells. Evidence has emerged that MSCs suppress the alloreaction of both CD4[+] and CD8[+], naïve and memory T cells in a dose-dependent manner. Blocking of proliferation, decrease in cytokine production, and reduced cytolytic activity of T cells triggered by cellular or non-specific mitogenic stimuli in autologous, allogeneic or even xenogeneic settings were still apparent by delayed MSC addition *(16,17,24–40)*. CD4[+] and CD8[+] T cells are equally

affected, although differences appear among studies, depending on the experimental setting *(24,34,38,39)*. In addition, allogeneic MSCs were shown to
exhibit stronger immunosuppressive effect than autologous MSCs *(38)*.

The majority of studies indicate a direct suppressive effect of MSCs on
T cells, mediated by soluble factors secreted following co-culture of MSCs with
the T cells. However, there is great confusion concerning their identity. MSC
immunomodulative ability was initially ascribed to TGF-β and HGF *(24)*, but a
number of subsequent studies dispute this finding *(25,30–32,34)*. Tryptophan-
deprivation of the culture medium by IDO, whose activity in MSCs is
induced by the IFN-γ produced by T cells, was identified as another potential
mechanism *(27,28,40)*, yet, again, to be argued by other investigators *(30,35)*.
Most recently, Aggarwal et al. *(35)* identified prostaglandin E2 (PGE2) as the
potential factor mediating the suppressing properties of MSCs *(35)*, but there
are studies contradicting this result *(25,30)*. Rasmusson et al. *(26)* supported the
participation of PGE2 in mitogen assays but not in mixed lymphocyte reactions
(MLRs). Differences regarding MSC isolation, resulting in diverse populations, as well as the detection system for immunosuppression (i.e., lymphocyte
population used, type of stimulus, and timing of analysis) could account for the
apparent discrepancies. Furthermore, diverse factors may participate at different
time points (such as PGE2 around the third day and IDO activity after the
fifth) or in different experimental protocols. To this end, Rasmusson et al. *(26)*
demonstrated that different mechanisms mediate the suppression of MLRs and
mitogen-stimulating assays. Specifically, in MLRs, MSC addition resulted in
increased IL-2, soluble IL-2 receptor and IL-10 levels, with the latter having
an activating role, and in PMA-stimulated cultures, MSC addition resulted in
a decrease of IL-2 levels, whereas soluble IL-2 receptor and IL-10 remained
unaffected.

By contrast, two murine studies suggested that cellular contact is essential
for T-cell suppression, the first pointing out that MSCs hinder T-cell contact
with antigen-presenting cells (APCs) in a non-cognate transient manner *(17)*
and the other indicating programmed death 1 (PD-1) molecule interaction
with its ligands as the potential mechanism *(37)*. The species-specific differences of MSCs, particularly between human and murine MSCs, could justify
those results. Besides, in the former study, a minor contribution of soluble
factors was actually observed. T-cell suppression is generally not attributed
to apoptosis *(24,30)*, although there is a report supporting that MSCs induce
apoptosis of activated T cells, having no effect on resting T cells *(28)*. After
MSC removal and restimulation of T cells, normal proliferation is detected *(24)*.
Upon co-culture with MSCs, activating receptors such as CD25, CD38, and

CD69 are down-regulated *(34)*, as regards human MSCs at least, because in a mouse setting, CD25 and CD69 remained unaffected *(39)*. In the latter report, it was demonstrated that MSCs do not interfere with early T-cell activation, allowing the initial protein synthesis, but blocking subsequent DNA synthesis, thus resulting in an arrest in the G_0/G_1 phase, inducing division arrest anergy. When MSCs are removed and T cells are stimulated with the cognate peptide, IFN-γ production is induced but not T-cell proliferation, in contrast to the reversibility of inhibition in the human or other primate systems *(24,29)*. Besides, Maccario et al. *(38)* using human MSCs showed that upon MSC removal, proliferation is resumed only for CD4$^+$ but not for CD8$^+$ cells.

An important subtype of T cells consists of the CD4$^+$CD25^{bright} regulatory T (T_{regs}), naturally arising, thymus-derived cells, that effectively control immune responses. When PB mononuclear cells (PBMCs) are cultured together with MSCs, there is an increase in the percentage of CD4$^+$CD25^{bright} Fox-P3 and/or CTLA-4-expressing T_{regs} *(35,38)*. However, the increased percentage of T_{regs} may represent only an additional means for MSC-mediated T-cell suppression, because their removal from the co-cultures does not affect suppression *(17,23)*. Additionally, Djouad et al. *(31)* implied the involvement of regulatory CD8$^+$ T or natural killer (NK)T cells as a potential mechanism for MSC immunosuppressive activity.

The complexity of lymphocyte stimulation, involving activating and inhibitory signals, contributes to the generation of the above contradictory results. Although the underlying mechanisms are not yet clarified, overall, MSCs suppress T-cell proliferation, inhibit IFN-γ and tumor necrosis factor-α (TNF-α) production and induce an increase in IL-4 and IL-10 levels, thus shifting the immune response from a pro-inflammatory toward an anti-inflammatory/tolerant state. This is further fortified by the alteration of the dendritic cell (DC)-profile from a DC1 to a DC2 *(35,38)* and the favoring of T_{reg} development.

3.2. Modulation of DC Properties

DCs are the most potent antigen-presenting cells (APCs), playing a crucial role in antigen uptake, transport, and presentation and thus in the initiation of an effective immune response. Their life span consists of an immature and a subsequent mature stage characterized by high efficiency in antigen uptake and processing the first and by potent antigen presentation the latter *(41)*. Whereas mature DCs induce the development of effector T cells, immature DCs induce peripheral tolerance either by T-cell deletion or T_{reg} expansion; although this concept was challenged recently *(42)*. The two subtypes DC1 and DC2, through polarizing mediators, trigger Th1 and Th2 responses, respectively. They can also interact with B and NK cells. Thus, modulation of DC properties is a potent mechanism to orchestrate cellular and humoral T-cell responses, giving rise to different final yields of effector (Th1 or Th2), memory, and regulatory T cells *(43)*.

MSCs are able to affect DC properties and function at all major stages of their life cycle, that is, differentiation, maturation, and activation. Specifically, they inhibit up-regulation of CD1a, CD40, CD80, CD86, and HLA-DR expression during differentiation and CD40, CD83, CD86, and MHC class I expression during maturation. Moreover, endocytosis capability and IL-2, IL-12, and TNF-α production are reduced, accompanied by increased secretion of IL-10 *(23,35,44,45)*. This denotes DC arrest in the immature stage, corresponding with a regulatory APC type, thus directing the immune reaction from inflammatory to anti-inflammatory and inhibiting T-cell activation. Using another experimental setting, namely, examining the effect of MSCs on DCs in MLR using unpurified PBMCs instead of isolated populations, Maccario and colleagues *(38)* showed that there is a sharp decrease in the total number of DCs in the cultures containing MSCs, and this outcome concerns the DC1 subpopulation.

Regarding the underlying mechanisms of MSC-mediated DC suppression, small differences within protocols, such as the maturation agents used [lipopolysaccharide (LPS) *(23,45)* or TNF-α *(44)*] or the MSC:DC ratios (varying from 1:1 to 1:40), result in a discrepancy whether the phenomenon is cell-to-cell contact dependent or not. This inconsistency could possibly be attributed to the higher concentrations of soluble factors in the microenvironment around MSCs. To this end, Jiang and colleagues found that IL-6 and M-CSF secreted by MSCs exhibit a role in retaining the DC immature immunophenotype. However, their neutralization failed to produce CD1a$^+$ DCs, whereas, upon removal of MSCs, functional DCs were generated *(45)*. Besides, Groh and colleagues *(36)* demonstrated a reverse phenomenon, where PB monocytes activate MSCs through soluble factors including IL-1β to secrete the inhibitory molecules mediating T-cell suppression.

MSC-mediated modulation of DC properties results in DC lock at an immature stage, suggesting an alternative means of modulating the immune system during the initial immune response, thereby adding to their direct impact on T cells. This may be of therapeutic significance in autoimmune diseases and graft versus host disease (GVHD), where regulatory DCs exhibit a central role.

Apart from the modulation of DC functions, MSCs themselves were recently ascribed APC function. Specifically, IFN-γ-treated mouse and human MSCs, strictly in syngeneic settings, were able to process exogenous antigens, subsequently activate T cells in vitro and induce antigen-specific immune responses in vivo. Up-regulation of the B7-H1 molecule was proposed to hold a role in MSC-mediated antigen presentation *(18)*. This work was supported to some extent by a recent study illustrating that at low concentrations of IFN-γ, MSCs robustly up-regulate MHC-class II and exert APC functions, although solely to recall antigens, whereas at elevated IFN-γ levels, there is a gradual decrease of those properties *(46)*. The dual role of MSC as APCs and suppressors of the immune response may have significant ramifications in autologous MSC-based therapies.

3.3. Modulation of B-Cell Properties

B lymphocytes represent an important component of adaptive immunity, which, upon activation, differentiate into Ab-producing plasma cells or memory cells. Their development takes place in the bone marrow (BM) where interaction of B cells with BM stroma determines their survival and proliferation *(47)*.

Very few studies have focused on the immunomodulatory effects of MSCs on humoral immunity and B cells. Using murine splenocytes and the combination of anti-CD40 antibody and IL-4, which exclusively stimulate B cells, MSCs were shown to inhibit B-cell proliferation *(39)*. In another murine study, proliferation of B cells within total splenocyte cultures activated with pokeweed mitogen, which stimulates both T and B cells, was inhibited through MSC-derived soluble factors *(37)*. Moreover, using BXSB mice, a model of systemic lupus erythematosus, Deng and colleagues *(48)* demonstrated that MSCs have an inhibitory effect on the proliferation, activation, and IgG production of the hyperactive B cells, whereas they slightly induce typical CD40 and suppress ectopic CD40L expression. Both these results indicate the inhibition of B-cell development into plasma cells.

Regarding human B cells, a recent study investigated the in vitro interaction of human MSCs with PB-derived B cells *(49)*. Using an effective model of B-cell activation, Corcione and colleagues demonstrated that following incubation with MSCs, proliferation, differentiation to Ab-producing cells and chemotaxis of B cells are significantly impaired. Specifically, B cells do not undergo apoptosis but are arrested in the G_0–G_1 phases of the cell cycle, similar to what occurs with T cells. Production of IgM, IgG, and IgA is inhibited. Moreover, CXCR4, CXCR5, and CCR7 expression are down-regulated: the first even at low MSC-to-B cell ratios, leading to the decrease of B-cell chemotaxis in response to CXCL12 and CXCL13, thus possibly eliminating migration to the secondary lymphoid organs. All these effects of MSCs are mediated by soluble factors, which, however, were not characterized. On the contrary, TNF, IFN-γ, IL-4, and IL-10 production by B cells is not impaired, and surface expression of HLA-DR, CD40, CD80, and CD86 is not down-regulated, indicating that APC-function of B cells may not be affected.

In another study, however, MSCs are able to suppress B-cell proliferation only in the presence of exogenously added IFN-γ *(40)*. The discrepancy possibly reflects the difference of the activation protocol, because the latter study uses CpG-containing DSP30F oligodeoxynucleotide in the presence of irradiated allogeneic T-cell-depleted PBMCs.

MSC-mediated B-cell suppression suggests the potential usage of MSCs in cellular therapeutic protocols for autoimmune diseases, where robust B-cell activation plays the major role.

3.4. Modulation of NK-Cell Properties

NK cells are the major effector cells of innate immunity, displaying a crucial role in early host defense against infected and tumor cells. They exert their function through immunoregulatory cytokine/chemokine secretion, cytolysis, and Ab-dependent cellular cytotoxicty mediation. NK-cell function is regulated by a balance of signals transduced by activating and inhibitory receptors inter-acting with specific surface molecules on the target cells *(50)*.

Using freshly isolated NK cells from PB, initial studies concluded that MSCs escape lysis by alloreactive NK cells *(15,32)*. However, after triggering NK cells with IL-2, inhibition of IFN-γ secretion was reported *(35)*. In another experimental setting, using PBMCs in conventional MLRs, the number of NK cells was found to be reduced when MSCs were added to the culture *(38)*.

More detailed information on the cellular interactions between MSCs and NK cells was published following the investigation of the phenotypical and functional properties of isolated CD56$^+$CD3$^-$ cells. As MSCs exhibit surface expression of ligands for DNAX accessory molecule-1 (DNAM-1) (such as PDGF- and VEGF- related receptor (PVR) and Nectin-2) and NK group 2D (NKG2D) (such as UL16-binding proteins (ULBPs) and MHC class I-related chain A (MIC-A)) and produce regulatory factors such as IL-15, TGF-β1, and PGE2, they are able to interact with NK cells and affect their proliferation and function *(51–53)*.

When no triggering factor is added in the culture, MSCs induce activation of NK cells, which is expressed by up-regulation of the activation antigen CD69 and secretion of IFN-γ and TNF-α. This effect is cell contact dependent and requires LFA/ICAM-1 interaction, whereas NKp30 partakes in cytokine production *(51)*. However, when NK cells are activated by IL-2 or IL-15, there is an inhibition of proliferation and reduction of cytokine secretion and cytolytic capacity *(40,52,53)*. Inhibition of NK-cell proliferation mainly affects the CD56^{bright} subpopulation and is mediated by soluble factors (including TGF-β1 and PGE2) induced to MSCs upon interaction with NK cells *(52)*. Cytotoxic ability is also reduced upon co-culture with MSCs *(40,52)*, possibly through down-regulation of γ$_c$-chain expression induced by PGE2 *(52)* or IDO production *(40)*. Production of IFN-γ, TNF-α, and IL-10 is also inhibited, yet again as regards the CD56^{brigt} cell subset *(52)*.

Despite the MSC-mediated NK cell-suppression and the initial finding that fresh NK cells do not lyse MSCs, studies using cytokine-activated NK cells have altered the perception that MSCs are neglected by the former. To this end, after activation with IL-2, IL-15, IL-12, and/or IL-18, NK cells exhibit potent cytolytic activity against both allogeneic and autologous MSCs *(51–53)*. The responsible activating receptors include NKp30, NKG2D, and DNAM-1 *(53)* or LFA1 *(51)*. Furthermore, for MSC-targeted cytolytic activity, both calcium and perforins are involved *(51)*. Surprisingly, although IFN-γ mediated up-regulation of HLA-class I molecules renders them less susceptible to NK cytotoxic activity *(53)*, masking HLA-class I molecules has no impact on their lysis *(51,53)*.

Together, all these findings suggest that the outcome of MSC/NK cell inter-action relies on NK-cell activation status and should be taken into consideration when estimating the therapeutic benefit of MSCs in tissue regeneration and of activated NK cells in cancer immunotherapy, to avoid elimination of donor cells by resident NK cells or tumor stroma, respectively.

4. MSCs Are Not Immunogenic and Avoid Allogeneic Rejection

MSC surface phenotype reflects poor recognition by both T and NK cells, because of the absence of MHC class II and co-stimulatory molecules, as well as the presence of MHC class I molecules, respectively. There are several studies demonstrating that MSCs escape recognition by allogeneic lympho-cytes *(17,28–30,54,55)*. MSCs fail to induce allogenic T-cell responses even after differentiation into osteocytes, adipocytes, or chondrocytes, indicating the safety of transplantation for tissue regeneration studies *(54,56)*. However, this cannot be attributed solely to the absence of HLA-class II molecules, because their IFN-γ-induced up-regulation does not result in recognition *(30,54)*, nor to co-stimulation deficiency, because MSCs transduction with B7-1 and B7-2 *(25)* as well as anti-CD3 Ab addition to the culture *(30)* do not elicit significant lymphoproliferation. Conversely, Beyth et al. *(23)* used MSCs to efficiently activate highly purified T cells, an effect attenuated by APC addition, whereas in another study, MSCs constitutively expressing HLA-class II elicited T-cell reaction *(16)*. Interestingly, studies illustrating the suppressive role of MSCs indicated that, in very low concentrations, they are capable of stimulating T-cell reactions *(33,55)*. Nevertheless, Kushnekova et al. *(25)* implied that MSCs initiate alloreactive T-cell activation but do not elicit T-cell responses because of active suppressive mechanisms.

In agreement with the in vitro data, several in vivo studies have shown that infusion of MHC-mismatched or even xenogeneic MSCs are not rejected by the host, irrespective of immunocompressive treatment, thus suggesting the potential use as universal donor MSCs *(29,57–63)*. The avoidance of rejection of MSCs cannot be ascribed solely to their lack of immunogenicity. We should thus hypothesize that their immune-privileged phenotype together with their immunosuppressive properties accounts for this distinctive attribute.

In contrast to the preceding results, recent work in the murine system indicated that MSCs may trigger immune responses in non-immunocompromised recipients *(64,65)*. Eliopoulos et al. *(64)* demonstrated that, in complete MHC class I and II mismatched setting, transplanted MSCs are rejected, outcome amplified through repeated challenge. Besides, Nauta et al. *(65)* illustrated that allogeneic MSCs not only induce memory T-cell responses but also increase rejection of allogeneic BM cells. Moreover, xenogeneic transplantation of human MSCs in rat myocardium resulted in rejection in immunocompetent but not in athymic mice *(66)*. The discrepancy in these studies could partially be attributed to the species specificity. Besides, Eliopoulos et al. used high numbers of transplanted MSCs after several population doublings (passages 12–17 for control MSCs and even greater for transduced MSCs), which may lead to chromosomal instability *(67)*. Nevertheless, these data should be an important consideration when planning clinical trials with allogeneic MSCs.

5. Therapeutic Application of MSC-Mediated Immunosuppression

MSC immunoregulatory properties observed in vitro, together with the fact that they are well tolerated in an allogeneic setting, led to their in vivo use for the prolongation of graft survival, as well as for the treatment of autoimmune diseases and GVHD. To this end, Bartholomew et al. *(29)* demonstrated in a baboon model that a single dose of MSCs was able to emit a modest but significant prolongation of skin graft survival. Regarding HSC transplantation, which represents the treatment of choice for numerous disorders, MSCs have been used to significantly increase the survival rate after MHC-mismatched HSC transplantation *(68)*, to ameliorate engraftment of multiple cord blood transplantations, possibly through elimination of graft-versus-graft reaction (**69**) and to allow engraftment of limited numbers of human HSCs in NOD/SCID mice (**55**). In the human setting, MSCs were used to improve survival and eliminate the risk of GVHD in hematopoietic malignancy patients (**62**), as well as to prevent GVHD in a patient receiving haplotype-mismatched CD34⁺ cells (**70**). An attractive therapeutic modality for MSCs would be the treatment of ongoing GVHD. As a proof of concept, Le Blanc et al. successfully treated a pediatric patient with refractory GVHD with two doses of haploidentical "third party" MSCs. After MSC infusion, patient's lymphocytes efficiently responded to in vitro induction with donor lymphocytes, indicating in vivo immunosuppression, excluding the establishment of tolerance (**61**).

In autoimmune diseases, MSCs can be used for both immunosuppression and repair of tissue damage caused by chronic inflamation. In a murine model of experimental autoimmune encephalomyelitis, MSCs were shown to ameliorate symptoms in both preventive and therapeutic protocols but not in chronic disease (**71,72**). T-cell anergy (**71**) or reduction of inflammatory infiltrates and demyelination and oligedendrogenesis stimulation (**72**) were identified as the potential mechanisms. Furthermore, in a study using unfractionated BM for the treatment of systemic lupus erythematosus, stromal cells have been presumed to exert the major role (**73**). However, in a murine model of rheumatoid arthritis, MSCs failed to produce any benefit, intensifying Th1 responses, a fact reversed in vitro by TNF-α addition (**74**).

Although still at the beginning, these studies encourage the prospect for immune modulation using MSCs. However, some of the contradictory results indicate that considerable work is required before MSCs can be used as cellular immunosuppressive agents in organ-transplant recipients and patients with severe autoimmune diseases, either alone or in combination with other immunosuppressive modalities.

6. MSCs and Cancer

The tolerogenic effects of MSCs together with their ability to produce angiogenic factors may have a role in cancer development, if tumor micromasses are present in the host. Djouad et al. (**31**) demonstrated that allogeneic melanoma cell lines, which are normally rejected, develop tumors when co-injected with MSCs. The targeting of microscopic tumors by MSCs and their proliferation and contribution to the tumor stroma was confirmed by in vivo imaging (**75**), whereas enrichment of angiogenesis of the developed tumors and elevated capability of proliferation and highly metastatic ability of the cancer cells within were also evidenced (**76**). Conversely, in a xenogeneic model of Kaposi's sarcoma, human MSCs clearly exhibited essential antitumor effect through the inhibition of Akt activation within sarcoma cells *(77)*.

On the contrary, the fact that MSCs are attracted by growth factors secreted by the tumor, become incorporated as components of the tumor stroma, and contribute as a significant proportion of it enables their use as "Trojan Horses," namely as cellular vehicles to locally deliver anti-tumor agents at the tumor site. MSCs are easily genetically modified, with high gene-transfer efficacy and long-term transgene expression. MSCs transduced with IFN-β *(78,79)* or IL-2 *(80)* have already been successfully used for tumor growth inhibition and increased patient survival.

In addition to the potential enhancement of tumor growth, a few studies indicated MSCs as precursors for tumorigenic cells *(67,81)*. Although cellular life span is normally under strict control and spontaneous transformation is rare, stem cells, because of their longevity, may have higher probability of accumulating mutations and consequently of malignant transformation. Likewise, Rubio et al. *(81)*, using adipose tissue-derived cells, demonstrated that while MSC-expansion was safe for 2 months, long-term (4–5 months) in vitro culture resulted in spontaneous transformation and elevated potential for in vivo tumor formation *(81)*. This was also confirmed for BM-derived murine MSC and was attributed to accumulated chromosomal abnormalities, amplified c-Myc expression and elevated telomerase activity *(67)*. However, applying the same experimental setting in the human system, the latter team showed that human BM-derived MSCs even at the latest stage of their life span are still capable of maintaining their chromosomal stability *(67)*. However, following transduction with the telomerase *hTERT* gene and after numerous population doublings, MSCs were able to form tumors *(82)*. Nevertheless, those results raise considerable concerns for the use of extensively expanded MSCs.

7. Conclusion

MSC use as universal "off the self" immunosuppressive elements looks very appealing and could have potential in clinical instances with high risk of engraftment failure (HLA-mismatched donors or cord blood transplantations), in the treatment of GVHD and autoimmune disorders, as well as for the support of hematopoietic recovery following HSC transplantation. Their generally recognized ability to avoid rejection may render them candidates for application in tissue engineering across HLA barriers. However, further research is required to elucidate fundamental questions regarding the mechanisms of manipulation of immune responses by MSCs, before realization of their therapeutic use in cell and gene therapy.

References

1. Conheim, J. (1867) Ueber Entzundung und Eiterung. *Pathol. Anat. Physiol. Klin. Med.* **40**, 1–79.
2. Friedenstein, A. J., Chailakhyan, R. K., Latsinik, N. V., Panasyuk, A. F. and Keiliss-Borok, I. V. (1974) Stromal cells responsible for transferring the microenvironment of the hemopoietic tissues. Cloning in vitro and retransplantation in vivo. *Transplantation* **17**, 331–340.
3. Deng, W., Obrocka, M., Fischer, I. and Prockop, D. J. (2001) In vitro differentiation of human marrow stromal cells into early progenitors of neural cells by conditions that increase intracellular cyclic AMP. *Biochem. Biophys. Res. Commun.* **282**, 148–152.
4. Toma, C., Pittenger, M. F., Cahill, K. S., Byrne, B. J. and Kessler, P. D. (2002) Human mesenchymal stem cells differentiate to a cardiomyocyte phenotype in the adult murine heart. *Circulation* **105**, 93–98.
5. Lee, K. D., Kuo, T. K., Whang-Peng, J., Chung, Y. F., Lin, C. T., Chou, S. H., Chen, J. R., Chen, Y. P. and Lee, O. K. (2004) In vitro hepatic differentiation of human mesenchymal stem cells. *Hepatology* **40**, 1275–1284.
6. Dezava, M., Ishikawa, H., Itokazu, Y., Yoshihara, T., Hoshino, M., Takeda, S., Ide, C. and Nabeshima, Y. (2005) Bone marrow stromal cells generate muscle cells and repair muscle degeneration. *Science* **309**, 314–317.
7. Kuznetsov, S. A., Mankani, M. H., Gronthos, S., Satomura, K., Bianco, P. and Gehron-Robey, P. (2001) Circulating skeletal stem cells. *J. Cell Biol.* **153**, 1133–1139.
8. Wagner, W., Wein, F., Seckinger, A., Frankhauser, M., Wirkner, U., Krause, U., Blake, J., Schwager, C., Eckstein, V., Ansorge, W. and Ho, A. D. (2005) Comparative characteristics of mesenchymal stem cells from human bone marrow, adipose tissue, and umbilical cord blood. *Exp. Hematol.* **33**, 1402–1416.
9. Horwitz, E. M., Le Blanc, K., Dominici, M., Mueller, I, Slaper-Cortenbach, I., Marini, F. C., Deans, R. J., Krause, D. S. and Keating A. (2005) Clarification of the nomenclature for MSC. The International Society for Cellular Therapy position statement. *Cytotherapy* **7**, 393–395.
10. Haynesworth, S. E., Baber, M. A. and Caplan, A. I. (1992) Cell surface antigens on human marrow-derived mesenchymal cells are detected by monoclonal Abs. *Bone* **13**, 69–80.
11. Barry, F., Boynton, R., Murphy, M., Haynesworth, S. and Zaia, J. (2001) The SH-3 and SH-4 antibodies recognize distinct epitopes on CD73 from human mesenchymal stem cells. *Biochem. Biophys. Res. Commun.* **289**, 519–524.
12. Cognet, P. A. and Minguell, J. J. (1999) Phenotypical and functional properties of human bone marrow mesenchymal progenitor cells. *J. Cell. Physiol.* **181**, 67–73.
13. Majumdar, M. K., Keane-Moore, M., Buyaner, D., Hardy, W. B., Moorman, M. A., McIntosh K. R. and Mosca, J. D. (2002) Characterization and functionality of cell surface molecules on human mesenchymal stem cells. *J. Biomed. Sci.* **10**, 228–241.

14. Wynn, R. F., Hart, C. A., Corradi-Perini, C., O'Neil, L., Evans, C. A., Wraith, J. E., Fairbairn, L. J. and Bellantuono, I. (2004) A small proportion of mesenchymal stem cells strongly expresses functionally active CXCR4 receptor capable of promoting migration to bone marrow. *Blood* **104**, 2643–2645.

15. Le Blanc, K. (2003) Immunomodulatory effects of fetal and adult mesenchymal stem cells. *Cytotherapy* **5**, 485–489.

16. Potian, J. A., Aviv, H., Ponzio, N. M., Harrison J. S. and Rameshwar P. (2003) Veto-like activity of mesenchymal stem cells: functional discrimination between cellular responses to alloantigen and recall antigens. *J. Immunol.* **171**, 3426–3434.

17. Krampera, M., Glennie, S., Dyson, J., Scott, D., Laylor. R., Simpson, E. and Dazzi, F. (2003) Bone marrow mesenchymal stem cells inhibit the response of naïve and memory antigen-specific T cells to their cognate peptide. *Blood* **101**, 3722–3729.

18. Stagg, J., Pommey, S., Eliopoulos, N. and Galipeau J. (2006) Interferon-γ-stimulated marrow stromal cells: a new type of nonhematopoietic antigen-presenting cell. *Blood* **107**, 2570–2577.

19. Bruder, S. P., Jaiswal, N. and Haynesworth, S. E. (1997) Growth kinetics, self-renewal, and the osteogenic potential of purified human mesenchymal stem cells during extensive subcultivation and following cryopreservation. *J. Cell. Biochem.* **64**, 278–294.

20. Haynesworth, S. E., Baber, M. A. and Caplan, A. I. (1996) Cytokine expression by human marrow-derived mesenchymal progenitor cells in vitro: effects of dexamethasone and IL-1α. *J. Cell. Physiol.* **166**, 585–592.

21. Majumdar, M. K., Thiede M. A., Mosca, J. D., Moorman. M. and Gerson, S. L. (1998) Phenotypic and functional comparison of cultures of marrow-derived mesenchymal stem cells (MSCs) and stromal cells. *J. Cell. Physiol.* **176**, 57–66.

22. Kim, D. H., Yoo, K. H., Choi, K. S., Choi, J., Choi, S. Y., Yang S. E., Yang, Y. S., Im, H. J., Kim, K. H., Jung, H. L., Sung, K. W. and Koo, H. H. (2005) Gene expression profile of cytokine and growth factor during differentiation of bone marrow-derived mesenchymal stem cell. *Cytokine* **31**, 119–126.

23. Beyth, S., Borovsky, Z., Mevorach, D., Liebergall, M., Gazit, Z., Aslan, H., Galun, E. and Rachmilewitz, J. (2005) Human mesenchymal stem cells alter antigen-presenting cell maturation and induce T-cell unresponsiveness. *Blood* **105**, 2214–2219.

24. Di Nicola, M., Carlo-Stella, C., Magni, M., Milanesi, M., Longoni, P. D., Matteucci, P., Grisanti, S. and Gianni, A. M. (2002) Human bone marrow stromal cells suppress T-lymphocyte proliferation induced by cellular or nonspecific mitogenic stimuli. *Blood* **99**, 3838–3843.

25. Klyushnekova, E., Mosca, J. D., Zernetkina, V., Majumdar, M. K., Beggs, K. J., Simonetti, D. W., Deans, R. J. and McIntosh, K. R. (2005) T cell responses to allogeneic human mesenchymal stem cells: immunogenicity, tolerance, and suppression. *J. Biomed. Sci.* **12**, 47–57.

26. Rasmusson, I., Rindgen, O., Sundberg, B. and Le Blanc, K. (2005) Mesenchymal stem cells inhibit lymphocyte proliferation by mitogens and alloantigens by different mechanisms. *Exp. Cell Res.* **305**, 33–41.

27. Meisel, R., Zibert, A., Laryea, M., Gobel, U., Daubener, W. and Dilloo, D. (2004) Human bone marrow stromal cells inhibit allogeneic T-cell responses by indoleamine-2,3-dioxygenase-mediated tryptophan degradation. *Blood* **103**, 4619–4621.

28. Plumas, J., Chaperot, L., Richard, M. J., Molens, J. P., Bensa, J. C. and Favrot, M. C. (2005) Mesenchymal stem cells induce apoptosis of activated T cells. *Leukemia* **19**, 1597–1604.

29. Bartholomew, A., Sturgeon, C., Siatskas, M., Ferrer, K., McIntosh, K., Patil, S., Hardy, W., Devine, S., Ucker, D., Deans, R., Moseley, A. and Hoffman, R. (2002) Mesenchymal stem cells suppress lymphocyte proliferation in vitro and prolong skin graft survival in vivo. *Exp. Hematol.* **30**, 42–48.

30. Tse, W. T., Pendleton, J. D., Beyer, W. M,. Egalka, M. C. and Gionan, E. C. (2003) Suppression of allogeneic T-cell proliferation by human bone marrow stromal cells: implications in transplantation. *Transplantation* **75**, 389–397.

31. Djouad, F., Plence, P., Bony, C., Tropel, P., Apparailly, F., Sany, J., Noel, D. and Jorgensen, C. (2003) Immunosuppressive effect of mesenchymal stem cells favors tumor growth in allogeneic animals. *Blood* **102**, 3837–3844.

32. Rasmusson, I., Rindgen, O., Sundberg, B. and Le Blanc, K. (2003) Mesenchymal stem cells inhibit the formation of cytotoxic T lymphocytes, but not activated cytotoxic T lymphocytes or natural killer cells. *Transplantation* **27**, 1208–1213.

33. Le Blanc, K., Tammik, L., Sundberg, B., Haynesworth, S. E. and Ringden, O. (2003) Mesenchymal stem cells inhibit and stimulate mixed lymphocyte cultures and mitogenic responses independently of the major histocompatibility complex. *Scand. J. Immunol.* **57**, 11–20.

34. Le Blanc, K., Rasmusson, I., Gotherstrom, C., Seidel, C., Sundberg, B., Sundin, M., Rosendahl, K., Tammik, C. and Ringden, O. (2004) Mesenchymal stem cells inhibit the expression of CD25 (interleukin-2 receptor) and CD38 on phytohaemagglutinin-activated lymphocytes. *Scand. J. Immunol.* **60**, 307–315.

35. Aggarwal, S. and Pittenger, M. F. (2005) Human mesenchymal stem cells modulate allogeneic immune cell responses. *Blood* **105**, 1815–1822.

36. Groh, M. E., Maitra, B., Szekely, E. and Koc, O. (2005) Human mesenchymal stem cells require monocyte-mediated activation to suppress alloreactive T cells. *Exp. Hematol.* **33**, 928–934.

37. Augello, A., Tasso, R., Negrini, S. M., Amateis, A., Indiveri, F., Cancedda, R. and Pannesi, G. (2005) Bone marrow mesenchymal progenitor cells inhibit lymphocyte proliferation by activation of the programmed death 1 pathway. *Eur. J. Immunol.* **35**, 1482–1490.

38. Maccario, R., Podesta, M., Moretta, A., Cometa, A., Comoli, P., Montagna, D., Daudt, L., Idatici, A., Piaggio, G., Pozzi, S., Frassoni, F. and Locatelli, F. (2005)

Interaction of human mesenchymal stem cells with cells involved in alloantigen-specific immune response favors the differentiation of CD4$^+$ T-cell subsets expressing a regulatory/suppressive phenotype. *Haematologica* **90**, 516–525.

39. Glennie, S., Soeiro, I., Dyson, P. J., Lam, E. W. F. and Dazzi, F. (2005) Bone marrow mesenchymal stem cells induce division arrest anergy of activated T cells. *Blood* **105**, 2821–2827.

40. Krampera, M., Cosmi, L., Angeli, R., Pasini, A., Liotta, F., Andreini, A., Santarlasci, V., Mazzinghi, B., Pizzolo, G., Vinante, F., Romagnani, P., Maggi, E., Romagnani, S. and Annunziato, F. (2006) Role of IFN-γ in the immunomodulatory activity of human bone marrow mesenchymal stem cells. *Stem Cells* **24**, 386–398.

41. Banchereau, J., Briere, F., Caux, C., Davoust, J., Lebecque, S., Liu, Y.J., Pulendran, B. and Palucka, K. (2000) Immunobiology of dendritic cells. *Annu. Rev. Immunol.* **18**, 767–811.

42. Rutella, S., Danese, S. and Leone, G. (2006) Tolerogenic dendritic cells. Cytokine modulation comes of age. *Blood* **108**, 1435–1440.

43. Steinman, R. M., Pack, M. and Inaba, K. (1997) Dendritic cells in the T-cell areas of lymphoid organs. *Immunol. Rev.* **156**, 25–37.

44. Zhang, W., Ge, W., Li, C., You, S., Liao, L., Han, Q., Deng, W. and Zhao, R. C. H. (2004) Effects of mesenchymal stem cells on differentiation, maturation, and function of human monocyte-derived dendritic cells. *Stem Cells Dev.* **13**, 263–271.

45. Jiang, X. X., Zhang, Y., Liu, B., Zhang, S. X., Wu, Y., Yu, X. D. and Mao, N. (2005) Human mesenchymal stem cells inhibit differentiation and function of monocyte-derived dendritic cells. *Blood* **105**, 4120–4126.

46. Chan, J. L., Tang, K. C., Patel, A. P., Bonilla, L. M., Pierobon, N., Ponzio, N. M. and Rameshwar, P. (2006) Antigen presenting property of mesenchymal stem cells occurs during a narrow window at low levels of interferon-γ. *Blood* **107**, 4817–4824.

47. Ollila, J. and Vihinen, M. (2005) B cells. *Int. J. Biochem. Cell Biol.* **37**, 518–523.

48. Deng, W., Han, Q., Liao, L., You, S., Deng, H. and Zhao, R. C. H. (2005) Effects of allogeneic bone marrow-derived mesenchymal stem cells on T and B lymphocytes from BXSB mice. *DNA Cell Biol.* **24**, 458–463.

49. Corcione, A., Benvenuto, F., Ferretti, E., Giunti, D., Cappiello, V., Cazzanti, F., Risso, M., Gualandi, F., Mancardi, G. L., Pistoia, V. and Ucceli, A. (2006) Human mesenchymal stem cells modulate B-cell functions. *Blood* **107**, 367–372.

50. Papamichail, M., Perez, S. A., Gritzapis, A. D. and Baxevanis, C. N. (2004) Natural killer lymphocytes: biology, development, and function. *Cancer Immunol. Immunother.* **53**, 176–186.

51. Poggi, A., Prevosto, C., Massaro, A. M., Negrini, S., Urbani, S., Pierri, I., Saccardi, R., Gobbi, M. and Zocchi, M. R. (2005) Interaction between human NK cells and bone marrow stromal cells induces NK cell triggering: role of NKp30 and NKG2D receptors. *J. Immunol.* **175**, 6352–6360.

52. Sotiropoulou, P. A., Perez, S. A., Gritzapis, A. D., Baxevanis, C. N. and Papamichail. M (2006) Interactions between human mesenchymal stem cells and natural killer cells. *Stem Cells* **24**, 74–85.

53. Spaggiari, G. M., Capobianco, A., Beccheti, S., Mingari, M. C. and Moretta, L. (2006) Mesenchymal stem cell-natural killer cell interactions; evidence that activated NK cells are capable of killing MSCs, whereas MSCs can inhibit IL-2-induced NK-cell proliferation. *Blood* **107**, 1484–1490.

54. Le Blanc, K., Tammik, C., Rosendahl, K,. Zetterberg, E. and Ringden, O. (2003) HLA expression and immunologic properties of differentiated and undifferentiated mesenchymal stem cells. *Exp. Hematol.* **31**, 890–896.

55. Maitra, B., Szekely, E., Gjini, K., Laughlin, M. J., Dennis, J., Haynesworth, S. E. and Koc, O. N. (2004) Human mesenchymal stem cells support unrelated donor hematopoietic stem cells and suppress T-cell activation. *Bone Marrow Transplant.* **33**, 597–604.

56. Liu, H., Kemeny, D. M., Heng, B. C., Ouyang, H. W., Melendez, A. J. and Cao, T. (2006) The immunogenicity and immunomodulatory function of osteogenic cells differentiated from mesenchymal stem cells. *J. Immunol.* **176**, 2864–2871.

57. Horwitz, E. M., Gordon, P. L., Koo, W. K., Marx, J. C., Neel, M. D., McNall, R. Y., Muul, L. and Hofmann, T. (2002) Isolated allogeneic bone marrow-derived mesenchymal cells engraft and stimulate growth in children with osteogenesis imperfecta: implications for cell therapy of bone. *Proc. Natl. Acad. Sci. U.S.A.* **99**, 8932–8937.

58. Koc, O. N., Day, J., Nieder, M., Gerson, S. L., Lazarus, H. M. and Krivit, W. (2002) Allogeneic mesenchymal stem cell infusion for treatment of metachromatic leukodystrophy (MLD) and Hurler syndrome (MPS-IH). *Bone Marrow Transplant.* **30**, 215–222.

59. Saito, T., Kuang, J. Q., Bittira, B., Al-Khaldi, A. and Chiu, R. C. (2002) Xenotransplant cardiac chimera: immune tolerance of adult stem cells. *Am. Thorac. Surg.* **74**, 19–24.

60. Devine, S. M., Cobbs, C., Jennings, M., Bartholomew, A. and Hoffman, R. (2003) Mesenchymal stem cells distribute to a wide range of tissues following systemic infusion into nonhuman primates. *Blood.* **101**, 2999–3001.

61. Le Blanc, K., Rasmusson, I., Sundberg, B., Gotherstrom, C., Hassan, M., Uzunel, M. and Ringden, O. (2004) Treatment of severe acute graft-versus-host disease with third party haploidentical mesenchymal stem cells. *Lancet* **363**, 1439–1441.

62. Lazarus, H. M., Koc, O. N., Devine, S. M., Curtin, P., Maziarz, R. T., Holland, H. K., Shpall, E. J., McCarthy, P., Atkinson, K., Cooper, B. W., Gerson, S. L., Laughlin, M. J., Loberiza, F. R. Jr., Moseley, A. B. and Bacigalupo, A. (2005) Cotransplantation of HLA-identical sibling culture-expanded mesenchymal stem cells and hematopoietic stem cells in hematologic malignancy patients. *Biol. Blood marrow Transplant.* **11**, 389–398.

63. Rasulov, M. F., Vasil'chenkov, A. V., Onishchenko, N. A., Krasheninnikov, M. E., Kravchenko, V. I., Gorshenin, T. L., Pidtsan, R. E. and Potapov, I. V. (2005) First experience in the use of bone marrow mesenchymal stem cells for the treatment of a patient with deep skin burns. *Bull. Exp. Biol. Med.* **1**, 141–144.

64. Eliopoulos, N., Stagg, J., Lejeune, L., Pommey, S. and Galipeau, J. (2005) Allogeneic marrow stromal cells are immune rejected by MHC class I- and class II- mismatched recipient mice. *Blood* **106**, 4057–4065.

65. Nauta, A. J., Westerhuis, G., Kruisselbrink, A. B., Lurvink, E. G., Willemze, R. and Fibbe, W. E. (2006) Donor-derived mesenchymal stem cells are immunogenic in an allogeneic host and stimulate donor graft rejection in a non-myeloablative setting. *Blood* **108**, 2114–2120.

66. Grinnemo, K. H., Mansson, A., Dellgren, G., Klingberg, D., Wardell, E., Drvota, V., Tammik, C., Holgersson, J., Ringden, O., Sylven, C. and Le Blanc, K. (2004) Xenoreactivity and engraftment of human mesenchymal stem cells transplanted into infracted rat myocardium. *J. Thorac. Cardiovasc. Surg.* **127**, 1293–1300.

67. Miura, M., Miura, Y., Padila-Nash, H. M., Molinolo, A. A., Fu, B., Patel, V., Seo, B. M., Sonoyama, W., Zheng, J. J., Baker, C. C., Chen, W., Ried, T. and Shi, S. (2006) Accumulated chromosomal instability in murine bone marrow mesenchymal stem cells leads to malignant transformation. *Stem Cells* **24**, 1095–1103.

68. Chung, N. G., Jeong, D. C., Park, S. J., Choi, B. O., Cho, B., Kim, H. K., Chun, C. S., Won, J. H. and Ham, C. W. (2004) Cotransplantation of marrow stromal cells may prevent lethal graft-versus-host disease in major histocompatibility complex mismatched murine hematopoietic stem cell transplantation. *Int. J. Hematol.* **80**, 370–376.

69. Kim, D. W., Chung, Y.J., Kim, T. G., Kim, Y. L. and Oh, I. H. (2004) Cotransplantation of third-party mesenchymal stromal cells can alleviate single-donor predominance and increase engraftment from double cord transplantation. *Blood* **103**, 1941–1948.

70. Lee, S. T., Jang, J. H., Cheong, J. W., Kim, J. S., Meamg, H. Y., Hahn, J. S., Ko, Y. W. and Min, Y. H. (2002) Treatment of high-risk acute myelogenous leukemia by myeloablative chemoradiotherapy followed by co-infusion of T cell-depleted haematopoietic stem cells and culture-expanded marrow mesenchymal stem cells from a related donor with one fully mismatched leukocyte antigen haplotype. *Br. J. Haematol.* **118**, 1128–1131.

71. Zappia, E., Casazza, S., Pedemonte, E., Benvenuto, F., Bonanni, I., Gerdoni, E., Giunti, D., Ceravolo, A., Cazzanti, F., Frassoni, F., Mancardi, G. and Ucceli, A. (2005) Mesenchymal stem cells ameliorate experimental autoimmune encephalomyelitis inducing T cell anergy. *Blood* **106**, 1755–1761.

72. Zhang, J., Li, Y., Chen, J., Cui, Y., Lu, M., Elias, S. B., Mitchell, J. B., Hammill, L., Vanguri, P. and Chopp, M. (2005) Human bone marrow stromal cell treatment improves neurological functional recovery in EAE mice. *Exp. Neurol.* **195**, 16–26.

73. Ishida, T., Inaba, M., Hisha, H., Sugiura, K., Adachi, Y., Nagata, N., Ogawa, R., Good, R. A. and Ikehara, S. (1994) Requirement of donor-derived stromal cells in the bone marrow for successful allogeneic bone marrow transplantation. Complete prevention of recurrence of autoimmune diseases in MRL/MP-Ipr/Ipr mice by transplantation of bone marrow plus bones (stromal cells) from the same donor. *J. Immunol.* **152**, 3119–3127.

74. Djouad, F., Fritz, V., Apparailly, F., Louis-Plence, P., Bony, C., Sany, J., Jorgensen, C. and Noel, D. (2005) Reversal of the immunosuppressive properties of mesenchymal stem cells by tumor necrosis factor α in collagen-induced arthritis. *Arthritis Rheum.* **52**, 1595–1603.

75. Hung, S.-C., Deng, W.-P., Yang, W. K., Liu, R.-S., Lee, C.-C., Su, T.-C., Lin, R.-J., Yang, D.-M., Chang, C.-W., Chen, W.-H., Wei, H.-J. and Gelovani, J. G. (2005) Mesenchymal stem cell targeting of microscopic tumors and tumor stroma development monitored by noninvasive *in vivo* positron emission tomography imaging. *Clin. Cancer Res.* **11**, 7749–7756.

76. Zhu, W., Xu, W., Jiang, R., Qian, H., Chen, M., Hu, J., Cao, W., Han, C. and Chen, Y. (2006) Mesenchymal stem cells derived from bone marrow favor tumor cell growth in vivo. *Exp. Mol. Pathol.* **80**, 267–274.

77. Khakoo, A. Y., Pati, S., Anderson, S. A., Reid, W., Elshal, M. F., Rovira, I. I., Nguygen, A. T., Malide, D., Combs, C. A., Hall, G., Zhang, J., Raffeld, M., Rogers, T. B., Stetler-Stevenson, W., Frank, J. A., Reitz, M. and Finkel, T. (2006) Human mesenchymal stem cell exert potent antitumorigenic effects in a model of Kaposi's sarcoma. *J. Exp. Med.* **203**, 1235–1247.

78. Studeny, M., Marini, F. C., Champlin, R. E., Zompetta, C., Fidler, I. J. and Andreeff, M. (2002) Bone marrow-derived mesenchymal stem cells as vehicles for interferon-β delivery into tumors. *Cancer Res.* **62**, 3603–3608.

79. Studeny, M., Marini, F. C., Dembinski, J. L., Zompetta, C., Cabreira-Hansen, M., Bekele, B. N., Champlin, R. E. and Andreef, M. (2004) Mesenchymal stem cells: potential precursors for tumor stroma and targeted-delivery vehicles for anticancer agents. *J. Natl. Cancer Inst.* **96**, 1693–1603.

80. Nakamura, K., Ito, Y., Kawano, Y., Kurozumi, K., Kobune, M., Tsuda, H., Bizen, A., Monmou, O., Nutsu, Y. and Hamada, H. (2004) Antitumor effect of genetically engineered mesenchymal stem cells in a rat glioma model. *Gene Ther.* **11**, 1155–1164.

81. Rubio, D., Garcia-Castro, J., Martin, M. C., de la Fuente, R., Cigudosa, J. C., Lloyd, A. C. and Bernad, A. (2005) Spontaneous human adult stem cell transformation. *Cancer Res.* **65**, 3035–3039.

82. Serakinci, N., Guldberg, P., Burns, J. S., Abdallah, B., Schrodder, H., Jensen, T. and Kassem, M. (2004) Adult human mesenchymal stem cell as a target for neoplastic transformation. *Oncogene* **23**, 5095–5098.

17

Clinical Grade Expansion of Human Bone Marrow Mesenchymal Stem Cells

Panagiota A. Sotiropoulou, Sonia A. Perez, and Michael Papamichail

Summary

Mesenchymal stem cells or marrow stromal cells (MSCs) represent a multipotent adult cellular population with immunomodulatory functions. In the adult human body, they are present in various niches, but their main source is bone marrow (BM). The regeneration capability of MSCs, their ease to undergo gene modification, as well as their immunosuppressive capacity render them as popular candidates for tissue engineering, gene therapy, and immunotherapy. They exhibit a unique in vitro expansion capacity, which however does not always compensate for the large number of cells required for cellular therapeutic applications. Unfortunately, to date, a uniform worldwide approach to MSC-culture is not available. Thus, in this chapter, we try to describe the optimal conditions for the successful isolation and ex vivo expansion of human MSCs from BM, to be used in all types of cellular therapeutic approaches. Moreover, we describe the methods for identification of their quality in terms of both multilineage potentiality and immunosuppressive ability. Detailed protocols for fetal calf serum selection and other culture parameters that affect the final outcome are described.

Key Words: Mesenchymal stem cells; Ex vivo expansion; Clinical trials; Tissue engineering; Immunosuppression; Multilineage potential.

1. Introduction

Mesenchymal stem cells or marrow stromal cells (MSCs) are rare multipotent stem cells, residing mainly in bone marrow (BM), but have also been isolated from other tissues, such as umbilical cord blood, adipose tissue, and mobilized peripheral blood. However, to date, BM remains the main source

From: *Methods in Molecular Biology, vol. 407: Stem Cell Assays*
Edited by: M. C. Vemuri © Humana Press, Totowa, NJ

for clinical applications. MSCs are capable of differentiating along multiple pathways including bone, cartilage, cardiac and skeletal muscle, neural cells, tendon, adipose, and connective tissue. Systemic administration of MSCs in injury models has been reported to lead to specific migration to the site of damage, followed by tissue-specific differentiation patterns. The multilineage differentiation ability of MSCs, together with their relatively easy isolation from BM and their extensive capacity for in vitro expansion, led to the discovery of novel approaches for utilizing MSCs in tissue engineering as well as for gene therapy, for various congenital and acquired diseases, such as facilitation of hematopoietic recovery in hematopoietic stem cell transplantation, osteogenesis imperfecta, metabolic diseases, amyotrophic lateral sclerosis, stroke, and myocardial infarction *(1–3)*.

MSCs are not inherently immunogenic and seem to be natural immunosuppressive elements *(4)*. Administration of MSCs has already been used for the effective treatment of acute graft-versus-host disease and prolongation of histoincompatible graft survival in humans and animal models, respectively *(5)*. The ability of MSCs to modulate immune responses implies their potential role in cellular immunoregulatory therapy, whereas their susceptibility to gene modification enables them to serve as delivery vehicles for intratumoral production of anticancer agents, such as interferon-β (IFN-β) and interleukin-2 (*see* Chapter 15 for more information).

MSCs comprise a mere 0.0001%–0.01% of total BM-nucleated cells, depending on age and varying extensively among individuals, even within the same individual at different times *(3)*. The very low proportion of MSCs in the BM necessitates their in vitro expansion, to acquire vast numbers, sometimes exceeding 10^9 cells, which are needed for clinical therapeutic protocols. Whereas MSCs are currently a target of scientific research worldwide, a broadly accepted culture system has not yet been established for their isolation and in vitro expansion. This causes extensive protocol variability among laboratories, resulting in the generation of diverse MSC populations and consequently impeding comparisons of resulting outcomes.

Standard culture techniques for MSC in vitro expansion involve MSC attachment to the plastic surface of a culture vessel and, after an initial lag phase, rapid division until density intensification, when they enter a stationary phase. During the lag and log phases, MSCs are spindle-shaped and fibroblast-like *(6,7)*. However, if they are cultured in high densities or after numerous population doublings, they tend to adopt a flattened morphology in parallel with their entry into the stationary phase *(8)*.

Essential parameters for MSC culture comprise the type of basal medium, the batch of fetal calf serum (FCS), the initial and passaging cell-density, and the addition of growth factors. Various basal media have been used for MSC isolation and culture. A recent study by our group, in which we compared several media, has suggested that Modified Eagle Medium alpha (a-MEM) is the optimal type of basal medium, whereas, as a more general rule, stable glutamine and low glucose concentration favor MSC-isolation and in vitro expansion *(9)*.

The key component of MSC culture is FCS. Although, ideally, cells to be used in clinical procedures should be expanded in serum-free media to avoid potential hazards, this has not been applicable so far in MSC culture. A recent publication however suggests that the serum substitute ULTROSER® is suitable for supplementing a-MEM basal medium for clinical scale expansion of human MSCs *(10)*. If those promising results get confirmed by other laboratories as well, this might signify a new era for MSC-based therapies.

The optimum alternative to FCS is autologous serum. In this regard, there are publications indicating that the use of autologous serum provides equivalent or even superior results to FCS. For clinical protocols involving low numbers of MSCs, it is feasible to obtain the appropriate volume of peripheral blood. However, most types of cellular therapy require vast numbers of MSCs, which in turn necessitate large amounts of culture media and subsequently forbiddingly large volumes of peripheral blood. Moreover, the quantity of peripheral blood needed cannot be overcome by the use of pooled human serum, because allogeneic serum causes MSC growth arrest and death *(11)*.

To date, all reported clinical trials have used FCS for the culture of MSCs. However, the essential nutrients and growth factors provided by FCS appear to vary from lot to lot, thus affecting, to the highest degree, the quantity and quality of cells produced *(12)*. Thus, extensive screening of numerous batches should be performed to select the one resulting in optimum MSC growth (*see* **Subheading 2.3.**). The FCS-screening procedure should be planned at least 6 months before the FCS stock is depleted. The serum lot that produces cells with the appropriate phenotype (*see* **Subheading 2.4.**) provides the best results in terms of cell numbers and differentiation potential (*see* **Subheadings 2.6.1–2.6.6.**), and shows immunosuppressive capacity (*see* **subheading 2.7.**) should be selected. This outcome should not be far from that given by a control medium.

Another critical condition for MSC growth is culture density. MSCs appear to reduce their doubling period when seeding density is lowered, in both the initial BM mononuclear cell (MNC) plating and the passages to follow (ideally 1000 BM MNCs/cm^2 and 50 MSCs/cm^2, respectively). Given the high proliferative capacity of MSCs, especially during the first few passages, low initial plating density appears to be critical for obtaining final numbers, especially when limited culture time is required *(9,13)*. Unfortunately, the population of MSCs needed in clinical trials often involves large numbers of cells, thus making the optimum MSC culture density unrealistic. Therefore, depending on the clinical protocol to be used, one should compromise between laboratory capabilities and optimal conditions. In the **Subheading 2.2.**, we describe an easy method to handle MSC culture protocol that finishes with large numbers of high-quality MSCs, even from limited volumes of BM.

Several growth factors are known to augment MSC proliferation ability, either without affecting their multipotentiality or by favoring the selection and/or expansion of cells with differentiation potential toward a specific lineage. The most common growth supplement for MSC culture is basic fibroblast growth factor (bFGF). bFGF exhibits a dose-dependent excessive mitogenic function but also favors osteogenic and, to a lower extent, adipogenic differentiation potential, while limiting neurogenic potential and eliciting HLA-class I upregulation and HLA-class II induction. Still, in vitro immune responses to allogeneic MSCs are very weak compared to the ones elicited by allogeneic peripheral blood MNCs (PBMCs) *(9,14)*. Along these lines, bFGF addition to the MSC culture system depends on the clinical approach to be followed (i.e., target disorder and allogeneic/autologous transplants).

As MSCs comprise a heterogeneous population, confirmation of culture-expanded MSC quality requires various phenotypical and functional characterizations. In contrast to most cellular populations that are uniformly characterized by a unique surface molecule, no such marker has yet been determined for MSCs. Thus, MSC identification relies on a combination of positively and negatively expressed markers; such an expression profile allows us to identify them among other cellular subsets. Briefly, MSCs express on their surface CD29, CD41, CD44, CD73 (SH3, SH4), CD90, CD105 (SH2), CD106, CD120a, CD124, MAB-1470, STRO-1, and low levels of HLA-class I molecules, whereas they are negative for CD14, CD34, CD45, and HLA-DR *(1,7,9)*. For phenotypical characterization of MSCs, refer to **Subheading 2.4**.

Culture-expanded MSC populations incorporate various clones with different proliferative potential. The expandability of MSCs is of great importance for their in vivo regeneration ability, because the numbers homing to the damaged tissue might not suffice for the required regeneration process. Growth rates in culture are not necessarily predictive of MSC expandability, because MSCs represent a heterogeneous population consisting of diverse clones. The detection of the percentage of cells with high replicating potential in a MSC culture can be identified by means of a colony-forming assay *(15)* (*see* **Subheading 2.5.**).

Clinical protocols involving MSCs target either tissue regeneration or immunosuppression of the host. Thus, multipotential differentiation ability and immunosuppressive capacity need to be investigated. With regard to the former, osteogenic, adipogenic, and chondrogenic potential are generally enough to corroborate the multilineage differentiation capacity of MSCs *(9,16–18)*. However, evidence for their neurogenic or cardiomyogenic potential may be considered necessary depending on clinical relevance *(9,19,20)* (*see* **Subheadings 2.6.1–2.6.6.**). Moreover, MSCs are immunoprivileged, being able to escape immunologic responses, whereas they are capable of limiting T-cell responses when added as third party ingredients *(4,9)*. The immunological properties of culture-expanded MSCs can be estimated by adding them, as a third party or as stimulators, to mixed lymphocyte reactions (*see* **Subheading 2.7.**).

2. Materials

2.1. Isolation of BM MNCs by Ficoll–Hypaque Density Gradient Centrifugation

1. Ficoll–Hypaque (cat. no. L6115, Biochrom AG, Berlin, Germany).
2. Dulbecco's Phosphate Buffer Saline (DPBS; cat. no. 14190, Life Technologies, Paisley, Scottland).
3. Trypan blue (cat. no. T0776, Sigma-Aldrich, St. Louis, CA, USA): A final concentration of 0.2% trypan blue is added to the cell suspension 5–10 min before counting with a hemacytometer. A stock solution of 2% in PBS is suggested.

2.2. Culture Expansion of BM MSCs

1. Complete a-MEM [a-MEM with Glutamax (cat. no. 32571, Life Technologies), 10% FCS (lot selected for the optimal MSC-expansion); different sources and batches should be tested *see* **Subheading 2.3.**) and 50 µg/ml gentamicin (cat. no. 15710, Life Technologies].
2. DPBS.
3. Tyrode's Salt solution (cat. no. T2397, Sigma).
4. Medium M199 (cat. no. 31150, Life Technologies).
5. Human serum albumin.
6. Saline.

For MSC in vitro expansion, regular culture flasks may be used. To this end, Falcon flasks Becton-Dickinson (Mountain View, CA, USA) have shown to induce higher proliferative rates compared with flasks from Nunc (Nunc A/S, Kamstrup, Denmark), Greiner (Greiner Bio-One, Frickenhausen, Germany), and Corning (Corning, NY, USA). However, given the extensive dimensions of surface required to achieve the optimal cellular density for MSCs, large-scale culture systems, such as Cell Factories (Nunc) and cellSTUCK (Corning) appear to be more suitable.

2.3. Selection of the Optimal FCS Lot for MSC-Growth a-MEM

1. Gentamicin.
2. FCS of different lots (heat inactivated for 30 min at 56 °C).
3. Hank's Balanced Salt Solution (HBSS; cat. no. 14170, Life Technologies).
4. 0.05% Trypsin/EDTA (cat. no. 15200, Life Technologies).
5. Control medium [a-MEM supplemented with 10% FCS preselected for optimal MSC growth or an MSC-specific medium, such as MSCGM (cat. no. PT-3001, Cambrex BioScience, Nottingham, UK), Mesencult (cat. no. 05401 and 05402, Stem Cell Technologies, Vancouver, Canada), or NH MSCM (cat. no. 130-091-680, Miltenyi Biotec, Gladbach, Germany)].

2.4. Phenotypical characterization of MSCs

1. Wash buffer 1: PBS supplemented with 1% BSA.
2. Wash buffer 2: 0.01% Tween-20 (cat. no. P7949, Sigma) in PBS.
3. 1% Paraformaldehyde-A (PFA; cat. no. P6148, Sigma) in PBS.
4. Monoclonal antibodies (Abs) against human CD14, CD29, CD34, CD44, CD45, CD90, CD73, CD105, CD106, HLA-ABC, and HLA-DR, and fluorochrome-labeled anti-mouse immunoglobulins, as well as the appropriate isotype controls. For the Ab combinations described herein, fluorescein isothiocyanate (FITC)-conjugated Abs against CD105 (cat. no. MCA1557F) can be purchased from Serotec (Oxford, UK). PerCP- or PE-Cy5-conjugated Abs against CD45 (cat. no. IM2653) and HLA-DR (cat. no. IMMU357) and FITC-conjugated Abs anti- HLA-ABC (cat. no. 1838) can be obtained from Immunotech (Beckman Coulter, Paris, France). PE-conjugated Abs anti- CD34 (cat. no. 348057), CD44 (cat. no. 553135) and CD14 (cat. no. 30545X), FITC-conjugated Ab against CD106 (cat. no. 551146) and APC-conjugated Abs against CD29 (cat. no. 30869X) and CD90 (cat. no. 559869), as well as all isotype controls can be obtained from BD Pharmingen (Mountain View, CA, USA). For the determination of CD73 expression, culture supernatant derived from the hybridoma cell line SH2 (cat. no. ATCC-HB-10743) or SH3 (cat. no. ATCC-HB-10744) can be obtained from the Developmental Studies Hybridoma Bank (Iowa, IA, USA). Goat secondary Abs against mouse immunoglobulins conjugated with PE (cat. no. R0480) and aqueous mounting medium (cat. no. S3025) for immunofluorescence can be obtained from DACO A/S (Glostrup, Denmark).

2.5. Colony-Forming Assay

1. Complete a-MEM.
2. PBS.
3. Trypsin.
4. 0.5% crystal violet (cat. no. 1.15940, Merck & Co, Whitehouse, NJ, USA) in methanol (cat. no. 632546, Sigma).

2.6. Differentiation Assays

2.6.1. Osteogenic Differentiation

1. Osteogenic differentiation medium: DMEM with 1000 g/l Glucose and Glutamax (cat. no. 21885, Life Technologies), 10% FCS, 10^{-7} M dexamethasone (Decadron; Merck), 10 mM b-glycerol phosphate (cat. no.50020, Fluka, Buchs, Switzerland), 50 μM L-ascorbic acid-2 phosphate (cat. no. A8960, Sigma), and 50 μg/ml gentamicin.
2. Formalin buffer: 10% formalin (cat.no. F1635, Sigma) in PBS.
3. Sodium carbonate formaldehyde: 5% sodium carbonate (cat. no. S7795) in 25% formalin in distilled water.
4. 2.5% silver nitrate (cat.no. S8157, Sigma) in distilled water.
5. Alkaline phosphatase substrate: 0.002 g Naphthol AS MX-PO$_4$ (cat. no. N5000), 0.015 g red violet LB salt (cat. no. F1625), 100 μl N,N-dimethylformamide (DMF, cat. no. D4551) and 25 ml Tris–HCl 0.1 M, pH 8.3 (all from Sigma).

2.6.2. Adipogenic Differentiation

1. Adipogenic differentiation medium: DMEM with 1000 g/l Glucose and Glutamax, 2% FCS, 0.5 μM dexamethasone, 0.5 mM isobutylmethylxanthine (IBMX; cat. no. I5879, Sigma), 60 μM indomethacin (cat. no. I7378, Sigma) and 50 μg/ml gentamicin.
2. PBS.
3. 10 % formalin in PBS.
4. Oil Red O stain [60% stock solution (0.5% in isopropanol; cat.no. O0625, Sigma) in distilled water, freshly prepared and filtered].

2.6.3. Chondrogenic Differentiation

1. Chondrogenic differentiation medium: DMEM with 4500 g/l Glucose and Glutamax (cat. no. 31966, Life Technologies), 500 ng/ml bone morphogenetic protein-6 (cat. no. 507-BP, R&D Systems Abington, UK), 10 ng/ml transforming growth factor-β3 (cat. no. 243-B3, R&D Systems), 10^{-7} M dexamethasone, 50 μg/ml L-ascorbic acid-2-phosphate (cat. no. A8960, Sigma), 40 μg/ml proline (cat. no. P5607, Sigma), 100 μg/ml pyruvate (cat. no. 11360, Life Technologies) and 50 mg/ml ITS+ Premix [cat. no. 354350, Beckton Dickinson; this can be replaced with 6.25 μg/ml insulin (Humulin Regular, Lilly, France), 6.25 μg/ml transferrin (cat. no. T3309, Sigma), 6.25 ng/ml selenous acid (cat. no. 211176, Sigma), 1.25 mg/ml BSA, and 5.35 mg/ml linoleic acid (cat. no. L5900, Sigma)].
2. 4 % PFA in PBS.
3. Alcian blue stain [75% alcian blue stain (cat. no. A3157, Sigma) in 25% glacial acetic acid in ethanol].
4. Mouse anti-human collagen type II Ab (cat. no. MAB8887, Chemicon, Temecula, CA, USA).

5. PE-conjugated goat-anti-mouse immunoglobulins Ab.
6. Methanol.
7. 25 mg/ml hyaluronidase (cat. no. H3506, Sigma) in PBS.
8. BSA.

2.6.4. Cardiomyogenic Differentiation

1. Cardiomyogenic differentiation medium 1: DMEM with 1000 g/l Glucose and Glutamax, 2% FCS, 3 μM 5-azacytidine (cat. no. A1287, Sigma), and 50 μg/ml gentamicin.
2. Cardiomyogenic differentiation medium 2: DMEM with 1000 g/l Glucose and Glutamax, 2% FCS, and 50 μg/ml gentamicin.
3. Abs against human ANP (cat. no. A5050, Sigma), desmin (cat. no. MAB1698, Chemicon), sarcomeric α-actinin (cat. no. A2172, Sigma), troponin (cat. no. MAB3438, Chemicon), and β-myosin heavy chain (cat. no. MAB1548, Chemicon), PE-conjugated Ab anti-mouse immunoglobulins.
4. 4% PFA in PBS.
5. Triton X-100 (cat.no. 93443, Sigma).
6. Wash buffer: 0.01% Tween-20 in PBS.
7. BSA.
8. 4-well or 8-well chamber slides (cat.no. 177437 and 177445, respectively, Nunc).

2.6.5. Neurogenic Differentiation

1. Complete a-MEM.
2. Neurogenic differentiation medium: DMEM with 1000 g/l Glucose and Glutamax, 2% FCS, 10^{-7} M dexamethasone, 0.5 μM linoleic acid, 10 ng/ml platelet-derived growth factor (cat. no. 120-HD, R&D Systems), 10 ng/ml epidermal growth factor (cat. no. 236-EG, R&D Systems), and 50 μg/ml gentamicin.
3. Abs against human neurofilament M (cat. no. MAB1592), synaptophysin (cat. no. MAB8368), tubulin (cat. no. MAB1637), and galactoserebroside (cat. no. AB142, all from Chemicon), PE-conjugated goat Ab anti-mouse immunoglobulins.
4. Methanol.
5. Wash buffer: 0.01% Tween-20 in PBS.
6. BSA.
7. 4-well or 8-well chamber slides (Nunc).

2.7. Immunological Assays

1. Complete a-MEM.
2. HBSS.
3. Phytohemagglutinin (PHA; cat. no. L9017 Sigma).
4. [^{3}H]TdR (cat. no. TRK418, Amersham Pharmacia Biotech, Buckinghamshire, UK).

3. Methods (see Notes 1 and 2)

3.1. Isolation of BM Mononuclear Cells by Ficoll–Hypaque Density Gradient Centrifugation

1. Place BM (heparinized, typically obtained from iliac crest under local anesthesia) into 50-ml conical tubes up to 25 ml/tube.
2. Dilute 1:2 in DPBS.
3. Place 25 ml of Ficoll–Hypaque in 50-ml conical tubes.
4. Slowly overlay the BM/DPBS mixture on the top of Ficoll–Hypaque.
5. Spin at 400 g for 25 min, at 20 °C, without brake.
6. Remove the upper layer and transfer the white/hazy mononuclear cell layer to another 50-ml tube (use one new tube for each three Ficoll–Hypaque tubes).
7. Add to the mononuclear cells 50 ml of DPBS and spin at 600 g for 10 min, at 20 °C with low brake.
8. Remove supernatant, wash twice in DPBS and count cells, determining viability by Trypan Blue dye exclusion (*see* **Note 3**).

3.2. Culture Expansion of BM MSCs

1. Adjust BM MSC suspension to 7000–175,000/ml in complete a-MEM (cellular concentration depends on the total number of cells to be cultured and the facilities available).
2. Distribute 25 ml cell-suspension per 175-cm^2 flask or 90 ml/large-scale culture system tray.
3. Incubate at 37 °C for 48 h, wash with Tyrode's salt solution to remove non-adherent cells, and add the initial quantity of fresh medium.
4. Change medium twice weekly until colonies consisting of approximately 50 cells are formed (cells within colonies should not have very close contact and, consequently, colonies should not be compact). Then perform passage 1.
5. Passage 1: Wash cells once with Tyrode's salt solution, add medium M199 and incubate for 1 h at 37 °C. Discard medium M199 and add trypsin/EDTA (5 ml per 175-cm^2 flask and 20 ml per large scale culture system tray). Incubate for 5 min at 37 °C, until cells are detached. Add complete a-MEM at four times the volume of trypsin/EDTA, spin, and resuspend in complete a-MEM. Count cells, adjust density to 350–1400 cells/m and distribute 25 ml cell-suspension per 175-cm^2 flask or 90 ml/large-scale culture system tray.
6. Repeat **steps 4** and **5** at weekly intervals for subsequent passages, changing medium twice weekly.
7. Before transplantation, wash cells once in Tyrode's salt solution, add medium M199, and incubate for 1 h at 37 °C. Discard medium M199 and add trypsin/EDTA (5 ml per 175-cm^2 flask and 20 ml per large-scale culture system tray). Incubate for 5 min at 37 °C, until cells are detached. Add M199 supplemented with 1% human serum albumin at four times the volume of trypsin/EDTA and count cells.

Wash three times in M199/1% human serum albumin. Resuspend (care should be taken to have a single-cell suspension) in the appropriate volume of saline and transfer to the operation room immediately (*see* **Note 4**).

3.3. Selection of the Optimal FCS lot for MSC-Growth

1. Add 50 μg/ml Gentamicin to the a-MEM medium, mix well and distribute 45 ml/tube in 50-ml tubes.
2. Add 5 ml FCS to each tube (six tubes per FCS lot to be tested are required). Mix well and store at 4 °C.
3. If an MSC-specific medium is used as a control, prepare according to manufacturer's instructions, dispense in 50 ml aliquots and store at the appropriate temperature.
4. Isolate BM MNCs and make a final cell-suspension of 8×10^6/ml in serum-free a-MEM.
5. Add 0.5 ml cell suspension to 25 ml of each medium and place in a 175-cm² flask.
6. After a 48-h incubation, discard non-adherent cells and add 25 ml of fresh medium (of the respective type).
7. Culture for 8 days changing the medium twice weekly.
8. On day 10, wash cells twice in HBSS, add 5 ml of trypsin/EDTA, and incubate for 5 min at 37 °C, until cells are detached.
9. Inactivate trypsin with 20 ml of the respective medium and spin.
10. Count cells, transfer 17,500 cells to a new tube and spin.
11. Resuspend in 25 ml of the respective medium and place in a 175-cm² flask (thus having for passage 1 one flask/type of medium).
12. Repeat **steps 8–11** at weekly intervals for passages 2 and 3.
13. Calculate the total number of passage 3 cells derived from 10^6 BM MNC-seeded and test for phenotype, colony-forming ability and adipogenic, osteogenic, and chondogenic differentiation potential (*see* **Subheadings 3.4.–3.7.**).

3.4. Phenotypical Characterization of MSCs

3.4.1. Phenotypical Characterization of Cells in Suspension Using Direct Immunofluorescent Assay

1. Count MSCs and resuspend in wash buffer 1 at approximately 100,000/ml. Incubate for 1 h at room temperature to block non-specific Ab binding.
2. To FACS tubes, add saturating concentrations of labeled Abs at the appropriate combinations (for combinations, refer to **Note 5**).
3. Spin cell suspension, resuspend in wash buffer 1 at approximately 2×10^6/ml and distribute 50 μl/tube to the FACS tubes prepared in **step 2**.
4. Mix gently and incubate for 30 min in an ice bath (*see* **Note 6**).
5. Wash cells twice with 2 ml cold wash buffer 1.
6. Add 0.4 ml/tube 1% PFA in PBS and keep on ice, protected from light, until analysis.

3.4.2. Phenotypical Characterization of Cells in Suspension Using Indirect Immunofluorescent Assay

1. Count MSCs and resuspend in buffer at approximately 100,000/ml. Incubate for 1 h at room temperature to block non-specific Ab binding.
2. To FACS tubes, add saturating concentrations of primary (unlabeled) Abs.
3. Spin cell suspension, resuspend in buffer at approximately 2×10^6/ml and distribute 50μl/tube to the FACS tubes prepared in **step 2**.
4. Mix gently and incubate for 30 min in an ice bath.
5. Wash cells twice with 2 ml cold wash buffer 1.
6. Add saturating concentration of the respective secondary Ab (labeled Ab against the immunoglobulins of the host species of the primary Ab; *see* **Note 6**).
7. Repeat **steps 4** and **5**.
8. Add 0.4 ml/tube 1% PFA in PBS and keep on ice, protected from light, until analysis (*see* **Note 7**).

3.4.3. Immunocytochemistry/Immunohistochemistry Using Fluorescent Abs

1. Prepare tissue sections or cultured cells on slides and fix as described in the respective Subheadings.
2. Block non-specific Ab binding by 1-h incubation with 2% BSA in PBS.
3. Cover slides with primary Abs diluted at the appropriate concentration in PBS/0.5%BSA or add solely PBS/0.5%BSA as a negative control and incubate 30 min at room temperature or overnight at 4 °C.
4. Wash twice with wash buffer 2.
5. Add PE-conjugated anti-mouse immunoglobulins Ab, diluted 1/20 in 0.5% BSA in PBS over all slides and incubate for 30 min at room temperature (*see* **Note 6**).
6. Wash three times with PBS.
7. Mount with an aqueous mountant and observe under fluorescence microscope.

3.5. Colony-Forming Assay

1. Count MSCs, resuspend in complete a-MEM at 50 cells/ml and distribute 2 ml/dish in 35-mm Petri dishes.
2. Culture for 2 weeks changing medium twice weekly.
3. On day 14, remove medium and wash once with PBS.
4. Stain for 5 min at room temperature with 2.5 ml/dish crystal violet.
5. Wash twice with distilled water and leave the dishes to dry.
6. Count colonies with a diameter greater than 2 mm under a stereomicroscope or inverted microscope and estimate the number of colonies formed per 100 cells seeded.

3.6. Differentiation Assays

3.6.1. Osteogenic Differentiation Culture Protocol

1. Count MSCs and resuspend in complete a-MEM at 50,000/ml.
2. Distribute cell suspension at 2 ml/well in 6-well culture plates.
3. The following day decant culture medium and replace with an equal volume of osteogenic differentiation medium (test cultures) or complete a-MEM (control cultures).
4. Culture cells for 3 weeks replacing culture supernatant with the respective medium twice weekly and proceed to the evaluation assay.

3.6.1.1. EVALUATION (ALKALINE PHOSPHATASE/VON KOSSA STAINING)

1. Wash wells once with cold PBS.
2. Fix by adding 2ml/well cold formalin buffer and incubate for 15 min.
3. Wash three times with distilled water, leaving the water for 15 min during the last wash.
4. While waiting, prepare the substrate.
5. Add 2 ml/well substrate and incubate for 45 min at room temperature.
6. Repeat **step 3**.
7. Add 2 ml/well silver nitrate and incubate at room temperature for 30 min.
8. Repeat **step 3**.
9. Add sodium carbonate formaldehyde for 15 s to 2 min. When the color of the mineralized nodules starts to deepen, immediately wash carefully in tap water. Fill with tap water and leave at room temperature for 30 min.
10. Discard water and leave the dish to dry.
11. Estimate the percentage of the mineralized (calcified) area (brown color, stained by Von Kossa) and the area with alkaline phosphatase activity (red color) in the total culture area. Osteogenic differentiated areas are positive for both stains.

3.6.2. Adipogenic Differentiation Culture Protocol

1. Count MSCs and resuspend in complete a-MEM at 50,000/ml.
2. Distribute cell suspension at 2 ml/well in 6-well culture plates.
3. The following day, decant culture medium and replace with equal volume of adipogenic differentiation medium (test wells) or complete a-MEM (control wells).
4. Culture cells for 3 weeks replacing culture supernatant with fresh medium of the respective type, twice weekly, and proceed to the evaluation assay.

3.6.2.1. EVALUATION (OIL RED O LIPID STAINING)

1. Wash wells once with PBS.
2. Fix by adding 2 ml/well 10% formalin and incubate at room temperature for 10 min.
3. Discard formalin, add 2 ml/well fresh Oil Red O stain and incubate at room temperature for 15 min.

4. Wash three times with distilled water.
5. Estimate the percentage of adipocytes in the culture by counting the number of cells having lipid droplets (red stained) in a total of 500 cells.

3.6.3. Chondrogenic Differentiation

1. Count MSCs, place 200,000 in 15-ml polypropylene conical tubes and spin at 450 g for 10 min.
2. Add to the pellet 0.5 ml chondogenic differentiation medium (test cultures) or complete aMEM (control culture) and incubate at 37 °C with 5% CO_2. After 24 h of incubation, test cultures should have formed a spherical aggregate not adherent to the walls of the tube.
3. Incubate pellets for 3 weeks, replacing with fresh culture medium of the respective type, twice weekly and proceed to the evaluation assays.

3.6.3.1. EVALUATION (ALCIAN BLUE STAINING, IMMUNOHISTOCHEMISTRY)

Aggregates should be paraffin-embedded, cut into 5-mm sections, and deparaffinized/rehydrated before Alcian blue staining or immunohistochemistry.

3.6.3.1.1. Alcian Blue Staining

1. Fix sections with PFA for 5 min at room temperature.
2. Wash three times in water.
3. Cover sections with freshly made, filtered alcian blue stain and incubate for 30 min at room temperature.
4. Wash five times in water.
5. Observe chondrogenesis by detecting blue-stained mucopolysaccharide- regions.

3.6.3.1.2. Immunohistochemistry

1. Fix sections with a 10-min incubation in methanol at room temperature.
2. Digest sections with hyaluronidase for 30 min at 37 °C for optimal antigen retrieval.
3. Perform **steps 2–7** of **Subheading 4.2**, using as primary Ab mouse anti-human type II collagen, diluted at 1 µg/ml.

3.6.4. Cardiomyogenic Differentiation Culture Protocol

1. Count MSCs and resuspend in complete a-MEM at 25,000/ml Distribute cell suspension at 5,000 cells/cm² in chamber slides.
2. The following day, decant culture medium and replace with equal volume of cardiomyogenic differentiation medium 1 (test cultures) or complete a-MEM (control cultures).
3. After 24 h of incubation, remove supernatant and replace with equal volume of cardiomyogenic differentiation medium 2 (where cardiomyogenic differentiation medium 1 was used) or fresh complete a-MEM (in the control cultures).

4. Culture cells for 3 weeks replacing supernatant with fresh cardiomyogenic differentiation medium 2 or complete a-MEM, respectively, twice weekly and proceed to the evaluation assays.

3.6.4.1. Evaluation (Immunocytochemistry, see **Note** 8)

1. Remove chambers from the slides and wash slides twice with PBS.
2. Fix by adding PFA for 10 min at room temperature.
3. Permeabilize with 0.1% Triton X-100 for 15 min at room temperature.
4. Wash once with wash buffer 2.
5. Perform **steps 2–7** of **Subheading 4.2**, using as primary Abs mouse anti-human β-myosin heavy chain diluted at 1/10 and anti-human ANP, desmin, sarcomeric α-actinin and troponin diluted at 1/200.

3.6.5. Neurogenic Differentiation Culture Protocol

1. Count MSCs and resuspend in complete a-MEM at 20,000/ml.
2. Distribute cell suspension at 10,000 cells/cm^2 in chamber slides.
3. The following day, decant culture medium and replace with equal volume of neurogenic differentiation medium (test cultures) or complete a-MEM (control cultures).
4. Culture cells for 2 weeks replacing culture medium with fresh medium of the respective type twice weekly and proceed to the evaluation assays.

3.6.5.1. Evaluation (Immunocytochemistry)

1. Remove chambers from the slides and wash slides twice in wash buffer 2.
2. Fix cells by incubating with methanol at −20 °C for 10 min.
3. Wash three times with PBS.
4. Perform **steps 2–7** of **Subheading 4.2**, using as primary Ab mouse anti-human galactoserebriside diluted 1/50 and anti-human neurofilament M, synaptophysin and tubulin, 1/200.

3.7. Immunological Assays

MSCs do not elicit T-cell responses and furthermore, when added as a third party, they suppress T-cell responses against alloantigens or non-specific mitogenic stimuli. Thus, mixed lymphocyte reactions using MSCs as stimulators or as regulatory/immunosuppressive cells (utilizing additionally allogeneic PBMCs or PHA for stimulation) should be performed.

1. Irradiate MSCs at 25 Gy and wash once with HBSS to remove free radicals (*see* **Note 9**).
2. Count MSCs and dispense into three centrifuge tubes 1.4×10^6, 1.4×10^5, and 1.4×10^4 cells, respectively.
3. Spin and resuspend in 1.4 ml complete a-MEM.

Table 1
Template for Mixed Lymphocyte Reaction

	1	2	3	4	5	6	7	8	9	10	11	12
A	100,000 MSCs	100,000 MSCs	100,000 MSCs	100,000 MSCs 100,000 PBMC-1 100,000 PBMC-2	100,000 MSCs 100,000 PBMC-1 100,000 PBMC-2	100,000 MSCs 100,000 PBMC-1 100,000 PBMC-2	100,000 MSCs 100,000 PBMC-1 PHA	100,000 MSCs 100,000 PBMC-1 PHA	100,000 MSCs 100,000 PBMC-1 PHA	100,000 MSCs 100,000 PBMC-1	100,000 MSCs 100,000 PBMC-1	100,000 MSCs 100,000 PBMC-1
B	10,000 MSCs	10,000 MSCs	10,000 MSCs	10,000 MSCs 100,000 PBMC-1 100,000 PBMC-2	10,000 MSCs 100,000 PBMC-1 100,000 PBMC-2	10,000 MSCs 100,000 PBMC-1 100,000 PBMC-2	10,000 MSCs 100,000 PBMC-1 PHA	10,000 MSCs 100,000 PBMC-1 PHA	10,000 MSCs 100,000 PBMC-1 PHA	10,000 MSCs 100,000 PBMC-1	10,000 MSCs 100,000 PBMC-1	10,000 MSCs 100,000 PBMC-1
C	1,000 MSCs	1,000 MSCs	1,000 MSCs	1,000 MSCs 100,000 PBMC-1 100,000 PBMC-2	1,000 MSCs 100,000 PBMC-1 100,000 PBMC-2	1,000 MSCs 100,000 PBMC-1 100,000 PBMC-2	1,000 MSCs 100,000 PBMC-1 PHA	1,000 MSCs 100,000 PBMC-1 PHA	1,000 MSCs 100,000 PBMC-1 PHA	1,000 MSCs 100,000 PBMC-1	1,000 MSCs 100,000 PBMC-1	1,000 MSCs 100,000 PBMC-1

(Continued)

Table 1
(Continued)

	1	2	3	4	5	6	7	8	9	10	11	12
D	100,000 PBMC-1	100,000 PBMC-1	100,000 PBMC-1									
E	100,000 PBMC-2	100,000 PBMC-2	100,000 PBMC-2									
F	100,000 PBMC-1 100,000 PBMC-2	100,000 PBMC-1 100,000 PBMC-2	100,000 PBMC-1									
G	100,000 PBMC-1 PHA	100,000 PBMC-1 PHA	100,000 PBMC-1 PHA									
H	–	–	–	–	–	–	–	–	–	–	–	–

4. Distribute 100 μl/well from each concentration to 12 wells of 96-well flat-bottomed plates, thus having wells with 10,000, 1000, and 100 cells/well.
5. After 5–12 h, once the MSCs have attached to the plastic surface and formed a cellular stroma, isolate PB MNCs from 2 HLA-mismatched donors (*see* **Note 10**).
6. Irradiate the PBMCs of one donor (PBMCs-2; used as stimulators) at 20 Gy and wash once with HBSS to remove free radicals (*see* **Note 9**).
7. Count PBMCs of both donors and resuspend at 10^6/ml in complete a-MEM.
8. Remove supernatant from the wells with MSC stroma.
9. Without allowing cells to dry, add 100 μl/well cell suspension from one or both donors or complete a-MEM, without or with 10 μg/ml PHA, as presented in **Table 1**, thus having triplicates for each condition (*see* **Note 11**).
10. Adjust culture volume to 200 μl/well for all wells.
11. On day 5, add 50 μl/well of 20 μCi/ml [^{3}H]TdR (1 μCi/well final).
12. Incubate for 18 h.
13. Harvest cells on glass-fiber filters (*see* **Note 12**).
14. Measure [^{3}H]TdR uptake in a β-counter.

4. Notes

1. The protocols described in this chapter involve live human cells, animal derivatives, and hazardous agents; thus, appropriate biosafety procedures should be followed. Furthermore, when the following protocols are used for clinical purpose, they should be performed under GMP/GLP conditions, according to the recommendations of the respective government law regarding cellular therapy.
2. All centrifugations are performed at 700 g for 10 min, unless otherwise mentioned.
3. BM MNCs can also be isolated by Percoll density gradient centrifugation.
4. At regular time intervals and before infusion, MSC- cultures should be tested for bacteria/fungi and mycoplasma contamination, as well as for endotoxin levels.
5. Suggested combinations for phenotypical characterization of MSCs are as follows:

   Tube 1: Isotype-FITC/isotype-PE/isotype-PE-Cy5/isotype-APC.
   Tube 2: CD105-FITC/CD14-PE/CD45-PE-Cy5/CD29-APC.
   Tube 3: CD106-FITC/CD34-PE/HLA-DR-PE-Cy5/CD90-APC.
   Tube 4: HLA-ABC-FITC/CD73+anti-mouse immunoglobulins-PE/CD44-PE-Cy5.

6. Incubations with fluorescent Abs should be performed in the dark.
7. When a combination of labeled and unlabeled Abs derived from the same species is used (such as tube 4 mentioned in **Note 5**), first perform indirect staining without the final step of PFA-addition and then carry out direct staining. Thus, the secondary Ab will bind only to the unlabeled Abs and not to the labeled ones as well.
8. Instead of or parallel to immunocytochemistry, mRNA expression for the indicated proteins may be tested. In this case, cultures should be performed in 25-cm^2 flasks to acquire sufficient numbers of cells.

9. If a blood irradiator is not available on the facilities, irradiation can be replaced by mitomcin C treatment [45 min incubation with 100 µg/ml mitomycin C (Kyowa, Tokyo, Japan) in serum-free medium, extensive wash and 1 h rest in complete a-MEM].

10. Instead of allogenic PBMCs (PBMC-2 on the table), as stimulators HLA-mismatched monocytes, dendritic cells or lymphoblastoid cell lines (LCL) can also be used, depending on the availability in each laboratory. In this case, PBMC from only one donor should be isolated (PBMC-1 on the table).

11. For allogeneic stimulation, test cultures are wells with PBMCs-1/PBMCs-2/MSCs and positive controls PBMCs-1/PBMCs-2; for mitogen stimulation, test cultures are wells with PBMCs-1/PHA/MSCs and positive controls PBMCs-1/PHA; when MSC are tested for lymphocyte-stimulation ability, test cultures are PBMCs-1/MSCs.

12. At this point, depending on the type and settings of the cell harvester, water may overflow because of the large size of MSCs. In this case, carry out a freeze-thaw cycle on the plate before harvesting.

Acknowledgements

Authors would like to thank Dr. Nike Cacoullos for critical review of the manuscript.

References

1. Barry, F. P. and Murphy J. M. (2004) Mesenchymal stem cells: clinical applications and biological characterization. Int. J. Biochem. *Cell Biol.* 36, 568–584.

2. Minguell J. J., Erices A. and Cognet P. (2001) Mesenchymal stem cells. *Exp. Biol. Med.* 226, 507–520.

3. Haynesworth S. E., Reuben D. and Caplan A. I. (1998) Cell-based tissue engineering therapies: the influence of whole body physiology. *Adv. Grug Deliv. Rev.* 33, 3–14.

4. Aggarwal S. and Pittenger M. F. (2005) Human mesenchymal stem cells modulate allogeneic immune cell responses. *Blood* 105, 1815–1822.

5. Ryan J. M., Barry F. P., Murphy J. M. and Mahon B. P. Mesenchymal stem cells avoid allogeneic rejection. *J. Inflamm.* 26, 8–18.

6. Bruder S. P., Jaiswal N. and Haynesworth S. E. (1997) Growth kinetics, self-renewal, and the osteogenic potential of purified human mesenchymal stem cells during extensive subcultivation and following cryopreservation. *J. Cell. Biochem.* 64, 278–294.

7. Cognet P. A. and Minguell J. J. (1999) Phenotypical and functional properties of human bone marrow mesenchymal progenitor cells. *J. Cell. Physiol.* 181, 67–73.

8. Sekiya I., Larson B. L., Smith J. R., Poshampally R., Cui J. G. and Prockop D. J. (2002) Expansion of human adult stem cells from bone marrow stroma: conditions that maximize the yields of early progenitors and evaluate their quality. *Stem Cells* 20, 530–541.

9. Sotiropoulou P. A., Perez S. A., Salagianni M. and Papamichail M. (2006) Characterization of the optimal culture conditions for clinical scale production of human mesenchymal stem cells. *Stem Cells* 24, 462–471.

10. Meuleman N., Tondreau T., Delforge A., Dejeneffe M., Massy M., Libertalis M., Bron D. and Lagneaux L. (2006) Human marrow mesenchymal stem cell culture: serum-free medium allows better expansion than classical a-MEM medium. *Eur. J. Haematol.* 76, 309–316.

11. Sotiropoulou P. A., Perez S. A., Salagianni M. and Papamichail M. (2006) Cell culture medium composition and translational adult bone marrow-derived stem cell research. *Stem Cells*, 24, 1409–1410.

12. Lennon D. P., Haynesworth S. E., Bruder S. P., Jaiswal N. and Caplan A. I. (1996) Human and animal mesenchymal progenitor cells from bone marrow: identification of serum for optimal selection and proliferation. *In vitro Cell. Dev. Biol.* 32, 602–611.

13. Colter D. C., Class R., DiGirolamo C. M. and Prockop D. J. (2000) Rapid expansion of recycling stem cells in cultures of plastic-adherent cells from human bone marrow. *Proc. Natl. Acad. Sci. U.S.A.* 97, 3213–3218.

14. Solchaga l. A., Penick K., Porter J. D., Goldberg V. M., Caplan A. I. and Welter J. F. (2005) FGF-2 enhances the mitotic and chondrogenic potentials of human adult bone marrow-derived mesenchymal stem cells. *J. Cell. Physiol.* 203, 398–409.

15. DiGirolamo C. M., Stokes D., Colter D., Phinney D. G., Class R. and Prockop D. J. (1999) Propagation and senescence of human marrow stromal cells in culture: a simple colony-forming assay identifies samples with the greatest potential to propagate and differentiate. *Br. J. Haematol.* 107, 275–281.

16. Jaiswal N., Haynesworth S. E., Caplan A. I. and Bruder S. P. (1997) Osteogenic differentiation of purified, culture-expanded human mesenchymal stem cells in vitro. *J. Cell. Biochem.* 64, 295–312.

17. Colter D. C., Sekiya I. and Prockop D. J. (2001) Identification of a subpopulation of rapidly self-renewing and multipotential adult stem cells in colonies of human marrow stromal cells. *Proc. Natl. Acad. Sci. U.S.A.* 98, 7841–7845.

18. Pochampally R. R., Smith J. R. and Prockop D. J. (2004) Serum deprivation of human marrow stromal cells (hMSC) selects for a subpopulation of early progenitor cells with enhanced expression of OCT-4 and other embryonic genes. *Blood* 103, 1647–1652.

19. Reyes M. and Verfaillie C. M. (1999) Turning marrow into brain: generation of glial and neuronal cells from adult bone marrow mesenchymal stem cells. *Blood* 94 (Suppl 1), 377a.

20. Kadivar M., Khatami S., Mortazavi Y., Shokrgozar M. A., Taghikhani M. and Soleimani M. (2006) In vitro cardiomyogenic potential of human umbilical vein-derived mesenchymal stem cells. *Biochem. Biophys. Res. Commun.* 340, 639–647.

18

Adenoviral Transduction of Mesenchymal Stem Cells

Pablo Bosch and Steven L. Stice

Summary

Intact or genetically manipulated mesnechymal stem cells (MSCs) are being considered an important cell source for developing human cell-based therapeutic approaches. For applications in which transient, high-level expression of the transgene is necessary, adenovirus vectors have become increasingly popular gene-transfer vehicles. However, host range and cell-type tropism restrict the use of specific adenovectors, sometimes necessitating the lengthy development of vectors with appropriate cell specificity. Here, we present a versatile and inexpensive porcine MSC transduction procedure that can also be used on other cell types from various species, including human that are otherwise refractory to adenovirus infection.

Key Words: Mesenchymal stem cells; Viral vectors; Adenovirus; Flow cytometry; Transduction.

1. Introduction

Mesenchymal stem cells (MSCs) are attractive candidates for the development of cell-based therapy for various human diseases. MSCs are relatively easy for isolation from adult mesodermal tissues such as bone marrow (BM), and endowed with extensive ex vivo proliferative potential and pluripotent differentiation capacity both in vivo and in vitro *(1–3)*. For many therapeutic schemes, transient or permanent genetic modification of in vitro expanded MSCs is desirable. Transient expression of endogenous or foreign genes may be valuable for the manipulation of the differentiation fate of stem cells or induce immunologic tolerance. Moreover, MSCs engineered to transiently express selected molecules may serve as vehicles for the delivery of therapeutics into

From: *Methods in Molecular Biology, vol. 407: Stem Cell Assays*
Edited by: M. C. Vemuri © Humana Press, Totowa, NJ

target tissues *(4–6)*. Adenoviruses provide particularly attractive vectors for ex vivo gene transfer when high-level transgene expression for a limited period of time is required. Additional advantages of adenoviral vectors are their ability to infect different cell types, infect dividing and quiescent cells and accommodate large pieces of exogenous DNA *(7)*. However, a limitation of adenoviral gene transfer is the poor infection rate achieved with cells lacking the specific adenoviral cell surface receptor such as hematopoietic stem cells *(8,9)* and human MSCs *(10)*. Consequently, different approaches have been developed to overcome this limitation including the use of fiber-modified adenovectors *(11)* and Ad5 chimeric vectors in which the fiber protein involved in cell internalization is replaced by one from a different adenovirus serotype, thus changing the adenovector's cell tropism *(10,12)*. Another approach relies on complexing adenovirus with polycations or cationic lipids facilitating in vitro transduction of refractory cells *(13,14)*. Here, we describe a novel methodology, using the polyamine-based transfection reagent GeneJammer (Stratagene, La Jolla, CA), to efficiently infect cultured porcine MSCs [~90% of green fluorescence protein (GFP)-expressing cells] with replication-defective adenoviral vectors carrying the *GFP* gene *(15)*. We have also used this methodology to induce high levels of expression of GFP and bone morphogenetic protein (BMP; a protein with potential therapeutic effects) in human MSCs and murine stromal cell line *(16)*.

2. Materials

1. Preanesthesia: Atropine sulfate injection (Phoenix Pharmaceutical, St. Joseph, MO, USA). Anesthesia: ketamine hydrochloride (Ketaset, Fort Dodge Laboratories, Fort Dodge, IA, USA) xylazine hydrochloride (Rompun, Bayer, Shawnee Mission, KS, USA).
2. Bone marrow aspiration needle 11G 4″ (Medical Device Technologies, Gainesville, FL, USA).
3. Betadine surgical scrub (Purdue Pharma L.P., Stamford, CT, USA).
4. Povidone-iodine solution 10% (Betadine Solution, Purdue Pharma L.P.).
5. Disposable syringes (20 mL; BD, Franklin Lakes, NJ, USA).
6. Anticoagulant: Heparin sodium salt (300 USP units/mL of BM; cat. no. H9399, Sigma, St. Louis, MO, USA). Dissolve heparin at 30,000 USP per mL of Dulbecco's phosphate-buffered saline without calcium and magnesium (Ca^{2+}-Mg^{2+} free D-PBS; cat. no. 14190, Gibco, Carlsbad, CA, USA), filter sterilize and load into BM aspiration syringes; that is, to make uncoagulable 20 mL of BM, load 200 μL of heparin solution per 20-mL syringe before BM aspiration.
7. Chromic gut suture No 3 (CP Medical, Portland, OR, USA).

8. Antibiotic: Penicillin G procaine (6,500 units/Kg BW IM; Agri-Cillin, AgriLabs, St. Joseph, MO). Anti-inflammatory: Flunixin meglumine (1.5 mg/Kg BW, IM; Flunixamine, Fort Dodge Laboratories).
9. Histopaque 1077 (cat. no. 10771, Sigma).
10. Hank's Balanced Salt Solution (HBSS) (cat. no. 14175-095, Gibco).
11. Culture medium for pMSCs: Minimum Essential Medium Alpha (MEM alpha) medium (cat. no. 12000-022, Invitrogen Corporation, Carlsbad, CA, USA) supplemented with 10% fetal bovine serum (FBS) (Hyclone, Logan, UT, USA) and 1% antibiotic-antimycotic (cat. no. A5955, Sigma).
12. Plastic centrifuge tubes (15 mL; Corning, Corning, NY, USA).
13. Disposable pipettes (5, 10, and 25 mL; Corning).
14. Neubauer hematocytometer (Fisher, Pittsburgh, PA, USA).
15. Cell culture flasks (Corning).
16. Disposable syringes (5 mL; BD).
17. Stainless steel 18G 6″ pippeting needle (Popper & Sons, Inc. Lincoln, RI).
18. GeneJammer Transfection Reagent (Stratagene).
19. MEM alpha (cat. no. 12000-022, Invitrogen Corporation).
20. Cell culture 12-well plates (cat. no. 3513, Corning).
21. Trypsin-ethylenediamine tetraacetic acid (EDTA) solution (cat. no. T4049, Sigma).
22. Propidium iodide (PI; cat. no. 1348639, Roche, Indianapolis, IN, USA).
23. Round-bottom polypropylene tubes (12×75 mm) (cat. no. 352003, BD Biosciences, San Jose, CA, USA).
24. Round-bottom polypropylene tubes (12×75 mm) with cell-strainer cap (cat. no. 352235, BD Biosciences).

3. Methods

3.1. Porcine Bone Marrow Aspiration

1. Animals are continuously monitored throughout the process by a trained veterinary staff to ensure proper animal care and welfare.
2. Pigs (50–150 Kg BW) undergoing BM aspiration are not given water and food 6 and 12 h respectively before the surgical procedure.
3. Atropine sulfate (0.04 mg/kg BW) is administered intramuscularly 10–15 min before anesthesia is induced. The pig is anesthetized with a mixture of ketamine hydrochloride (5.5 mg/Kg BW) and xylazine hydrochloride (2.75 mg/Kg BW). Half of the total dose is administered intravenously (e.g., through ear vein) and the remaining intramuscularly. Combination of these anesthetics at the indicated dose usually provides 15–20-min surgical time enough to complete BM aspiration procedure (*see* **Note 1**).
4. Once the pig is in surgical plane of anesthesia, it is placed on dorsal recumbence on a v-shape stretcher, and the shoulder area is prepared for the procedure. The area where the epiphysis of the humerus is located is clipped, cleaned with Betadine

surgical scrub and disinfected with Betadine Solution or any other povidone-iodine solution.

5. The greater tubercle of the humerus is identified by palpation and a small skin incision ($\sim 5\,mm$) is made with a scalpel blade over the tubercle.

6. The BM aspiration needle is pushed through the muscle layer. Once the needle is in contact with the surface of the bone, it is introduced 1–1.5 cm into the spongy bone of the epiphysis of the humerus with a clockwise/counter-clockwise motion and forward pressure.

7. Once the BM aspiration needle is in place, the stylet is removed from the shaft of the BM aspiration needle and a 20-mL syringe containing heparin is coupled.

8. Then, 15–20 mL of BM is aspirated into the syringe previously loaded with heparin. The syringe is uncoupled from the aspiration needle and left in a secure place.

9. The stylet is reintroduced into the shaft of the aspiration needle. The BM aspiration needle is gently removed and sterile gauze is immediately applied over the incision to reduce bleeding. A horizontal mattress stitch of gut suture is used to close the skin incision.

10. The animal receives antibiotic and anti-inflammatory intramuscularly before it is moved to a clean, dry pen with access to water; feeding is withheld for 6–12 h after the surgical procedure. The pig is monitored for recovery.

3.2. Isolation and Culture of Porcine Bone Marrow MSCs

1. Once the heparinated BM sample is in the laboratory, it is diluted with the same volume of HBSS and thoroughly mixed. Diluted BM is transferred to 15-mL centrifuge tubes (8 mL/tube).

2. Using the stainless steel pipetting needle coupled to a 5-mL plastic syringe, transfer 3 mL of Histopaque (*see* **Note 2**) under the diluted BM. To achieve this, place the tip of the blunt-ended needle on the bottom of the tube and release the Histopaque under the diluted BM by gently pushing the syringe's plunger; Histopaque will displace the diluted BM column upward.

3. The tubes are centrifuged at $400\,g$ for 30 min at room temperature (RT; $\sim 23°C$).

4. The distinct layer at the plasma–Histopaque interface formed during centrifugation (containing mononuclear cells) is transferred to 15-mL centrifuge tubes with a Pasteur pipette coupled to a hand pipettor.

5. Mononuclear cells recovered from the interface are washed $3\times$ with HBSS by centrifugation ($400\,g$, 10 min) to remove any remaining Histopaque.

6. After the last centrifugation, the supernatant is discarded and the cell pellet resuspended in 5 mL of pMSC culture medium. Cell concentration is determined with hematocytometer (use 1:20 dilution, e.g., $5\,\mu L$ in $95\,\mu L$ of D-PBS).

7. Cells are plated at 600,000–800,000 mononuclear cells/cm^2 in 75-cm^2 culture flasks in pMSC culture medium, with first media change at 24 h (*see* **Note 3**); medium is replaced every 3–4 days afterward. Cultures are passaged by trypsinization (*see* **Note 4**) when they reach 80–90% confluence (~ 14 days after initial plating).

Subculture is performed at $5,000\,cells/cm^2$. Porcine MSCs grow well in an atmosphere of 5% CO_2 in air and maximum humidity; however, we have demonstrated that pMSCs proliferate faster in a low oxygen atmosphere (5% O_2, 5% CO_2, and 90% N_2) *(1)*.

3.3. Adenoviral Transduction of Porcine MSCs

1. Porcine MSCs are trypsinized and plated at a density of $43,000\,cells/cm^2$ per well of 12-well plates (150,500 cells/well). Viral infection starts 24 h after cell plating when the cultures have reached 80–90% confluence.
2. The method described here has been optimized for transduction with replication defective E1-E3 deleted first generation human type five adenovirus (Ad5) or a modified form in which the normal fiber protein is substituted for the human adenovirus type 35 fiber (Ad5F35). This adenovectors are constructed to contain the *GFP* gene in the E1 region of the virus *(17)* under the control of the immediate-early promoter of cytomegalovirus (CMV) (*see* **Note 5**). The recommended virus particles to plaque-forming units should be 200 or less. Aliquots of viral stock are kept at $-80\,°C$ until use.
3. Eight microliters of GeneJammer are added to $200\,µL$ of serum-free antibiotic-free MEM alpha and incubated 10 min at RT. (This is the amount for 1 well; scale up according to the number of wells to be transduced.)
4. Then, stock adenoviral suspension is added to the diluted GeneJammer transfection reagent and incubated at RT for 10 min. According to our experience, adenoviral multiplicity of infection (MOI, defined as pfu/cell) required to achieve high infection efficiency of pMSCs without compromising cell viability ranges between 10 and 14 MOI *(15)*.
5. The culture medium in each well is replaced by $300\,µL$ of fresh MEM alpha with 10% FBS, and $200\,µL$ of the GeneJammer-virus mixture prepared in **steps 2** and **3** is added drop-wise to achieve a final volume of $500\,µL$ per well. As a result of these adjustments, the final concentration of GeneJammer is $1.6\,µL/100\,µL$ of medium (1.6%; *see* **Fig. 1**).
6. Plates are incubated 6 h at $37\,°C$ in 5% CO_2 in air and $500\,µL$ of MEM alpha with 10% FBS are added in each well (final volume: 1 mL/well). Cultures are incubated at $37\,°C$ in 5% CO_2 in air until flow cytometric analysis 24–48 h after viral infection (*see* **Note 6**).

3.4. Flow Cytometric Analysis of GFP Expression

1. Medium is removed from each well and washed off with Ca^{2+}-Mg^{2+} free D-PBS. Immediately, $150\,µL$ of trypsin-EDTA solution are added to each well, and the plate is incubated for 4–5 min at $37\,°C$. To stop trypsin, 2 mL of MEM alpha with 20% FBS is transferred to each well, and the cells are dispersed by pipetting. The cell suspension from each well is transferred to individual round-bottom polystyrene

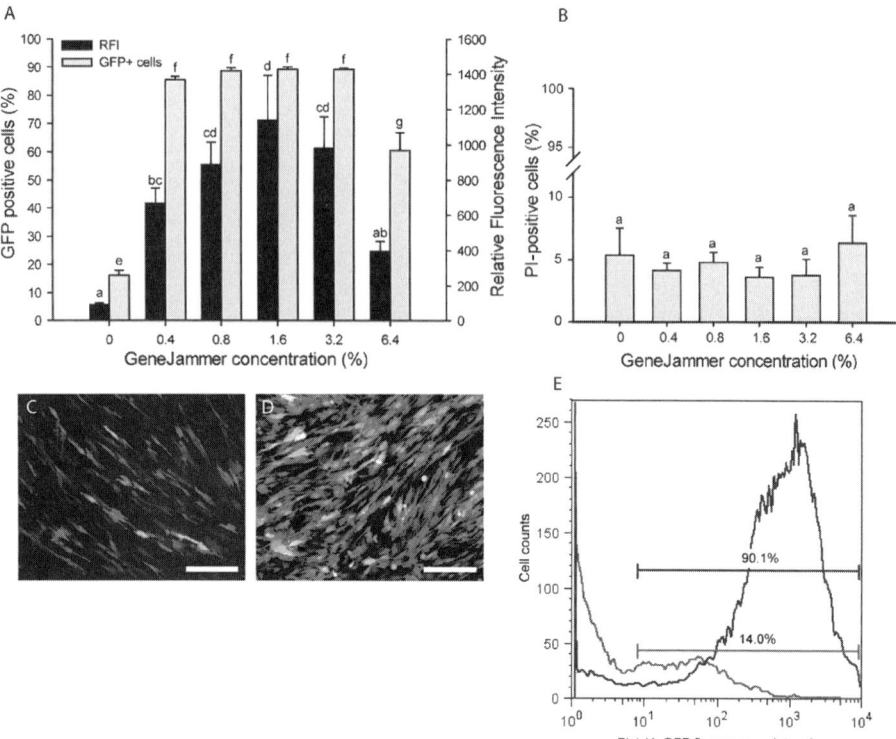

Fig. 1. Effect of GeneJammer on transduction efficiency of pMSCs with Ad5F35eGFP (13.8 MOI). pMSC cultures were transduced without (control) or with increasing concentrations of GeneJammer. Twenty-four hours after the induced infection, the percentage of pMSCs expressing GFP and RFI were analyzed through flow cytometry (**A**). GeneJammer treatment did not affect the percentage of non-viable cells (**B**). Photomicrographs of pMSC cultures 24 h after transduction in the absence (**C**) or presence (**D**) of GeneJammer (1.6%). Representative distribution of GFP-negative/positive cells from non-treated (red curve; 14.0% of positive cells) and GeneJammer treated cells (1.6 μL/100 μL; blue curve, 90.1% of positive cells) (**E**). Bars with different letters denote significant differences ($P < 0.05$; ANOVA) within each variable studied. Results are expressed as mean ± SEM from three independent replicates. Bar = 200 μm. (Reproduced from **ref. *15*** with permission of Wiley InterScience).

tubes and centrifuged at 400 g for 10 min. After discarding the supernatant, cells are resuspended in 0.5 mL of D-PBS with 2% FBS and kept on ice until flow cytometric analysis.

2. Immediately before running the sample through the cytometer, 50 μL of PI is added to each tube containing the cell suspension. The sample is transferred to a

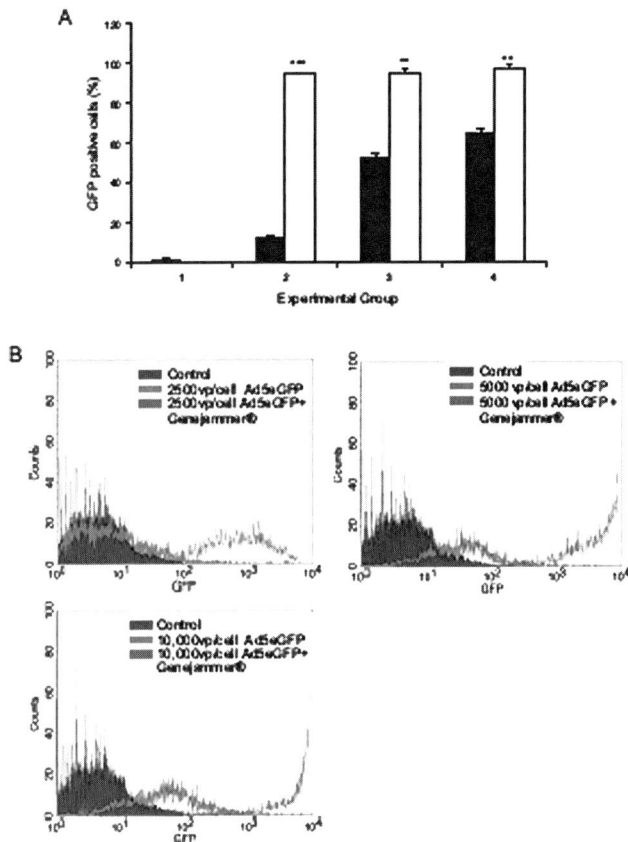

Fig. 2. **(A)** Flow cytometric quantification of GFP expression after transduction of hBM-MSCs with (i) Ad5-empty, (ii) Ad5eGFP 2,500 VP/cell, (iii) Ad5eGFP 5,000 VP/cell, or (iv) Ad5eGFP 10,000 VP/cell in the absence (solid column) or presence (open column) of GeneJammer. The percentage of GFP-positive cells is depicted as the average eGFP fluorescence, where $n = 3$. Columns and errors bars represent means ± standard deviation, respectively, for $n = 3$ experiments. ***$P < 0.001$ and **$P < 0.01$ (Student t test). **(B)** GFP fluorescence intensity shifts in the flow cytometry profiles of eGFP expression in the hBM-MSCs transduced with Ad5GFP at 2500, 5000, or 10,000 VP/cell, in the presence or absence of GeneJammer [shown in **(A)**]. In all samples, 100% of the cells were found to express eGFP. (Reproduced from **ref. *16*** with permission of Mary Ann Liebert).

round-bottom polypropylene tube (12×75 mm) with cell-strainer cap to remove any clump of cells normally present.

3. The samples are run through a flow cytometer (we routinely use a FACSCalibur cytometer, Becton Dickinson, San Jose, CA, USA) equipped with an argon-ion laser with an excitation wavelength of 488 nm. GFP fluorescence emission is collected by the photomultiplier tube FL1 after passing a 530/30 BP filter, whereas PI emission is collected by the photomultiplier tube FL3 after passing a 530/30 BP filter. At least 20,000 events are acquired for further data analysis, which can be performed with appropriate flow cytometry software such as FlowJo (Tree Star, Ashland, OR). The analysis is carried out to determine percentage of GFP-positive cells, relative fluorescence intensity (RFI) of the GFP-positive cell population and cell viability by exclusion of PI on a gated cell population (the gate is established to exclude debris and cell clumps). (Examples of results in pMSCs and human MSCs are shown in **Figs. 1** and **2**, respectively).

4. Notes

1. Other drug combinations, such as xylazine plus telazol, are also appropriate to perform short surgical procedures in pigs like the one described here. In addition, approval by local animal use committees is required and may require the use of alternative materials and methods, including additional anesthesia and analgesics.

2. Histopaque is stored at 4 °C. Histopaque must be brought to RT before use.

3. Porcine MSCs are enriched based on their ability to adhere and growth attached to plastic. Non-adherent cells, presumably hematopoietic cells, are removed out of the culture during first media exchange 24 h after culture initiation. Using this methodology, we have been able to consistently establish porcine cell lines that display morphology, surface antigen profile and pluripotency characteristics of MSCs *(1)*.

4. To trypsinize monolayers of pMSCs for passaging or other uses, the culture medium is removed and pMSC monolayer is washed once with Ca^{2+}-Mg^{2+} free D-PBS. Then, trypsin-EDTA solution is added (just enough to cover the cell monolayer). After incubation for 3–4 min at 37 °C, MEM alpha medium supplemented with 10% FBS is added to the cells to inhibit trypsin catalytic effect. Cell suspension is transferred to a 15-mL conical tube and washed $2\times$ with MEM alpha containing 10% FBS by centrifugation at 400 g, 10 min.

5. Laboratory work that involves the use of adenovectors may pose some potential risks for the personnel if strict biosafety guidelines are not followed. Recombinant adenoviral vectors have been classified as Class I (minimum risk) or Class II (potentially higher risk). Vectors expressing transgenes such as markers (e.g., LacZ, neomycin phosphotransferase, or chloramphenicol acetyl transferase) are grouped in Class I. On the contrary, vectors expressing toxic products or transgenes involved in the regulation of cell function should be grouped in Class II. All procedures involving Class I adenovectors are performed under Biosafety Level 2, whereas for those involving Class II vectors, Biosafety Level 2 with the addition of Biosafety

Level 3 practices apply. Adhere to specific biosafety guidelines for handling of adenovectors according to institutional policies before starting experimentation.

6. If flow cytometry analysis is performed at 48 h, medium containing adenovirus in each well is changed by fresh pMSC culture medium at 24 h. At 48 h after viral infection proceed with **step 3.4.1**.

Acknowledgments

We thank Julie Nelson from the Center for Tropical and Emerging Global Diseases, Flow Cytometry Facility at The University of Georgia for her assistance with flow cytometry analysis and funding from ViaGen.

References

1. Bosch, P., Pratt, S. L., and Stice, S. L. (2006) Isolation, characterization, gene modification and nuclear reprogramming of porcine mesenchymal stem cells. *Biol Reprod* **74**, 46–57.
2. Pittenger, M. F., Mackay, A. M., Beck, S. C., Jaiswal, R. K., Douglas, R., Mosca, J. D., Moorman, M. A., Simonetti, D. W., Craig, S., and Marshak, D. R. (1999) Multilineage potential of adult human mesenchymal stem cells. *Science* **284**, 143–7.
3. Woodbury, D., Schwarz, E. J., Prockop, D. J., and Black, I. B. (2000) Adult rat and human bone marrow stromal cells differentiate into neurons. *J Neurosci Res* **61**, 364–70.
4. Nakamizo, A., Marini, F., Amano, T., Khan, A., Studeny, M., Gumin, J., Chen, J., Hentschel, S., Vecil, G., Dembinski, J., Andreeff, M., and Lang, F. F. (2005) Human bone marrow-derived mesenchymal stem cells in the treatment of gliomas. *Cancer Res* **65**, 3307–18.
5. Studeny, M., Marini, F. C., Champlin, R. E., Zompetta, C., Fidler, I. J., and Andreeff, M. (2002) Bone marrow-derived mesenchymal stem cells as vehicles for interferon-beta delivery into tumors. *Cancer Res* **62**, 3603–8.
6. Studeny, M., Marini, F. C., Dembinski, J. L., Zompetta, C., Cabreira-Hansen, M., Bekele, B. N., Champlin, R. E., and Andreeff, M. (2004) Mesenchymal stem cells: potential precursors for tumor stroma and targeted-delivery vehicles for anticancer agents. *J Natl Cancer Inst* **96**, 1593–603.
7. Amalfitano, A. (2004) Utilization of adenovirus vectors for multiple gene transfer applications. *Methods* **33**, 173–8.
8. Thoma, S. J., Lamping, C. P., and Ziegler, B. L. (1994) Phenotype analysis of hematopoietic CD34+ cell populations derived from human umbilical cord blood using flow cytometry and cDNA-polymerase chain reaction. *Blood* **83**, 2103–14.
9. Rebel, V. I., Hartnett, S., Denham, J., Chan, M., Finberg, R., and Sieff, C. A. (2000) Maturation and lineage-specific expression of the coxsackie and adenovirus receptor in hematopoietic cells. *Stem Cells* **18**, 176–82.

10. Olmsted-Davis, E. A., Gugala, Z., Gannon, F. H., Yotnda, P., McAlhany, R. E., Lindsey, R. W., and Davis, A. R. (2002) Use of a chimeric adenovirus vector enhances BMP2 production and bone formation. *Hum Gene Ther* **13**, 1337–47.

11. Kawabata, K., Sakurai, F., Koizumi, N., Hayakawa, T., and Mizuguchi, H. (2006) Adenovirus vector-mediated gene transfer into stem cells. *Mol Pharm* **3**, 95–103.

12. Yotnda, P., Onishi, H., Heslop, H. E., Shayakhmetov, D., Lieber, A., Brenner, M., and Davis, A. (2001) Efficient infection of primitive hematopoietic stem cells by modified adenovirus. *Gene Ther* **8**, 930–7.

13. Dodds, E., Piper, T. A., Murphy, S. J., and Dickson, G. (1999) Cationic lipids and polymers are able to enhance adenoviral infection of cultured mouse myotubes. *J Neurochem* **72**, 2105–12.

14. Byk, T., Haddada, H., Vainchenker, W., and Louache, F. (1998) Lipofectamine and related cationic lipids strongly improve adenoviral infection efficiency of primitive human hematopoietic cells. *Hum Gene Ther* **9**, 2493–502.

15. Bosch, P., Fouletier-Dilling, C., Olmsted-Davis, E. A., Davis, A. R., and Stice, S. L. (2006) Efficient adenoviral-mediated gene delivery into porcine mesenchymal stem cells. *Mol Reprod Dev* **73**, 1393–403.

16. Fouletier-Dilling, C. M., Bosch, P., Davis, A. R., Shafer, J. A., Stice, S. L., Gugala, Z., Gannon, F. H., and Olmsted-Davis, E. A. (2005) Novel compound enables high-level adenovirus transduction in the absence of an adenovirus-specific receptor. *Hum Gene Ther* **16**, 1287–97.

17. Olmsted, E. A., Blum, J. S., Rill, D., Yotnda, P., Gugala, Z., Lindsey, R. W., and Davis, A. R. (2001) Adenovirus-mediated BMP2 expression in human bone marrow stromal cells. *J Cell Biochem* **82**, 11–21.

19

Directed Differentiation of Human Embryonic Stem Cells to Dendritic Cells

Maxim A. Vodyanik and Igor I. Slukvin

Summary

Embryonic stem cells represent a pluripotent population of cells capable of self-renewal, large-scale expansion, and differentiation in various cell lineages including cells of hematopoietic lineage. In this chapter, we describe a three-step cell culture method for directed differentiation of human embryonic stem cells (hESCs) to dendritic cells (DCs) that includes (1) hESC differentiation into hematopoietic progenitors by coculture with OP9 stromal cells, (2) expansion of myeloid DC precursors in suspension bulk cultures with granulocyte monocyte-colony stimulating factor (GM-CSF), and (3) differentiation of myeloid precursors to DCs in the serum-free medium with GM-CSF and interleukin-4 (IL-4). The method employs cell culture conditions selecting an almost pure population of myeloid DC precursors and does not require isolation of hematopoietic progenitors. With this method, hESCs can be differentiated to functional DCs within 30 days at an efficiency of at least four DCs per single undifferentiated hESC. Directed differentiation of DCs from hESCs could be useful for studying cellular and molecular mechanisms of DC development and potentially for the generation of antigen-presenting cells for cellular immunotherapy.

Key Words: Human embryonic stem cells; Dendritic cells; OP9; Hematopoiesis; Dendritic cell precursors; Myeloid progenitors.

1. Introduction

Dendritic cells (DCs) are powerful antigen-presenting cells, which play a key role in directing immune responses (*1*). For many years, functional studies of DCs were hindered by difficulties in obtaining a sufficient number of cells as DCs are rare in tissues and blood. Development of protocols for DC generation

From: *Methods in Molecular Biology, vol. 407: Stem Cell Assays*
Edited by: M. C. Vemuri © Humana Press, Totowa, NJ

from CD34$^+$ hematopoietic progenitors and monocytes significantly facilitated studies of DCs and made possible the production of DCs for immunotherapeutic purposes *(2–5)*. However, obtaining large numbers of human DC progenitors is still a laborious process and poses potential risks for donors. Human embryonic stem cells (hESCs) represent a unique population of cells capable of self-renewal and large-scale expansion as well as differentiation toward hematopoietic cells *(6–9)*. Therefore, hESCs can be seen as a novel source of DCs.

In this chapter, we describe the protocol for high-scale production of DCs from hESCs. Generation of hESC-derived DCs is performed in a three-step culture. Hematopoietic differentiation of hESCs is first induced through coculture with OP9 stromal cells. Myeloid DC precursors are then expanded under non-adherent conditions in the presence of granulocyte monocyte-colony stimulating factor (GM-CSF). Finally, myeloid DC precursors are induced to differentiate into DCs in serum-free medium containing GM-CSF and interleukin-4 (IL-4). The most critical step for successful DC generation is efficient induction of hematopoietic differentiation in hESC/OP9 coculture. OP9 cells induce multilineage hematopoietic differentiation in mouse and non-human primate ESCs *(10,11)* and have been used successfully to induce mouse ESC differentiation into myeloid, lymphoid, erythroid, megakaryocytic, and DC lineages *(12–15)*. We have shown that OP9 cells induce efficient hematopoietic differentiation of hESCs as well *(7)*. Hematopoietic differentiation of hESCs in coculture with OP9 cells proceeds very rapidly, and the first CD34$^+$ hematopoietic progenitors can be detected after 4–5 days of coculture. Recently, we demonstrated that hESC-derived CD34$^+$ cells are heterogeneous and include CD34$^+$CD43$^+$ hematopoietic progenitors, CD34$^+$CD43$^-$KDR$^+$CD31$^+$ endothelial cells, and CD34$^+$CD43$^-$KDR$^-$CD31$^-$ mesenchymal cells *(16)*. The major subpopulation of hematopoietic CD43$^+$ cells expresses glycophorin A (CD235a) and represents erythroid progenitors (CD43$^+$CD235a$^+$), whereas multilineage progenitors (CD43$^+$CD235a$^-$CD45$^{+/-}$Lin$^-$) comprise the minor subpopulation *(16)*. Because only the later cells give rise to the myeloid progeny, their sufficient number in hESC/OP9 cocultures ($> 2\%$ of total cells) was found to be critical for success of the following DC differentiation steps.

Selective expansion and subsequent isolation of a pure population of myeloid precursors is achieved by cultivation of total cells collected from hESC/OP9 coculture in the serum-containing medium supplemented with GM-CSF in flasks coated with anti-adhesive polymer–poly(2-hydroxyethyl methacrylate) (pHEMA). The differentiation potential changes significantly, and myeloid cell expansion could not be achieved when cells are allowed to adhere in regular flasks. During the "forced" suspension culture with GM-CSF, all adherent cells form large floating aggregates, whereas myeloid cells proliferate in suspension. Thus, a pure population of myeloid cells ($\sim$95%) could be obtained by removal of cell aggregates by filtration. The subsequent culture of myeloid precursors with GM-CSF and IL-4 in the serum-free medium induces their differentiation to functional DCs, which phenotypically resemble interstitial DCs. Using the described protocol, we typically produce at least 4×10^6 DCs from 10^6 initially plated hESC, and therefore, a sufficient number of DCs can easily be generated for functional studies or genetic manipulations. In addition, with this method, myeloid DC precursors are obtained at different stages of maturation accommodating studies of DC development.

2. Materials

2.1. Coating of Plastic Surface with Gelatin

1. 0.1% gelatin type A (cat. no. G1890, Sigma, St. Louis, MO, USA) solution in endotoxin-free reagent grade water prepared by autoclaving of gelatin slurry at 120°C for 45 min. Store at 4°C.
2. 6-well tissue culture plates (cat. no. 353046, BD Labware, Bedford, MA, USA); 100-mm tissue culture dishes (cat. no. 353003, BD Labware).

2.2. Coating of Plastic Surface with pHEMA

1. 10% pHEMA (cat. no. P3932, Sigma) coating solution. Add 4 g pHEMA to 40 ml 95% ethanol containing 10 mM NaOH in 50-ml polypropylene tube. Close tube and place on rotation mixer immediately, to prevent pHEMA crystals from precipitating at the bottom or tube walls. Dissolve by continuous rotation at 37°C overnight. Store at room temperature.
2. T75 tissue culture flasks (cat. no. 353134, BD Labware).
3. 5-ml glass serological pipettes.

2.3. Preparation of Feeder Plates for hESC Subculture

1. Frozen mouse CF-1 strain embryonic fibroblasts (MEF; passage no. 3, 2.5 $\times$ 10^6 cells/vial) prepared according to established protocol from WiCell Research Institute, Madison, WI, USA (protocol I, http://www.wicell.org).
2. MEF growth medium: Dulbecco's modified eagle medium (DMEM) (cat. no. 12100-046, Gibco-Invitrogen, Carlsbad, CA, USA) supplemented with 10% heat-inactivated fetal bovine serum (FBS; cat. 16000-044, Gibco-Invitrogen), and 1$\times$ non-essential amino acid solution (NEAA; cat. no. 11140-050, Gibco-Invitrogen).
3. Dulbecco's phosphate-buffered saline (PBS) without calcium and magnesium (cat. no. 21600-044, Gibco-Invitrogen).

4. 0.05% Trypsin–0.5 mM ethylenediaminetetraacetic acid (EDTA) solution (cat. no. 25300-054, Gibco-Invitrogen). Before use, pre-warm solution in the water bath at 37 °C for 15 min.
5. Gelatin-coated 6-well tissue culture plates.
6. 15-ml polypropelene centrifuge tubes, pipettes.
7. Cell counting supplies: hemacytometer (Bright-Line; cat. no. 1492, Hausser Scientific, Horsham, PA, USA), 0.4% trypan blue solution (cat. no. T8154, Sigma).
8. Gamma irradiator.

2.4. Culture of Undifferentiated hESCs

1. H1 (WA01) or H9 (WA09) hESC lines (WiCell Research Institute).
2. 6-well plates with pre-plated irradiated MEFs.
3. Sterile-filtered DMEM/F12 basal medium prepared from powder (cat. no. 12400-024, Gibco-Invitrogen).
4. Knockout serum replacer (KSR; cat. no. 10828-028, Gibco-Invitrogen). Store frozen at −20 °C in 50 ml aliquots.
5. 100× L-glutamine/2-mercaptoethanol (2-ME) solution. Dissolve 146 mg L-glutamine (cat. no. 21051-024, Gibco-Invitrogen) in 10 ml PBS, add 7 μl 2-ME (cat. no. M7522, Sigma), mix well, and store at 4 °C. Prepare fresh every week.
6. 100× NEAA 10 mM solution (cat. no. 11140-050, Gibco-Invitrogen).
7. Recombinant human basic fibroblast growth factor (bFGF; cat. no. 100-18B, PeproTech, Rocky Hill, NJ, USA). Prepare 500× solution (2 μg/ml) by dissolving 50 μg bFGF in 25 ml sterile-filtered PBS containing 2 mg/ml bovine serum albumin fraction V (BSA; cat. no. 15260-037, Gibco-Invitrogen). Store frozen at −80 °C in 0.5 ml aliquots.
8. hESC growth medium: 200 ml DMEM/F12, 50 ml KSR (20%), 2.5 ml NEAA (1×), 2.5 ml L-glutamine/2-ME (1×), 0.5 ml bFGF (4 ng/ml). Combine components in the upper chamber of 250 ml bottle top filter unit (cat. no. 5680020, Nalgene, Rochester, NY, USA) and sterilize by vacuum filtration. Prepare fresh every 2 weeks. Before use, pre-warm medium in the water bath at 37 °C for 15 min.
9. Sterile-filtered collagenase type IV (cat. no. 17104-019, Gibco-Invitrogen) solution (1 mg/ml in DMEM/F12). Store at 4 °C. Prepare fresh every week. Before use, pre-warm solution in the water bath at 37 °C for 15 min.
10. 15-ml polypropylene centrifuge tubes, 5- and 10-ml glass serological pipettes.

2.5. Culture of OP9 Cells

1. Mouse OP9 bone-marrow stromal cell line (cat. no. CRL-2749, ATCC, Manassas, VA, USA). OP9 cells used in the present method were originally obtained from Dr. Toru Nakano (Osaka University, Japan) (*see* **Note 1**).
2. Sterile-filtered α-MEM basal medium prepared from powder (cat. no. 12000-022, Gibco-Invitogen) (*see* **Note 2**).

3. OP9 growth medium: α-MEM supplemented with 20% non-heat-inactivated defined FBS (cat. no. SH30070.03, HyClone, Logan, UT, USA).
4. 0.05% Trypsin–0.5 mM EDTA solution.
5. PBS.
6. Gelatin-coated 100-mm tissue culture dishes.

2.6. Hematopoietic hESC Differentiation in OP9 Coculture

1. 100-mm dishes with overgrown OP9 cells on day 4–6 of post-confluence culture.
2. 6-well plates with undifferentiated hESCs on day 5–6 of culture.
3. 1000× monothioglycerol (MTG; cat. no. M6145, Sigma) solution (100 mM). Add 87 μl MTG to 10 ml endotoxin-free reagent grade water, mix well, sterilize by 0.2 μM filtration, and store frozen in 0.5 ml aliquots.
4. hESC differentiation medium: α-MEM supplemented with 10% non-heat-inactivated defined FBS (HyClone) and 1× MTG (100 μM) (*see* **Notes 2** and **3**).
5. Sterile-filtered collagenase type IV solution in DMEM/F12 (1 mg/ml).
6. 0.05% Trypsin–0.5 mM EDTA solution.
7. Cell counting supplies: hemacytometer and 0.4% trypan blue solution.

2.7. Expansion Culture of hESC-Derived Myeloid Progenitors

1. 100-mm dishes with hESC/OP9 cocultures on day 9–10 of differentiation.
2. Recombinant human GM-CSF (Leukine® liquid 500 μg/ml, Berlex Laboratories, Richmond, CA, USA). Prepare 1000× solution (100 μg/ml) by 1/5 dilution in sterile-filtered PBS containing 2 mg/ml BSA. Store frozen at −80°C in 100 μl aliquots.
3. Expansion medium: α-MEM supplemented with 10% non-heat-inactivated defined FBS (HyClone), 1× MTG (100 μM), and 100 ng/ml GM-CSF. Add GM-CSF before use.
4. Sterile-filtered solution of collagenase type IV in DMEM/F12 (1 mg/ml). Before use, prepare fresh and pre-warm in the water bath at 37°C for 15 min.
5. 0.05% Trypsin–0.5 mM EDTA solution.
6. Sterile 70 μM nylon filters (Cell strainer; cat. no. 352350, BD Labware).
7. Sterile-filtered cell wash buffer: PBS containing 5% FBS.
8. pHEMA-coated T75 tissue culture flasks.
9. Cell counting supplies: hemacytometer and 0.4% trypan blue solution.

2.8. Differentiation of Myeloid Precursors to DC

1. T75 flasks with myeloid precursors on day 9–12 of GM-CSF expansion culture.
2. Recombinant human IL-4 (cat. no. 200-04, PeproTech). Prepare 1000× solution (100 μg/ml) by dissolving 0.5 mg IL-4 in 5 ml sterile-filtered PBS containing 2 mg/ml BSA. Store frozen at −80°C in 100 μl aliquots.
3. DC differentiation medium: StemSpan serum-free expansion medium (SFEM; cat. no. 09650, Stem Cell Technologies, Vancouver, Canada) supplemented with 1/500

dilution of Ex-Cyte growth enhancement supplement (cat. no. 81-129-1, Serologicals Proteins, Kankakee, IL, USA), 100 ng/ml GM-CSF, and 100 ng/ml IL-4. Prepare fresh before use.

4. 25% Percoll solution. Prepare 100% Percoll by mixing 36 ml Percoll (cat. no. P1644 Sigma) and 4 ml 10× PBS (cat. no. 70013-032, Gibco-Invitrogen) in 50-ml polypropylene tube. Dilute 1/4 in PBS to obtain 25% Percoll solution.
5. Sterile 70 μM nylon filters.
6. Sterile-filtered cell wash buffer: PBS containing 5% FBS.
7. Cell counting supplies: hemacytometer and 0.4% trypan blue solution.

2.9. Fluorescence-Activated Cell Sorting Analysis

1. Fluorescence-activated cell sorting (FACS) buffer: PBS containing 2% FBS, 2 mM EDTA, and 0.05% sodium azide.
2. Normal mouse serum (cat. no. M5904, Sigma). Store frozen in 0.5 ml aliquots.
3. Staining buffer: FACS buffer containing 1% normal mouse serum.
4. Fluorochrome-conjugated monoclonal antibodies (*see* **Table 1**).

3. Method

Differentiation of hESC to DCs includes a three-step protocol of successive differentiation cultures for hESC hematopoietic induction, myeloid cell expansion, and differentiation of myeloid DC precursors to DCs. Monitoring

Table 1
Monoclonal Antibodies for FACS Analysis

Antigen	Format	Clone	Isotype[a]	Dilution[b]	Supplier, cat. no.[c]
CD43	FITC	1G10	IgG1	1:10	BD PharMingen, 555437
CD45	APC	HI30	IgG1	1:10	BD PharMingen, 555485
CD235a	PE	CLB-ery-1	IgG1	1:50	Caltag, MHCGLA04
CD11b	FITC	VIM12	IgG1	1:50	Caltag, CD11b01
CD11c	PE	S-HCL-3	IgG2b	1:20	BDIS, 340713
DC-SIGN	FITC	DCN46	IgG2b	1:10	BD PharMingen, 551264
CD1a	PE	VIT6B	IgG1	1:50	Caltag, MHCD1a04
CD14	FITC	M5E2	IgG2a	1:10	BD PharMingen, 555397
HLA-DR	PE	TÜ36	IgG2b	1:50	Caltag, MHLDR04

FITC, fluorescein isothiocyanate; PE, phycoerythrin; APC, allophycocianin.
[a] All isotype-matched control mAbs were from BD PharMingen.
[b] Optimal dilution for staining up to 5×10^5 cells in 100 μl incubation volume.
[c] BD Immunocytometry Systems (BDIS), San Jose, CA, USA; BD PharMingen, San Diego, CA, USA; Caltag Laboratories-Invitrogen, Carlsbad, CA, USA.

of differentiation is based on the analysis of relevant cell populations in each differentiation step by flow cytometry.

In the *first step*, hESC are induced to differentiate into hematopoietic progenitors by coculture with OP9 cells. The efficiency of differentiation is assessed by enumeration of $CD43^+$ multilineage hematopoietic progenitors ($CD235a^-CD45^{-/+}$) by flow cytometry (*see* **Fig. 1**). Cells from hESC/OP9 cocultures containing > 2% of $CD43^+CD235a^-CD45^{-/+}$ cells give rise to myeloid cell expansion in the *second step* bulk cultures with GM-CSF under non-adherent conditions in pHEMA-coated flasks. During GM-CSF culture, adherent cells form floating cell aggregates, whereas myeloid cells grow as single-cell suspension. Removal of aggregates and dead cells by $70\,\mu M$ filtration followed by Percoll separation results in the isolation of an almost pure population of $CD45^+$ myeloid progenitors (~95%) expressing early myeloid markers (CD15, CD33, CD11b, CD11c), but lacking markers of mature myeloid and DCs (HLA-DR, CD68, CD66b, CD1a) *(9)*. CD45- cells detected in second step cultures are mostly erythroid cells. Commitment to DC lineage in GM-CSF expansion cultures is assessed by the detection of $CD11b^+CD11c^+$ cells (*see* **Fig. 2**). GM-CSF-expanded cells containing > 60% of $CD11b^+CD11c^+$ cells effectively differentiate to DCs when cultured in the serum-free medium with GM-CSF and IL-4 (*third step*). DCs can be identified as large cells with high light-scatter profile, numerous cytoplasmic dendrites, and specific phenotype ($CD1a^+$, DC-SIGN$^+$, HLA-DR$^+$) (*see* **Fig. 3**). DCs generated by this method are fully functional and are capable of antigen processing, triggering naive T cells in mixed lymphocyte reaction, and presenting antigens to specific T-cell clones through the MHC class I pathway *(9)*.

3.1. Gelatinization of Tissue Culture Plastic

Cover plastic surface with gelatin solution (6 ml per 100-mm dish; 2 ml per well of 6-well plate) and incubate at least overnight in CO_2 incubator. Dishes and plates filled with gelatin solution can be stored in incubator for several days. Before use, aspirate gelatin solution and add cell growth medium.

3.2. Preparation of Feeder Plates for hESC Subculture

1. Prepare two gelatin-coated 100-mm dishes and 15-ml tube filled with 10 ml cold (4 °C) MEF growth medium (for 1 vial of thawed MEFs).
2. Thaw a frozen vial of MEFs (2.5×10^6 cells in 1ml) in the water bath at 37 °C. Add cell suspension to a 15-ml tube with 10 ml medium and centrifuge at $300\,g$ for 5 min. Aspirate supernate and resuspend cells in 2 ml growth medium. Add MEFs to 100-mm dishes with 10 ml growth medium (1 ml/dish). Distribute cells evenly by gentle agitation and place dishes in CO_2 incubator.

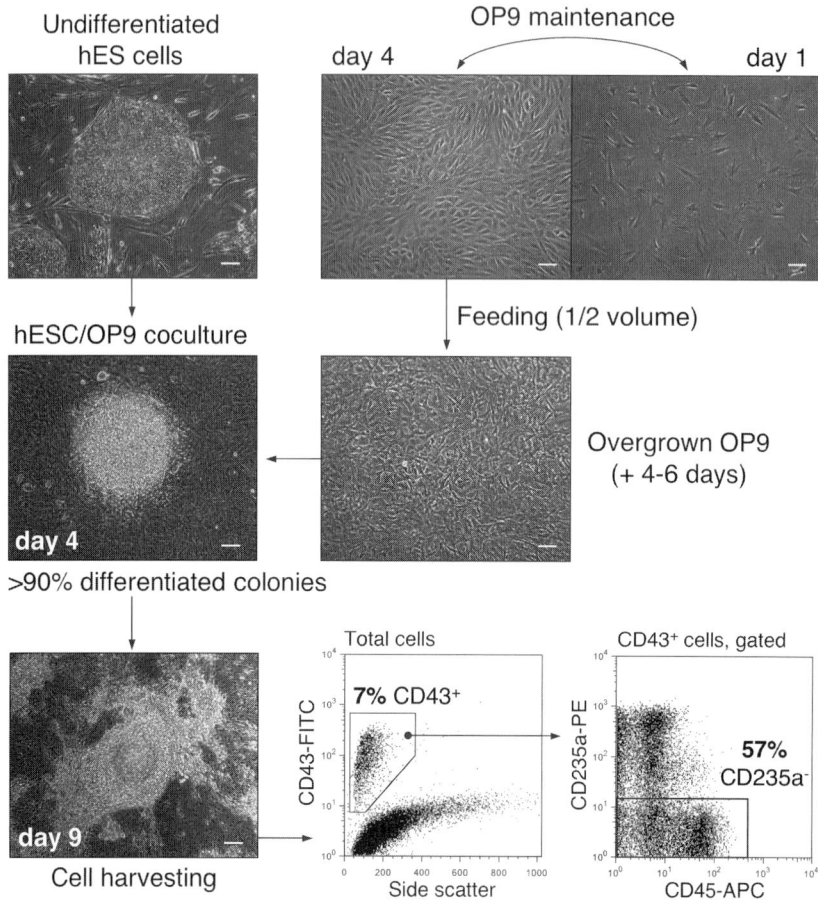

Fig. 1. Hematopoietic differentiation of hESCs in OP9 coculture (**step 1**). Photographs demonstrate major steps associated with preparation of hESC/OP9 cocultures and typical hESC/OP9 cocultures on days 4 and 9 of differentiation. In undifferentiated cultures, hESCs form typical flat colonies with well-defined borders. In the maintenance culture and at overgrowth (before hESC plating), OP9 cells do not demonstrate adipogenic differentiation. On day 4 of hESC/OP9 coculture, greater than 90% of hESC colonies are differentiated. On day 9 of hESC/OP9 coculture, cells are harvested and hematopoietic progenitors with myeloid potential ($CD43^+CD235a^-CD45^{-/+}$) are detected by flow cytometry (for success of subsequent differentiation steps, these progenitors should comprise > 2% of total cells in hESC/OP9 cocultures). For FACS analysis, $CD43^+$ cells are first gated using CD43/side scatter (SSC) dot-plot (left), and minimum 5000 CD43-gated events are acquired. Gated $CD43^+$ cells are then analyzed using CD235a/CD45 dot-plot (right) to determine proportion of $CD235a^-CD45^{+/-}$

3. Grow MEFs to confluence for 3–4 days.
4. Wash MEF monolayer by adding and removing 10 ml PBS. Add 4 ml of 0.05% trypsin–0.5 mM EDTA solution and incubate in CO_2 incubator for 2 min. MEFs detach quickly and completely. Collect cells by pipette and transfer in a 15-ml tube containing 5 ml MEF growth medium. Centrifuge at 300 g for 5 min. Aspirate medium and resuspend cells in 2 ml MEF growth medium.
5. Irradiate tube with 5000 rads of gamma irradiation.
6. Pellet cells by centrifugation (3 min at 300 g). Resuspend cells in 1 ml MEF growth medium and take aliquot (10 μl) for counting.
7. Count cells and prepare MEF suspension at 2×10^5 cells/ml in MEF growth medium. This suspension should be dispensed at 1 ml per well of 6-well plate. Prepare corresponding number of gelatin-coated 6-well plates pre-filled with 2 ml/well of MEF growth medium. Add MEFs to 6-well plates and incubate plates at least 24 h before hESC plating (*see* **Note 4**).

3.3. Culture of Undifferentiated hESCs

1. Have prepared MEF feeder plates for hESC subculture, optimally on 2–3 days after MEF plating. Aspirate MEF growth medium, wash wells with 2 ml PBS, and fill plate with 1.5 ml/well of hESC growth medium. Place MEF plates in CO_2 incubator.
2. hESC cultures where hESC colonies occupy 60–70% of growth surface area are usually split at 1/6 ratio (one well per one 6-well plate), typically on 6–7th day of hESC culture. For hESC passage, aspirate growth medium, wash wells quickly with 2 ml PBS and fill with 2 ml/well collagenase IV solution.
3. Incubate plates for 10–15 min in CO_2 incubator until colonies begin to detach from plate. Using a 5-ml glass pipette, dislodge hESC colonies with repetitive gentle washing of plastic surface with collagenase solution (*see* **Note 5**). Transfer suspension in 15-ml tube and centrifuge at 200 g for 5 min.
4. Aspirate medium and resuspend cells in 4 ml of hESC growth medium. By pipetting cells up and down against the bottom of tube, break up hESC colonies into a fine suspension of small cell aggregates. Centrifuge at 200 g for 5 min. Resuspend cells in 6 ml hESC growth medium and dispense suspension in 6-well MEF plate at 1 ml/well using a vertically positioned glass serological pipette.

Fig. 1. cells in CD43$^+$ population. The formula for calculation of the percent of CD43$^+$CD235a$^-$CD45$^{-/+}$ cells in hESC/OP9 coculture is: (% of CD43$^+$ cells in total cell population) $\times$ (% of CD235a$^-$ cells in CD43$^+$ (gated) population)/100. In the depicted experiment, CD43$^+$CD235a$^-$CD45$^{-/+}$ cells comprise 4% of total cells in H1/OP9 coculture ((7×57)/100). Representative experiment with H1 hESCs is shown; scale bar is 100 μM.

GM-CSF suspension culture

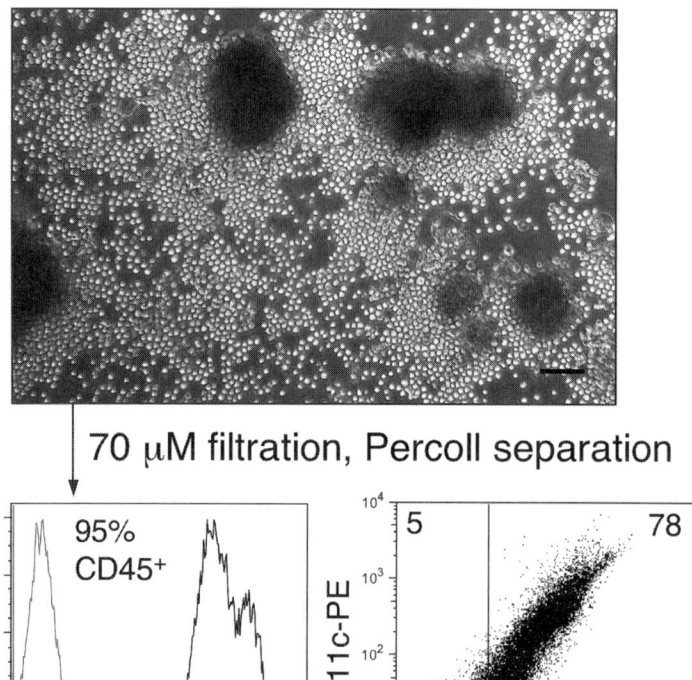

Fig. 2. Myeloid cell expansion in GM-CSF suspension cultures (**step 2**). Photograph shows H1-derived GM-CSF suspension culture on day 6 of expansion. Myeloid cells proliferate in suspension, while non-hematopoietic adherent cells form large floating ball-like aggregates (scale bar is $100\,\mu M$). As determined by flow cytometry, a single-cell suspension isolated from GM-CSF cultures by filtration and Percoll separation contains > 90% of CD45[+] hematopoietic cells (95% in the depicted experiment, left dot-plot). To achieve successful DC generation in subsequent differentiation cultures (**step 3**), myeloid DC precursors (CD11b[+]CD11c[+]) should comprise > 60% of cells isolated from GM-CSF cultures (78% in the depicted experiment, right dot-plot).

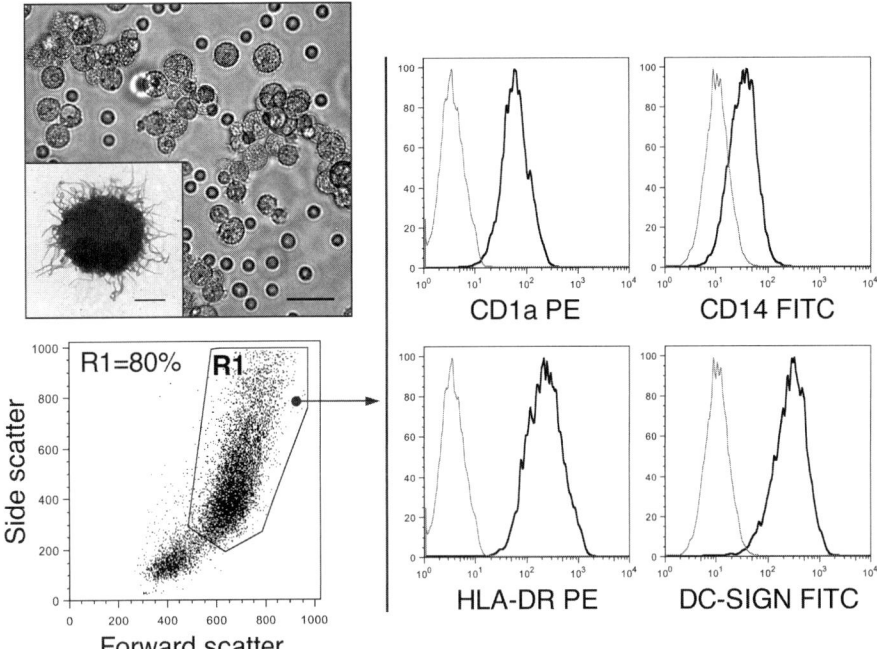

Fig. 3. DC differentiation culture (**step 3**). Photograph demonstrates culture of H1-derived DCs (scale bar is 40 μM). In culture, differentiated DCs can be distinguished from remaining undifferentiated myeloid precursors by their large size and numerous cytoplasmic dendrites (insert shows DC in the Wright-stained smear prepared from DC differentiation culture, scale bar is 10 μM). DCs can be clearly recognized by high light-scatter profile at FACS analysis (R1 gate, dot-plot). More than 95% of R1-gated cells have a DC-specific phenotype defined by CD1a^{high}, DC-SIGN^{high}, HLA-DR^{high}, and CD14^{low} expression (histograms). R1 cells typically comprise > 60% of total cells in DC differentiation cultures (80% in the depicted experiment).

5. Place plate on the shelf in CO_2 incubator and distribute hESC aggregates evenly throughout the growth surface by moving the plate back and forth 2–3 times. Avoid rotating motions as this will cause accumulation of hESC clamps in the middle of well. hESCs should attach to MEF monolayer during 24 h. Do not disturb plates during this time.
6. Feed hESC cultures daily by aspiration and adding 3 ml fresh hESC growth medium per well.

3.4. Culture of OP9 Cells

1. Aspirate growth medium and wash OP9 monolayer with 10 ml PBS. Add 5 ml of 0.05% trypsin–0.5 mM EDTA solution and incubate in CO_2 incubator for 10 min.
2. Resuspend cells by pipetting and add cell suspension to 15-ml tube containing 5 ml OP9 growth medium.
3. Centrifuge tube at 300 g for 5 min, aspirate supernate, and resuspend cells in 1 ml OP9 growth medium.
4. Add 1/7–1/10 volume of cell suspension to gelatin-coated 100-mm dish with 10 ml OP9 growth medium.
5. Grow OP9 cells to confluence and split as above, typically every 4 days (*see* **Note 6**).

3.5. Hematopoietic Differentiation of hESCs by Coculture with OP9 Stromal Cells

1. Plate OP9 cells for hESC differentiation in separate gelatin-coated 100-mm dishes as for regular passage (*see* **Subheading 3.4.**). After OP9 confluence on day 4, feed cultures by replacing 5 ml medium with fresh OP9 growth medium. Incubate OP9 dishes for additional 4–6 days.
2. Before hESC plating, aspirate growth medium from OP9 dishes completely and add 10 ml of hESC differentiation medium. Place OP9 dishes in CO_2 incubator.
3. Prepare undifferentiated hESCs in suspension of small cell aggregates as for regular hESC passage (*see* **Subheading 3.3.**), but perform washing step in differentiation medium and resuspend cells in 1 ml of differentiation medium per each well of hESCs collected.
4. To determine the absolute number of cells in suspension of hESC aggregates (*see* **step 3**), use one well of 6-well plate to prepare single-cell suspension for counting. Wash well with 2 ml PBS, add 1 ml 0.05% trypsin–0.5 mM EDTA solution, and incubate at 37 °C for 5 min. Dissociate hESC colonies by pipetting, add cell suspension to 15-ml tube with 2 ml differentiation medium, and centrifuge tube at 300 g for 5 min. Aspirate supernate, resuspend cells in 1 ml medium and take aliquot (10 μl) for counting. Calculate the total number of cells in 1 ml. This number is equal to cell concentration (per 1 ml) in suspension of hESC aggregates (*see* **step 3**).
5. Dilute hESC suspension to 1.5×10^6 cells/ml with differentiation medium (typically at 1.5–2× dilution) (*see* **Note 7**).
6. Plate hESC suspension on OP9 dishes at 1 ml/dish (plating day is day 0). Place dishes in CO_2 incubator and distribute hESC aggregates evenly throughout the OP9 monolayer by moving the dishes back and forth 2–3 times.
7. On the next day (day 1), agitate dishes and aspirate all medium containing non-attached hESC. Add 20 ml of fresh differentiation medium per dish.
8. Feed hESC/OP9 cocultures on day 4, 6 and 8 by replacing 10 ml culture medium with fresh differentiation medium. Inspect cultures on day 4 under the microscope.

Flat undifferentiated hESC colonies should not be detectable, and > 90% colonies should have a characteristic appearance of mesodermal colonies with an elevated central portions composed of tightly packed rounded cells (*see* **Fig. 1**).

9. Harvest hESC/0P9 cocultures on day 9–10 (*see* **Subheading 3.7.1**)

3.6. Coating of Plastic Surface with pHEMA

Pour 5 ml of pHEMA coating solution into T75 flask using glass serological pipette. Rotate the flask until plastic surface is covered by pHEMA solution. Put the flask in vertical position and allow the excess of pHEMA solution to drain into flask corner. Remove pHEMA solution completely by pipette and put flask in horizontal position immediately. Leave the flask opened in the laminar hood to air-dry at least overnight. Completely dried pHEMA-coated flasks can be then closed and stored at room temperature (*see* **Note 8**).

3.7. Expansion of Myeloid Progenitors in GM-CSF Bulk Cultures

1. Prepare single-cell suspension from hESC/OP9 cocultures by successive enzymatic treatment with collagenase–trypsin (*see* **Note 9**). Aspirate medium and wash hESC/OP9 dishes with 10 ml PBS. Add 5 ml/dish of collagenase IV solution and incubate at 37 °C for 20 min. Remove collagenase solution into collection tube (15- or 50-ml tube); keep this tube for collection of enzymatically digested cells. Add 5 ml/dish of 0.05% trypsin–0.5 mM EDTA solution and incubate for additional 20 min at 37 °C. Add 2 ml of PBS-5%FBS to stop trypsin digestion and resuspend cells by pipetting. Transfer cell suspension into collection tube, mix cells by inverting closed tube and centrifuge at 400 g for 5 min. Wash cells three times with PBS-5%FBS, including filtration through 70 µM nylon filter at last washing step. Resuspend cells in expansion medium (1 ml per 1 collected hESC/OP9 dish) and take aliquots for cell counting (5 µl) and subsequently for FACS analysis (10^6 cells). Calculate the total cell number and prepare cell suspension in expansion medium (with GM-CSF) at 2×10^6 cells/ml.

2. Dispense cell suspension into pHEMA-coated T75 flasks (20 ml/flask) and place flasks in CO_2 incubator.

3. Feed cultures on day 3, 6, and 8 by replacing half the volume (10 ml) of medium with fresh expansion medium. Some cells will be removed with medium. These cells should be pelleted, resuspended in fresh medium, and returned to the flask. If myeloid cells start a vigorous proliferation after feeding on day 3, double a cell culture volume by adding 20 ml of expansion medium on day 6.

4. Myeloid cultures are usually harvested on day 9–10; however, in case of delayed cell expansion (*see* **Note 10**), myeloid cells can be collected on day 12–14.

3.8. Differentiation of Myeloid Precursors to DC

1. Collect myeloid cultures in 50-ml tubes and centrifuge cells at 300 g for 10 min. Aspirate supernate and resuspend cell mass in 10 ml PBS-5%FBS using glass serological pipette. Be careful not to brake up cell aggregates as they contain irrelevant cell types; myeloid cells can be resuspended easily with several pipette fillings. Remove cell aggregates by filtration of cell suspension through 70 μM nylon filter into 15-ml tube. Pellet cell filtrate by centrifugation (5 min at 300 g) and resuspend cells in 5 ml PBS. Underlay cell suspension with 5 ml 25% Percoll solution using serological pipette. Centrifuge at 400 g for 15 min and discard PBS and Percoll phases by aspiration (PBS/Percoll interphase contains dead cells and fine cell aggregates). Wash cells twice in PBS-5%FBS and resuspend cells in 5 ml DC differentiation medium. Take aliquots for cell counting (10 μl) and subsequently for FACS analysis (10^6 cells). Calculate the total number of cells and prepare cell suspension in DC differentiation medium at 10^6 cells/ml.
2. Dispense cell suspension into pHEMA-coated T75 (20 ml) flasks and place flasks in CO_2 incubator.
3. Feed cultures on day 4 and 8 by replacing 10 ml of culture medium with fresh DC differentiation medium. Cells contained in the removed medium should be pelleted, resuspended in fresh medium, and returned to the flask.
4. Collect DC cultures on day 10–12 of differentiation.

3.9. Cell Staining with Fluorochrome-Conjugated mAbs

1. Prepare cells in FACS staining buffer at 5×10^6 cells/ml.
2. Add FITC, PE, and APC-labeled mAbs to the test tube and corresponding isotype-matched control mAbs to the control tube.
3. Add 100 μl of cell suspension to test and control tubes, mix well by shaking, and incubate at 4 °C for 40 min.
4. Wash cells once with 4 ml FACS buffer (5 min at 400 g) and resuspend cells in 0.4 ml FACS buffer without FBS. Store tubes at 4 °C before analysis.

3.10. Monitoring of Cell Differentiation by FACS Analysis

1. For cells collected from hESC/OP9 cocultures (step 1), use CD43−FITC, CD235a−PE and CD45-APC mAbs to determine $CD43^+CD235a^-CD45^{+/-}$ multilineage progenitors (*see* **Fig. 1**). Efficient expansion of myeloid cells and their subsequent differentiation into DCs can be achieved only if more than 2% of $CD43^+CD235a^-CD45^{+/-}$ cells are generated in hESC/OP9 coculture. As a rule, cultures with a lower number of $CD43^+CD235a^-CD45^{+/-}$ cells fail to expand and differentiate into DCs (*see* **Notes 11** and **12**).
2. For cells collected from GM-CSF expansion cultures (step 2), use CD11b−FITC, CD11c−PE, and CD45−APC mAbs to evaluate proportion of myeloid DC precursors ($CD45^+CD11b^+CD11c^+$). More than 90% of cells collected

from GM-CSF cultures are $CD45^+$ (*see* **Fig. 2**) and express early myeloid markers CD33 and CD15 *(9)*; however, the proportion of $CD11b^+CD11c^+$ cells may be variable. Myeloid cell populations containing more than 60% of $CD11b^+CD11c^+$ cells efficiently generate DCs in step 3 cultures with GM-CSF and IL-4.

3. DCs collected from step 3 differentiation cultures can be evaluated using $CD14-FITC/CD1a-PE$ and $DC-SIGN-FITC/HLA-DR-PE$ mAb combinations. At first, DCs can be distinguished from still undifferentiated myeloid cells by high light-scatter profile (see R1 gate in **Fig. 3**). R1-gated cells typically comprise more than 60% of total cells in culture and express specific DC phenotype ($CD1a^+DC-SIGN^+HLA-DR^+CD14^{-/+}$).

4. Notes

1. OP9 cells have recently been characterized as a preadipocyte cell line *(17)*. In fact, many researchers mention the massive adipogenesis in OP9 cultures shortly after confluence. OP9 cells used in our laboratory exhibit diminished adipogenic properties possibly because of continuous subculture on the gelatin-coated plastic that we initially found to be important for the selection of highly inductive OP9 cells. These cells, however, retain adipogenic potential because adipogenic differentiation can be induced by short-term culture in serum-free media. Less adipogenic OP9 cells allow us to obtain the high-density overgrown OP9 cultures, which are critical for efficient generation of hematopoietic progenitors in hESC/OP9 cocultures.

2. A quality of basal α-MEM medium is essential for OP9 cells. Freshly prepared α-MEM from powder formulation is more suitable than liquid commercial medium. We did not find an advantage of α-MEM formula supplemented with nucleosides for hematopoietic differentiation in hESC/OP9 cocultures, although this medium significantly increases proliferation of OP9 cells. One peculiarity of α-MEM formula is a high concentration of ascorbic acid (50 μg/ml). We found that additional supplementation of hESC differentiation medium (α-MEM-based medium for hESC/OP9 cocultures) with 50 μg/ml ascorbic acid increases the yield of $CD34^+$ and $CD43^+$ cells and may be recommended for poorly differentiating hESC lines.

3. Use of SH-agents in hESC differentiation medium is optional. Differentiation proceeds efficiently without addition of SH-agents. However, the yield of total cells in MTG-supplemented cultures is consistently higher, suggesting a favorable effect of SH-agents on cell survival and growth during differentiation. We found that 2-ME is suppressive for differentiation starting from 50 μM concentration, whereas MTG is permissive for up to 200 μM. We selected 100 μM MTG concentration as optimal for differentiation medium. Addition of MTG to GM-CSF expansion cultures may delay emergence of proliferating myeloid cells; however, its presence is essential for long-lasting growth of myeloid progenitors.

4. Semiconfluent MEF monolayers are optimal for hESCs culture. Over-crowding MEFs suppress growth of hESC colonies and may stimulate their spontaneous differentiation. MEFs from different batches and prepared using different lots of plastic and FBS may vary in cell size. Therefore, MEF plating density should be adjusted accordingly to ensure semiconfluent feeder layers for hESC subculture.

5. After treatment of hESC cultures with collagenase, nearly all hESC colonies should be dislodged easily by washing or very gentle scraping. Intensive scraping should be avoided to prevent excessive mechanical damage to cells and collection of the firmly attached colonies, which may contain differentiated cell types.

6. During OP9 maintenance, cultures should be split no later than next day after confluence. Because OP9 growth is largely influenced by FBS, a split ratio must be adjusted with each new lot of FBS. The same type of FBS is usually used for OP9 maintenance and differentiation in hESC/OP9 cocultures. We select FBS lots with minimal adipogenic effect on confluent OP9 cells after feeding with half the volume of fresh medium and prolonged culture for 4–6 days. We found that different lots of HyClone "defined" FBS (without heat inactivation) support efficient OP9 growth with minimal if detectable adipogenesis in overgrown OP9 cultures (*see* **Fig. 1**) and also provide a relatively stable hematopoietic differentiation in hESC/OP9 cocultures. Results with FBS from other suppliers were more variable, yet not systematically studied.

7. Optimal plating density of hESCs in OP9 cocultures may vary for different hESC lines. It is primarily dependent on intensity of hESC growth: hESC lines with a higher proliferation rate in undifferentiated cultures may require a lower plating density in OP9 cocultures. The density of the $1.5–2 \times 10^6$ cells/OP9 dish is optimal for H1 cells, although H9 cells differentiate more efficiently starting from a lower density $(1–1.5 \times 10^6$ cells/OP9 dish). Optimal plating density for other hESC lines should be established in preliminary experiments using an initial range of $0.5–2.5 \times 10^6$ cells/OP9 dish with 0.5 intervals.

8. Plastic coating with pHEMA-ethanol solution should be done quickly. Ethanol evaporates rapidly causing pHEMA concentration. It may lead to the formation of irregular excessive coating. Do not discard pHEMA solution after coating; pour it back into pHEMA storage tube. Coating of one T75 flask usually takes 1–2 min and consumes approximately 1 ml of pHEMA solution.

9. hESC/OP9 cocultures on day 9–10 of differentiation may form a plenty of extra-cellular matrix that withstands digestion with collagenase and trypsin. As a result, many cells may be lost because of clumping and mechanical damage during pipetting. Longer incubations with collagenase and trypsin (up to 30 min each) should be tried first to improve cell recovery. In addition, supplementation of collagenase solution with 0.1 mg/ml hyalouronidase IV-S (cat. no. H3884, Sigma) can be further used to improve dissociation of hESC/OP9 monolayers.

10. Because myeloid cell expansion in GM-CSF cultures is solely dependent on the number of hematopoietic progenitors generated in hESC/OP9 coculture, variable

efficiency of hESC/OP9 cocultures is a matter of subsequent variations in GM-CSF cultures. A burst-like proliferation of myeloid cells in GM-CSF cultures is usually detected on day 4–5 (after feeding on day 3), but intensity of proliferation in following days may vary significantly. The feeding of GM-CSF cultures, therefore, should be flexible to accommodate the optimal progression of myeloid cells. More frequent feeding is necessary for highly proliferating cultures, whereas less frequent feeding is optimal for slowly proliferating cultures. It is important to note that more than half of growth medium in GM-CSF cultures should not be changed, as excessive media replacement may block proliferation and induce differentiation in the myeloid population.

11. The first multilineage hematopoietic progenitors are detectable in hESC/OP9 cocultures on day 6 of differentiation *(16)*. Addition of GM-CSF (20 ng/ml) to hESC/OP9 cocultures beginning from day 6 favors commitment and expansion of the earliest myeloid progenitors and eventually will increase the yield of $CD45^+$ cells in hESC/OP9 cocultures. GM-CSF addition to hESC/OP9 cocultures can be used to improve efficiency of subsequent myeloid expansion cultures with GM-CSF (**step 2**), and is highly recommended for hESC lines generating a low ($< 2\%$) number of $CD43^+CD235a^-CD45^{-/+}$ cells.

12. Both, $CD45^-$ and $CD45^+$ cells in $CD43^+CD235a^-$ population contain multilineage hematopoietic progenitors. However, $CD45^+$ cells are highly enriched in myeloid colony-forming cells and represent more committed myeloid population *(16)*. Therefore, simultaneous detection of $CD45^+$ cells within $CD43^+CD235a^-$ population is essential to confirm an ongoing myeloid commitment in hESC/OP9 cocultures. On day 9–10, $CD45^+$ cells should predominate and comprise 60–80% of $CD43^+CD235^-$ cells.

Acknowledgments

We thank Dr. Toru Nakano for providing OP9 cells and Joan Larson for editing assistance. This work is supported by Defense Advanced Research Projects Agency (DARPA) grant to UWM and NIH grant P51 RR000167 to the National Primate Research Center, University of Wisconsin-Madison. Igor Slukvin is supported by NIH grant HD44067. NIH-approved stem cell lines were used for all DARPA-funded research.

References

1. Steinman, R. M. (1991) The dendritic cell system and its role in immunogenicity. *Annu Rev Immunol* **9**, 271–296.
2. Caux, C., Massacrier, C., Dezutter-Dambuyant, C., Vanbervliet, B., Jacquet, C., Schmitt, D., and Banchereau, J. (1995) Human dendritic Langerhans cells generated in vitro from CD34+ progenitors can prime naive CD4+ T cells and process soluble antigen. *J Immunol* **155**, 5427–5435.

3. Romani, N., Reider, D., Heuer, M., Ebner, S., Kampgen, E., Eibl, B., Niederwieser, D., and Schuler, G. (1996) Generation of mature dendritic cells from human blood. An improved method with special regard to clinical applicability. *J Immunol Methods* **196**, 137–151.

4. Sallusto, F., and Lanzavecchia, A. (1994) Efficient presentation of soluble antigen by cultured human dendritic cells is maintained by granulocyte/macrophage colony-stimulating factor plus interleukin 4 and downregulated by tumor necrosis factor alpha. *J Exp Med* **179**, 1109–1118.

5. Bender, A., Sapp, M., Schuler, G., Steinman, R. M., and Bhardwaj, N. (1996) Improved methods for the generation of dendritic cells from nonproliferating progenitors in human blood. *J Immunol Methods* **196**, 121–135.

6. Chadwick, K., Wang, L., Li, L., Menendez, P., Murdoch, B., Rouleau, A., and Bhatia, M. (2003) Cytokines and BMP-4 promote hematopoietic differentiation of human embryonic stem cells. *Blood* **102**, 906–915.

7. Vodyanik, M. A., Bork, J. A., Thomson, J. A., and Slukvin, II (2005) Human embryonic stem cell-derived CD34+ cells: efficient production in the coculture with OP9 stromal cells and analysis of lymphohematopoietic potential. *Blood* **105**, 617–626.

8. Zambidis, E. T., Peault, B., Park, T. S., Bunz, F., and Civin, C. I. Hematopoietic differentiation of human embryonic stem cells progresses through sequential hemato-endothelial, primitive, and definitive stages resembling human yolk sac development. *Blood* **106**, 860–870.

9. Slukvin, II, Vodyanik, M. A., Thomson, J. A., Gumenyuk, M. E., and Choi, K. D. (2006) Directed differentiation of human embryonic stem cells into functional dendritic cells through the myeloid pathway. *J Immunol* **176**, 2924–2932.

10. Umeda, K., Heike, T., Yoshimoto, M., Shiota, M., Suemori, H., Luo, H. Y., Chui, D. H., Torii, R., Shibuya, M., Nakatsuji, N., and Nakahata, T. (2004) Development of primitive and definitive hematopoiesis from nonhuman primate embryonic stem cells in vitro. *Development* **131**, 1869–1879.

11. Kitajima, K., Tanaka, M., Zheng, J., Sakai-Ogawa, E., and Nakano, T. (2003) In vitro differentiation of mouse embryonic stem cells to hematopoietic cells on an OP9 stromal cell monolayer. *Methods Enzymol* **365**, 72–83.

12. Nakano, T., Kodama, H., and Honjo, T. (1994) Generation of lymphohematopoietic cells from embryonic stem cells in culture. *Science* **265**, 1098–1101.

13. Nakano, T., Kodama, H., and Honjo, T. (1996) In vitro development of primitive and definitive erythrocytes from different precursors. *Science* **272**, 722–724.

14. Eto, K., Murphy, R., Kerrigan, S. W., Bertoni, A., Stuhlmann, H., Nakano, T., Leavitt, A. D., and Shattil, S. J. (2002) Megakaryocytes derived from embryonic stem cells implicate CalDAG-GEFI in integrin signaling. *Proc Natl Acad Sci USA* **99**, 12819–12824.

15. Senju, S., Hirata, S., Matsuyoshi, H., Masuda, M., Uemura, Y., Araki, K., Yamamura, K., and Nishimura, Y. (2003) Generation and genetic modification of dendritic cells derived from mouse embryonic stem cells. *Blood* **101**, 3501–3508.
16. Vodyanik, M. A., Thomson, J. A., and Slukvin, II (2006) Leukosialin (CD43) defines hematopoietic progenitors in human embryonic stem cell differentiation cultures. *Blood* **108**, 2095–2105.
17. Wolins, N. E., Quaynor, B. K., Skinner, J. R., Tzekov, A., Park, C., Choi, K., and Bickel, P. E. (2006) OP9 mouse stromal cells rapidly differentiate into adipocytes: characterization of a useful new model of adipogenesis. *J Lipid Res* **47**, 450–460.

20

Insulin-Producing Cells from Embryonic Stem Cells
Experimental Considerations

Enrique Roche, Roberto Ensenat-Waser, Nestor Vicente-Salar, Alfredo Santana, Martin Zenke, and Juan Antonio Reig

Summary

The main objective of cell bioengineering is to generate customized tissues that allow recovering the lost functions in the organism in the absence of immune rejection. Although the possibility of in vitro generation of entire organs is technically very complex, obtaining specific cell types for replacement therapies seems to be a more realistic goal at mean time. In this context, those pathologies affected by the dysfunction of a specific cell type, as it is the case of β-cell in diabetes, would be in principle candidates to benefit from cell transplantation protocols. Embryonic stem cells offer interesting possibilities in this context because they fulfill two important criteria: (i) High proliferation rate by symmetric cell division, overcoming the problem of biomass scarcity and (ii) Plasticity of differentiating to all cell types present in the adult organism, including the germ line. Different approaches have been developed in vitro to obtain insulin-producing cells from embryonic stem cells. Nevertheless, a definitive protocol does not exist yet. However, the experience accumulated in this field by the different laboratories has provided considering key points that would help to design a preferred protocol in the future.

Key Words: Embryonic stem cells; Insulin-producing cells; β-Cells; Diabetes.

1. Introduction

Diabetes mellitus is a degenerative pathology caused by the absence or low production of insulin by pancreatic β-cells. Insulin is a key hormone in controlling the uptake of circulating glucose and fatty acids by peripheral target tissues, such as skeletal muscle and adipose tissue (*1*). Therefore, diabetic people are obliged to

From: *Methods in Molecular Biology, vol. 407: Stem Cell Assays*
Edited by: M. C. Vemuri © Humana Press, Totowa, NJ

inject insulin daily to survive, because there are not compensatory hormones that can mimic exactly the function of this hormone. Diabetes is classically divided in two main groups: type 1 diabetes and type 2 diabetes. Type 1 diabetes is caused by an autoimmune destruction of pancreatic β-cells, leading therefore to total absence of insulin. Type 2 diabetes represents a more complex pathology that usually evolves to β-cell dysfunction caused by an excess of circulating fat and glucose. Recent evidences indicate that the persistently high concentrations of these nutrients could activate β-cell suicide programs, leading to apoptosis and insulin absence *(2)*. This is known as glucolipotoxicity, indicating that, at the end, type 2 diabetes displays as well as type 1, a significant reduced β-cell mass. Altogether, this information indicates that both type 1 and type 2 diabetes are candidate pathologies for cell therapy protocols.

The Edmonton protocol has been a key attempt in the treatment of diabetes by transplanting highly pure isolated islets from cadaveric donors *(3)*. The surgery protocol is minimally invasive for the patient and in this context represents a clear advantage versus the whole pancreas/kidney double transplant. However, this protocol has to face two main obstacles: to find a correct immunosuppressive regime and the scarcity of cadaveric pancreata, indicating that alternative sources for insulin-producing cells have to be considered *(4,5)*.

The proposed cell therapy strategies consider several alternatives including pancreas regeneration, isolation, expansion, and differentiation of pancreatic stem cells, transdifferentiation of adult stem cells from other organs, and bioengineering of embryonic stem cells (ESCs) *(5,6)*. ESCs are good candidates for in vitro strategies to obtain insulin-producing cells because they display a strong proliferation potential and broad plasticity. However, a detailed in vitro protocol is still missing. In this chapter, we plan to present the lastest version of the protocol we are currently using, pointing out the critical steps in which researches have to pay attention. Indeed, this is a quickly changing field, and most likely, this protocol will be improved in a near future. In any case, we intend to establish a rational working methodology that will help undoubtedly to design a definitive protocol.

2. Materials

2.1. Cell Culture

1. **Table 1** summarizes the composition of culture media used. Dulbecco's Modified Eagle's medium (DMEM) and chemicals are purchased from Gibco (cat. no. 32430-100, Invitrogen, Carlsbad, CA, USA). Fetal bovine serum (FBS) from selected batches was purchased from Biochrom, cat. no. 20S0115 (Berlin, Germany) and leukemia inhibitory factor (LIF) from Chemicon, cat. no. T01.ES61106 (Temecula, CA, USA).

Table 1
Culture Media Used with Undifferentiated and Differentiated ESCs

Reagent	Stock	Final concentration in the culture medium for undifferentiated ESCs	Final concentration in the culture medium for ESCs differentiation
DMEM 32430-027	$1\times$	$1\times$	$1\times$
FBS	100%	15%	15–10%[a]
Nonessential amino acids	100%	1%	$1\times$
β-Mercaptoethanol	50 mM	0.1 mM	0.1 mM
Penicillin Streptomycin	$100\times$	$1\times$	$1\times$
LIF	10^6 U/ml	10^3 U/ml	–

DMEM, Dulbecco's modified Eagle's medium; ESC, embryonic stem cell; LIF, leukemia inhibitory factor.

[a]3% of FBS is used in certain differentiation protocols *(10,11)*. See **Subheading 3.6** for more details.

2. Gelatin (cat. no. G-2625, Sigma, St Louis, MO, USA) solution (0.1%) is prepared in phosphate buffered saline (PBS) (Biochrom) and autoclaved.
3. Trypan blue solution in PBS is purchased from Gibco (cat. no. 15250-010, Invitrogen).
4. Trypsin ready-to-use solution containing 0.05% trypsin+0.002% Na$_4$ ethylendiaminetetraacetic acid (EDTA), was purchased from (cat. no. 25300-062, Gibco, Invitrogen, England).
5. Freezing medium consists of complete culture medium supplemented with 10% dimethyl sulfoxide (DMSO) (cat. no. D-2650, Sigma) and 25% FBS.

2.2. Control of the Differentiation State of ESCs

1. Blocking solution for stage-specific embryonic antigen-1 (SSEA-1) immunodetection consists in 10% goat serum (cat. no. G-9023, Sigma) in PBS supplemented with 1% bovine serum albumin (BSA) (Sigma).
2. Glycerin-gelatin (also named Glycerin Jelly) solution contains 7.63% gelatin, 53.83% glycerin, and 0.763% phenol, diluted in distilled water. Before adding glycerin and phenol, gelatin has to be dissolved in warm water to get a homogeneous solution and cooled down.

3. Alkaline Phosphatase Substrate, from Vector Labs, either Vector Blue (dark blue precipitate, cat. no. SK-5300) or Vector Red (red precipitate also fluorescent, cat. no. SK-5100) kit were used (Vector Red/Blue Alkaline Phosphatase Substrate Kit, Vector Laboratories, Burlingame, CA).

3. Methods

In general, the culture protocol is divided into three steps. First, ESCs are expanded as adherent monolayers in the presence of LIF, a cytokine of the interleukin-6 family that maintains the cells in an undifferentiated state. Second, differentiation programs are activated by transferring the cells to bacteriological plates, allowing the formation of floating cell aggregates called embryoid bodies (EBs) *(7)*. Under these conditions, cells start to express protein markers typical of the three embryonic layers: ectoderm, mesoderm, and endoderm. Insulin expression is evident at long-term incubation periods in EBs (21 days) *(8)*. Finally, cells are plated again in adherent dishes and incubated in the presence of different substances to increase insulin expression and hormone production. We have observed that the ESCs and EB culture conditions are critical and have an important influence in the phenotype of the final cell product. Therefore, the observation of some rules is instrumental in this context.

3.1. Culture of Undifferentiated Mouse ESCs

1. Mouse D3-ESCs obtained from American Type Culture Collection and R1-ESCs from Dr Andras Nagy (Mount Sinai Hospital, Toronto) were used.
2. Culture dishes (TPP, Trasadingen, Switzerland) are treated with 0.1% gelatin in PBS. The gelatin solution has to cover the culture surface. After 10-min incubation at room temperature, gelatin solution is discarded and the dish is ready to use (*see* **Note 1**).
3. Cell number is determined in Neubauer chamber after staining with 0.4% trypan blue solution in PBS. Living cells are determined according to their round and smooth morphology and trypan blue exclusion.
4. Cells are passaged in the corresponding culture plates or flasks (*see* **Table 2**) and cultured at 37°C in a 5% CO_2 incubator. When reaching 80% confluence, culture medium is discarded and cells are washed with the same volume of PBS. Once PBS is removed, a specific volume of trypsin solution is added (*see* **Table 2**), and the dish is incubated at 37°C for 1–5 min (longer periods of time are recommended for highly confluent cells but never longer than 5–10 min). When cells start to detach, add the same volume of culture medium (stopping volume in **Table 2**). Pipette up and down carefully to dissociate cell aggregates, minimizing always foam production. Transfer cell suspension to a sterile tube and wash the plate with the same volume of fresh culture medium as used to stop trypsin solution (washing volume in **Table 2**). Transfer again to the same tube and mix to obtain a

Table 2
Trypsin Volumes Used

Flask (cm^2) Plate (mm of diameter)	Surface (cm^2)	Trypsin volume	Stopping volume	Washing volume	Final volume of suspension
Flask (150) Plate (150)	150	4 ml	8 ml	8 ml	20 ml
Flask (75) Plate (100)	60–75	2 ml	4 ml	4 ml	10 ml
Flask (25) Plate (60)	25	1 ml	2 ml	2 ml	5 ml
Plate (24 multiwells)	1.76	250 μl	500 μl	500 μl	750 μl

homogeneous cell suspension (final volume in **Table 2**). Count cells according to **step 3.1.3** and plate at the appropriate concentration (*see* **Note 2**).

5. After trypsinization, cells could be frozen in 1–1.8 ml of freezing medium. To this end, the cell suspension obtained from **step 3.1.4** is centrifuged at 110 g for 3–5 min. Supernatant is discarded, and pellet is resuspended in freezing medium to obtain a homogeneous cell suspension ($3–5 \times 10^6$ cells/1–1.8 ml). Aliquots of 1–1.8 ml are transferred to cryovials (TPP, Trasadingen) that are introduced in a "Mr Frosty" (Nalgene, Rochester, NY) according to manufacturer instructions and then frozen at −80 °C for at least 4 h or overnight. Finally, cryovials are stored in a tank containing liquid N_2 at −196 °C.

6. Cell thawing is performed by quick transfer of the cryovial from liquid N_2 to a water bath at 37 °C. Cells are resuspended and mixed in a sterile tube with 6 ml of fresh culture medium. The cryotube is washed with additional 2 ml of culture medium. Cell suspension is centrifuged at 110 g for 3–5 min. The supernatant is discarded, and the pellet is resuspended in culture medium at the desired cell concentration. Cells are plated and the dishes are incubated at 37 °C in the CO_2 incubator overnight. To eliminate floating dead cells, it is recommended to change the medium next day after 24–36 h.

3.2. Control of the Undifferentiated State of ESCs

1. Cells are cultured routinely in adherent tissue culture dishes treated with gelatin. The undifferentiated state has to be controlled after certain number of passages by monitoring the colony morphology and the expression of specific markers by reverse transcriptase–polymerase chain reaction (RT–PCR) and immunostaining microscopy (*see* **Note 3**).

2. Specific markers of undifferentiated ESCs and ectoderm can be detected by RT–PCR, using the primers and conditions indicated in **Table 3**.
3. SSEA-1 (marker of undifferentiated ESCs) detection was performed by immunofluorescence according the following protocol (*see* **steps 4–8**).
4. Cells are washed three times and fixed with 100% methanol at −20 °C. After 10 min, cells are washed again three times with PBS.
5. At the end, cells are blocked with a 10% goat serum solution and incubated at room temperature in a humid chamber.
6. After 1 h, blocking solution is discarded and cells are covered with a PBS containing 1/100 anti-SSEA-1 (MC-480, developed by Dr Solter and obtained from Developmental Studies Hybridoma Bank, Iowa University).
7. Cells are incubated for 1 h at room temperature. After this period, cells are washed three times with PBS and quickly incubated with a PBS solution containing 1/600 goat anti-mouse immunoglobulin M (IgM) bound to Cy3 (Jackson Immunoresearch, West Grove, PA) and supplemented with 0.5% BSA. Cells are incubated with the secondary antibody for 45 min.
8. At the end, cells are washed three times, covered with glycerin-gelatin (Glycerin Jelly) solution and a coverslip and observed under microscope (inverted Nikon Eclipse TE200 microscope, Tokyo, Japan).
9. Alkaline phosphatase (marker of undifferentiated ESCs) activity is detected according to the **steps 10–12**.
10. Cells are washed once with PBS and immediately fixed with 4% paraformaldehyde solution in PBS (Sigma) at room temperature for 15 min. Alternatively, cells can be fixed with 100% cold methanol at −20 °C for 10 min.
11. Following fixation, the fixing reagent is discarded and cells are washed once with 100 mM Tris–HCl pH 8.3 (Sigma). After this, cells are incubated in the presence of alkaline phosphatase substrate (Vector Red) for 25 min. The processing of this substrate by the enzyme generates a red precipitate that can be detected by transmission as well as fluorescence microscopy.
12. Finally, cells are washed with 100 mM Tris–HCl pH 8.3, covered with glycerin-gelatin (Glycerin Jelly) solution and a coverslip and observed under microscope.

3.3. EB Formation by Massive Method

1. Trypsinized cells ($2.5–5 \times 10^5$ cells/ml) are transferred to a bacteriological plate (100 mm of diameter) in differentiation medium (*see* **Table 1**) to reach a final volume of 10 ml.
2. Cells are incubated at 37 °C in the CO_2 incubator for 2 days.
3. Medium is changed next by transferring the EB suspension to a tube with a conical bottom. EBs sediment by gravity in 10 min and supernatant is discarded. The medium is changed in the same plate from days 4 to 5.
4. EBs are then resuspended in fresh differentiation medium with no LIF and transferred to a new bacteriological Petri dish (*see* **Note 4**).

Table 3
List of Gene-Specific Primers (5′–3′) Used in PCR

Genes	Forward primer	Reverse primer	Product size (bp)	Tm/cycle number	GeneBank accession number
ESCs					
Oct3/4	AGGCCCGGAAGAGAAAGCGAACTA	TGGGGGCAGAGAGAAAGGATACAGC	265	67 °C/26	NM_013633
Nanog	AGGGTCTGCTACTGAGATGCTCTG	CAACCACTGGTTTTTCTGCCACCG	363	62 °C/29	AB093574
ESG-1 (Dppa5)	ATAAGCTTGATCTCGTCTTCC	CTTGCTAGGATGTAACAAAGC	501	55 °C/29	BC092354
Ectoderm					
Otx-2	CCATGACCTATACTCAGGCTTCAGG	GAAGCTCCATATCCTGGGTGGAAAG	211	67 °C/35	BC027104
N-200	GAGTGGTTCCGAGTGAGGTTGGAC	GACGTTGAGCAGGTCCTGGTACTC	343	67 °C/32	NM_010904
GFAP	CTGTTTGCCAGGCTCAGTTCCCAC	GGAAACTTCCAGCTCTGGCAACGG	297	67 °C/32	NM_010277
AChE	CCGGGTCTATGCCTACATCTTTGA	CACAGGTCTGAGCAGCGCTCCTGCTTGCTA	483	67 °C/35	BC046327
TH	AGTTCTCCCAGGACATTGGACTT	ACACAGCCCAAACTCCACAGT	100	67 °C/35	NM_009377
MBP	GTGCAGCTTGTTCGACTCCG	ATGCTCTCTGGCTCCTTGGC	153	67 °C/35	NM_010777
Nestin	CGGCCCACGCATCCCCCATCC	AGCGGCCTTCCAATCTCTGTTCC	258	67 °C/32	NM_016701
Mesoderm					
Brachyury	GCTCATCGGAACAGCTCTCCAACC	GGAGAACCAGAAGACGAGGACGTG	319	67 °C/30	NM_009309
α-MHC	CTGCTGGAGAGGTTATTCCTCG	GGAAGAGTGAGCGGCGCATCAAGG	301	64 °C/32	NM_010856
β-MHC	TGCAAAGGCTCCAGGTCTGAGGGC	GCCAACACCAACCTGTCCAAGTTC	205	64 °C/32	NM_080728
ANF	TGATAGATGAAGGCAGGAAGCCGC	AGGATTGGAGCCCGAAGTGGACTAGG	203	67 °C/29	NM_008725
Endoderm					
AFP	CCTTGGCTGCTCAGTACGACAAGG	CCTGCAGACACTCCAGCGAGTTTC	301	67 °C/26	NM_007423
Foxa2	GTTAAAGTATGCTGGGAGCCG	CGCCCACATAGGATGACATG	219	61 °C/30	NM_010446
Foxa3	CGGGCGAGGTGTATTCTCCA	GCCCAGTAGGAGGCCTTTGCC	520	62 °C/32	NM_008260
GATA4	GGCCCCTCATTAAGCCTCAG	CAGGACCTGCTGGCGTCTTA	249	61 °C/30	NM_008092

(Continued)

Table 3
(Continued)

Genes	Forward primer	Reverse primer	Product size (bp)	Tm/cycle number	GeneBank accession number
GATA5	GCGTCTGTCCTCATCCCGAA	CTGGAGGCCTGGGAGGTGATA	354	62°C/30	NM_008093
Amnionless[a]	ACTGCCTCCAACTGGAACCAGAAC	CGCAGAGGTCACAGCATTGTCCTT	667	62°C/30	BC087954.1
Glucagon[b]	CTTCCCAGACAGAAGCGCATGAGG	GTCCCTGGTGGCAAGATTGTCCAG	394	67°C/30	NM_008100
Neuroectoderm and β-cells					
Pax6	AGTGAATCAGCTTGGTGGTG	TCTGTCTCCGGATTTCCCAAG	294/336	60°C/32	AF443223
Isl1	CACTATTTGCCACCTAGCCAC	AAATCATGATTACACTCCGCAC	255	60°C/32	AJ132765
Pdx1	ACACAGCTCTACAAGGACCCGTGC	GCACAATCTTGCTCCGGCTCTTCG	655	60°C/32	NM_008814
Insulin	CCCACCCAGGCTTTTGTCAAACAGC	TCCAGCTGGTAGAGGGAGCAGATG	250	60°C/32	NM_008386/87
Positive control					
GAPDH	GCCATCAATGACCCCTTCATTG	CACCACCTTCTTGATGTCATCA	692	62°C/32	NM_008084
β-Actin	CCCTAGGCACCAGGGTGTGA	TCCCAGTTGGTAACAATGCCA	128	60°C/27	X03672

Endoderm markers are for primitive and definitive.
[a] Specific marker for primitive endoderm.
[b] Specific marker for definitive endoderm.

3.4. EB Formation by Hanging Drop Method

1. Trypsinized cells are prepared in a suspension of 600 cells/20 μl of differentiation medium (*see* **Note 5**).
2. Put the 20-μl drops containing the cell suspension on the upside down face of a cover of a bacteriological plate (75–150 drops/100 mm diameter cover).
3. Put 10–15 ml of PBS into the plate to create a humid atmosphere and avoid drop evaporation.
4. When everything is ready, take the cover containing the drops and turn quickly. The drops remain hanging in the upper surface of the cover.
5. Close the plate that contains PBS with the cover containing the hanging drops and incubate at 37 °C in CO_2 incubator for 3–5 days.
6. After this incubation period, EBs were pooled in 2 ml of differentiation medium (*see* **Table 1**) and transferred to a bacteriological plate of 100 mm of diameter. The EBs from two additional covers can be pooled in the same bacteriological plate bringing a 6 ml final volume.
7. EBs were then incubated in suspension, and the medium is changed as described in the massive method (*see* **steps 3.3.3** and **3.3.4**).

3.5. Additional Methodology to Use with ESCs

3.5.1. Transfection

1. All constructs must be linearized before transfection.
2. At the end of the digestion, DNA has to be precipitated under the hood in sterile conditions with 70–90% ethanol (2–2.5 vol) plus 3 M sodium acetate pH 5.2 (1/10 vol). Linear DNA precipitates at −80 °C overnight.
3. Next day, DNA is pelleted, washed with 70% ethanol and resuspended under the hood by adding sterile PBS.
4. Transfection is performed by electroporation. To this end, 3×10^7 ESCs were mixed with 50 μg (25 μg for 1.3×10^7 ESCs) of linear construct in PBS to a final volume of 0.7–1 ml.
5. The DNA/cell suspension were transferred to an electroporation cuvette and incubated on ice for 5–10 min.
6. Cells were electroporated with one single pulse of 0.8 KV and 3 μF using a Gene Pulser II® (Biorad, Hercules, CA).
7. After electroporation, cells are placed on ice, resuspended in culture medium (*see* **Table 1**) and transferred to 100-mm plates at 3×10^6 cells/plate.
8. Cells are cultured as described before (*see* **Subheading 3.1.**) for 2–3 days, changing the medium daily to eliminate floating dead cells (*see* **Note 6**).

3.5.2. Clonal Selection by Antibiotic Addition to the Culture Medium

Cell dilution could be used to select clones. Nevertheless, usually purest clones could derive from antibiotic selection after transfection, provided that the corresponding resistance gene is present in the DNA construction. In our laboratory, the DNA construct contains a cassette with the hygromycin resistance gene under the control of the constitutive promoter of the phosphoglycerate kinase gene *(9)*.

1. We add 400–800 μg/ml of hygromycin (GIBCO/Calbiochem, La Jolla, CA) to perform cell selection.
2. Clones are growing separately and are visible under the microscope. The dish has to be washed once with PBS and left with PBS. To pick up a certain clone, it is important to use a pipette adjusted to a final volume of 15 μl. The tip has to be blocked with FBS by repipetting several times. This avoids the adherence of the clone to the tip. The selected clone is pushed carefully with the tip. Once floating, the clone is picked up with the pipette and transferred to a Petri dish as a drop.
3. At the end, the plate contains several drops containing each one a different clone. Add to each drop 50 μl of 0.05% trypsin plus 0.02% EDTA.
4. Leave the plate containing the drops 5 min at 37 °C in the CO_2 incubator. This allows to cell dissociation in each drop.
5. After this, each clone is pipetted slowly to favor the total dissociation of the clone. The obtained cell suspension is brought to 750 μl final volume and transferred to well of 24 multiwell plate treated with gelatin as described in **step 3.1.2**. The plate is incubated at 37 °C for 2 days in a CO_2 incubator.
6. Medium is changed every 2 days, and incubation continues until the surface is totally covered by the cells.
7. When confluence is reached, each clone is trypsinized as described in **step 3.1.4** and transferred to a 60 mm of diameter plate treated with gelatin (*see* **step 3.1.2**) in a final volume of 5 ml. Change medium next day.
8. When cells reach the confluence, they can be expanded by transferring to subsequent Petri dishes (*see* **Table 2**). Alternatively, cells can be frozen in a cryovial as described in **step 3.1.5** (*see* **Note 7**).

3.5.3. Clonal Selection

This can be performed as well by cell sorting by using flow cytometry. To this end, cells have to be transfected with a construct codifying the expression of fluorescence marker, usually green fluorescence protein. For this type of analysis, we used the cytometer Vantage SE System with FACS separation (BD Biosciences, San Jose, CA, USA).

1. Cells are resuspended in PBS containing 1 mM EDTA to a final concentration of 10^6 cells/ml and kept on ice. The EDTA avoids cell aggregation.
2. Analysis is performed quickly on 10^4 cells, and then sorting can be carried out.
3. A different dissociation protocol before sorting has to be used with EBs (*see* **steps 4–9**).

4. Wash EBs with PBS containing 1 mM EDTA.
5. Cover the EBs with a solution containing 200–250 U/ml of collagenase (Sigma) in DMEM.
6. Incubate at 37 °C for 35–45 min in the CO_2 incubator.
7. At the end, add trypsin from a stock of 0.25% to reach a final concentration of 0.05%, directly to the collagenase solution. Let incubate for 15 min.
8. At the end of this incubation period, cells are carefully pipetted with a 5-ml tip to favor cell dissociation.
9. Viable cells are counted by trypan blue exclusion as described in **step 3.1.3**, quickly resuspended to 10^6 cells/ml in PBS plus 1 mM EDTA and proceeding for analysis.
10. After sorting, it is important to verify the purity of the culture by microscopy.

3.6. Differentiation Protocols

1. Differentiation starts in the EB culture and some manipulations can be introduced at this stage. For instance, EBs were cultured for 7 days in the presence of 3% FBS and different factors or conditioned media were added *(10,11)* (*see* **Note 8**).
2. If the culture medium is not manipulated, insulin gene expression seems to be evident in EBs after long incubation periods (21 days) *(8)*.
3. The resulting EBs were plated onto 100 mm of diameter tissue culture dishes and allowed to attach. Factors present during the EB incubation period or new factors can be added. FBS concentration in the medium is reduced to 10% at this stage *(10,11)* (*see* **Note 8**).
4. In cells transfected with a construct containing the neomycin resistance gene under the control of β-cell-specific promoters, the selection of insulin-producing cells can be achieved by adding 0.4–0.8 mg/ml G418 (Gibco, Invitrogen) *(10,11)*.
5. Before transplantation, cells are allowed to form aggregates that mimic the islet structure. To this end, cell aggregates are gently detached from the dish and transferred to 100-mm Petri dishes and further incubated with differentiation selection medium containing 10% FBS, 0.4–0.8 mg/ml G418, 10 mM nicotinamide and 5 mM glucose *(9)*, for their final maturation.

Notes

1. The treatment of the culture surfaces with gelatin can be performed at 4 °C from 1 to 24 h. However, we do not recommend this procedure because the additional manipulations increase the risk of contamination.
2. The number of cells in this phase is crucial for subsequent results. When cells are plated at high density (i.e., 7×10^6 cells/100 mm of diameter dish), we obtain commitment to definitive endoderm (according to the expression of specific markers) during the differentiation period in EBs. When the cells are expanded at low density (i.e., $1–2 \times 10^6$ cells/100 mm of diameter dish), commitment to definitive endoderm is observed during spontaneous differentiation, but at the

same time, markers of primitive endoderm are expressed as well, what might be a consequence of slight differences in the differentiation potential of undifferentiated ES cells depending on the culture conditions. This point is important because both primitive and definitive endoderm express insulin and many other markers that are found in mature β-cell and in β-cell precursors. If the final cell product has to mimic as much as possible the pancreatic β-cell phenotype, the culture protocol has to focus on the enrichment in definitive endoderm precursors.

3. We have observed that cells tend to differentiate to ectoderm when cultured in monolayer after several passages, even in the presence of LIF. Undifferentiated cells grow forming round shaped colonies and express specific markers such as Oct3/4, Nanog, and ESG-1 (now called Dppa5) (detected by RT–PCR), alkaline phosphatase activity (detected by microscopy), and SSEA-1 (detected by immunostaining). When cells start to commit to ectoderm, SSEA-1 expression is reduced, and the expression of typical ectodermal markers, such as neurofilament-200 (N-200) and glial fibrillary acidic protein (GFAP), starts to be evident. The cells change the morphology growing as a continuous monolayer with no colony formation (*see* **Fig. 1**). Karyotype alterations have been noticed at these late stages of culture.

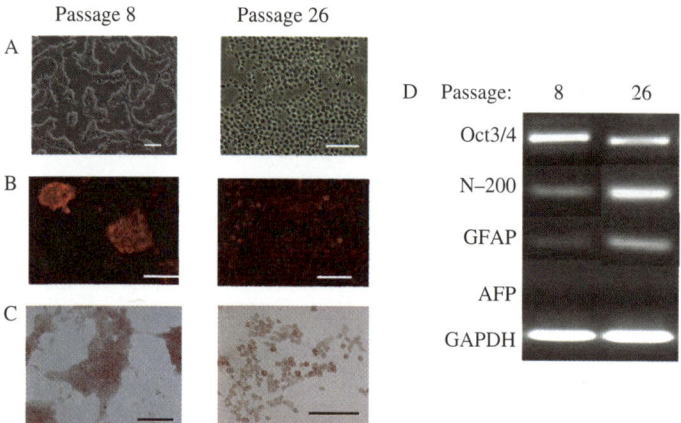

Fig. 1. (**A**) Phase-contrast image showing a typical culture of mouse R1-embryonic stem cells (ESCs) at passages 8 and 26. Bar: 100 μm. (**B**) Stage-specific embryonic antigen-1 (SSEA-1) immunofluorescence (rhodamine filter) for different passages (8 and 26) of mouse R1-ESCs. Bar: 100 μm. (**C**) Alkaline phosphatase staining of mouse R1-ESCs at passages 8 and 26. Bar: 100 μm. (**D**) Reverse transcriptase–polymerase chain reaction (RT–PCR) showing the increased expression of neuroectodermal genes [N-200 and [glial fibrillary acidic protein (GFAP)] at passage 26 respect to passage 8. Markers of endoderm [alphafetoprotein (AFP)] are not expressed. See **Table 3** for PCR conditions.

4. It is important to change the Petri dish because cellular debris tend to attach to the plastic surface, interfering with the normal growth of EBs. Markers (*see* **Table 3**) from the different embryonic layers can be detected at different times of EB culture. This kinetics is altered when EBs are obtained from late passages of ESCs (*see* **Fig. 2**).

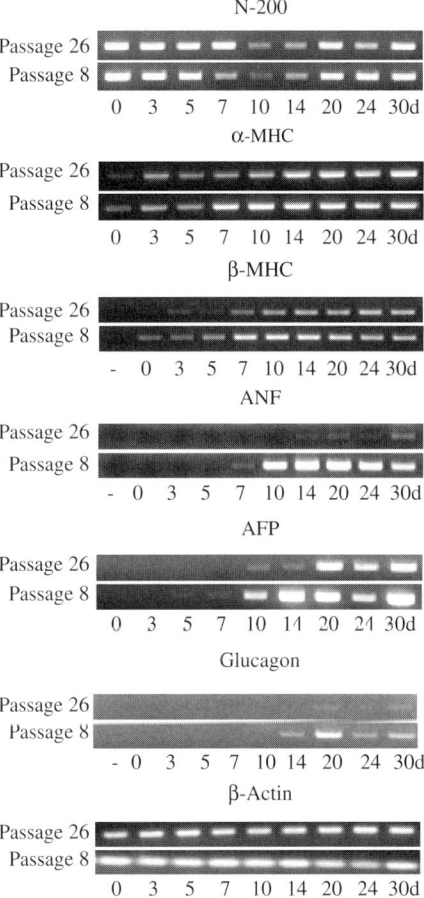

Fig. 2. Pattern of expression of different marker genes from ectoderm (N-200), mesoderm [α-myosin heavy chain (α-MHC), β-MHC, and atrial natriuretic peptide (ANF)], primitive and definitive endoderm (AFP) and definitive endoderm (glucagon) in embryoid body (EB) cultures from 0 to 30 days derived from mouse R1-embryonic stem cells (ESCs) at passages 8 and 26. β-Actin expression is used as unvariant control. See **Table 3** for PCR conditions.

5. The number of cells in this stage is critical for commitment into specific cell lineages: 200 cells/20 μl for neuroectoderm, 400 cells/20 μl for cardiac muscle, 600 cells/20 μl for pancreatic lineages and skeletal muscle and 800 cells/20 μl for smooth vascular muscle.

6. Electroporation works very well with ESCs, yielding around 10–30% of transfected cells. This yield depends on the cell passage, cell line, and the electroporator type. The best yields are obtained with square wave electroporators.

7. Usually 8×10^6 cells can be obtained from a 60 mm of diameter plate. Use $2-3 \times 10^6$ to extract RNA and analyze markers of pluripotentiality and freeze the rest. This is important to avoid saturation of the incubator because of the big number of clones that can be expanded and the high cost of the corresponding culture medium.

8. Eliminating LIF from the culture medium favors activation of differentiation programs. In addition, certain laboratories claim that different factors can be added to the medium to obtain insulin-producing cells. Some of them are 2 μg/ml of anti-sonic hedgehog (Developmental Studies Hybridoma Bank, Iowa) and conditioned medium obtained from incubation of mouse pancreatic rudiments isolated at embryonic day 16.5 *(10,11)*. At present, our laboratory has not fully tested these treatments yet, and several points still need to be clarified. For instance, the real origin of the obtained cells is not known. This can be performed in part just by checking the expression of insulin I gene (expressed in mature β-cells) or insulin II gene (expressed in neuroectoderm, primitive endoderm, and fetal liver). This has not been deeply addressed in these studies *(10,11)*. Furthermore, the final cell products still present high rates of BrdU incorporation, suggesting a potential tumoral risk in the therapeutical use of these cells. This can be checked by looking for tumor formation in transplanted animals. In these studies, transplanted animals are maintained for no longer than 14–16 days *(10,11)*, and in our laboratory, we have assessed that tumor formation is clearly detectable only 3 months after transplantation in similar experimental approaches. Other protocols based on nestin selection have been published *(12–15)*. It seems that the final cell product might have a neuroectodermal origin and thereby not be fully mimicking all functional β -cell features *(16)*. In conclusion, it is important to work in this stage of the protocol to improve the present methodology and obtain a reliable differentiation procedure.

References

1. DeFronzo, R. A., Ferrannini, E., Keen, H., and Zimmet, P. (eds.) (2004). *International Textbook of Diabetes Mellitus*. John Wiley and Sons, Chichester, UK.
2. Prentki, M., Joly, E., El-Assaad, W., and Roduit, R. (2002) Malonyl-CoA signalling, lipid partitioning, and glucolipotoxicity. Role in β-cell adaptation and failure in the etiology of diabetes. *Diabetes* **51 (Suppl 3)**, S405–S413.
3. Shapiro, A. M. J., Lakey, J. R. T., Ryan, E. A., Korbutt, G. S., Toth, E., Warnock, G. L., Kneteman, N. M., and Rajotte, R. V. (2000). Islet transplantation

in seven patients with type 1 diabetes mellitus using a corticoid-free immunosuppressive regime. *N. Engl. J. Med.* **343**, 230–238.

4. Ryan, E. A., Paty, B. W., Senior, P. A., Bigam, D., Alfadhli, E., Kneteman, N. M., Lakey, J. R., and Shapiro, A. M. (2005) Five-year follow-up after clinical islet transplantation. *Diabetes* **54**, 2060–2069.

5. Roche, E., Reig, J. A., Campos, A., Paredes, B., Isaac, J. R., Lim, S., Calne, R. Y., and Soria, B. (2005) Insulin-secreting cells derived from stem cells: clinical perspectives, hypes and hopes. *Transplant. Immunol.* **15**, 113–129.

6. Roche, E., Santana, A., Vicente-Salar, N., and Reig, J. A. (2005) From stem cells to insulin-producing cells: towards a bioartificial endocrine pancreas. *Panminerva Medica* **47**, 39–51.

7. Roche, E., Sepulcre, M. P., Enseñat-Waser, R., Maestre, I., Reig, J. A., and Soria, B. (2003) Bio-engineering insulin-secreting cells from embryonic stem cells: a review of progress. *Med. Biol. Eng. Comp.* **41**, 384–391.

8. Soria, B., Skoudy, A., and Martín, F. (2001) From stem cells to beta cells: new strategies in cell therapy of diabetes mellitus. *Diabetologia* **44**, 407–415.

9. Soria, B., Roche, E., Berna, G., León-Quinto, T., Reig, J. A., and Martín, F. (2000) Insulin-secreting cells derived from embryonic stem cells normalize glycemia in streptozotocin-induced diabetic mice. *Diabetes* **49**, 157–162.

10. Vaca, P., Martin, F., Vegara-Meseguer, J. M., Rovira, J. M., Berna, G., and Soria, B. (2006) Induction of differentiation of embryonic stem cells into insulin secreting cells by foetal soluble factors. *Stem Cells* **24**, 258–265.

11. Leon-Quinto, T., Jones, J., Skoudy A., Burcin, M., Soria, B. (2004) In vitro directed differentiation of mouse embryonic ítem cells into insulin-producing cells. *Diabetologia* **47**, 1442–1451.

12. Lumelsky, N., Blondel, O., Laeng, P., Velasco, I., Ravin, R., and McKay, R. (2001) Differentiation of embryonic stem cells to insulin-secreting structures similar to pancreatic islets. *Science* **292**, 1389–1394.

13. Hori, Y., Rulifson, I. C., Tsai, B. C., Heit, J. J., Cahoy, J. D., and Kim, S. K. (2002) Growth inhibitors promote differentiation of insulin-producing tissue from embryonic stem cells. *Proc. Natl. Acad. Sci. U.S.A.* **99**, 16105–16110.

14. Moritoh, Y., Yamato, E., Yasui, Y., Miyazaki, S., and Miyazaki, J. (2003) Analysis of insulin-producing cells during in vitro differentiation from feeder-free embryonic stem cells. *Diabetes* **52**, 1163–1168.

15. Blyszczuk, P., Czyz, J., Kania, G., Wagner, M., Roll, U., St-Onge, L., and Wobus, A. (2003) Expression of Pax4 in embryonic stem cells promotes differentiation of nestin-positive progenitor and insulin-producing cells. *Proc. Natl. Acad. Sci. U.S.A.* **100**, 998–1003.

16. Roche, E., Sepulcre, P., Reig, J. A., Santana, A., and Soria, B. (2005) Ectodermal commitment of insulin-producing cells derived from mouse embryonic stem cells. *FASEB J.* **19**, 1341–1343.

21

Efficient Generation of Dopamine Neurons from Human Embryonic Stem Cells

Chang-Hwan Park and Sang-Hun Lee

Summary

In this chapter, we introduce a co-culture protocol for human embryonic stem (hES) cell differentiation in which dopamine (DA) neurons with midbrain-specific markers are efficiently derived. Human ES cells on a feeder layer of stromal cells are induced to differentiate into neuroepithelial or neural precursor cells with embryonic midbrain precursor properties. The resulting neural precursor cells are then selectively expanded and serially passaged to obtain a large, homogeneous population of these cells. Under the conditions for terminal differentiation, the majority of hES-derived neural precursors differentiate into neuronal cells that are positive for DA neuronal markers such as tyrosine hydroxylase (TH) and function in vitro as presynaptic DA neurons.

Key Words: Dopamine neurons; Human embryonic stem cells; Tyrosine hydroxylase, Parkinson's disease.

1. Introduction

The derivation of specific cell fates from human embryonic stem (hES) cells represents the initial step for using these cells in developmental biology research as well as in regenerative medicine. Clinical transplantation experience using fetal midbrain tissues has suggested that cell replacement strategies may be a future therapeutic option for Parkinson's disease, which is caused by the specific degeneration of dopamine (DA) neurons in the substantia nigra within the midbrain *(1)*. Generation of DA neurons from stem cells is therefore of particular relevance in the use of hES cells as a renewable cell

From: *Methods in Molecular Biology, vol. 407: Stem Cell Assays*
Edited by: M. C. Vemuri © Humana Press, Totowa, NJ

source for the treatment of Parkinson's disease. Several protocols, including embryoid body (EB)-based lineage selection *(2)* and co-culture *(3)* methods, have been developed for the generation of DA neurons with midbrain cell properties from mouse ES (mES) cells. EB-based neural differentiation has yielded low numbers of DA cells from hES cells *(4,5)*. In contrast, the co-culture protocol, in which DA neuronal differentiation of ES cells is induced by signals derived from stromal cells [referred to as SDIA (stromal cell-derived inducing activity) in *(3)*], has recently been demonstrated to be applicable for DA differentiation of hES cells *(6–8)*. However, efficient DA differentiation of hES cells using these co-culture protocols requires a laborious step involving dissection of clusters of differentiated neural structures under a microscope *(6)*, thereby making mass production of DA cells unfeasible. Here, we introduce a substantially improved co-culture method in which hES cells synchronously and sequentially differentiate into midbrain neural precursors and DA neuronal cells. Repeated subcultures during the co-culture step produce a high yield of midbrain precursor clusters, rendering microscopic dissection unnecessary. The hES-derived midbrain precursor cell clusters in the subsequent step are dissociated into single cells that are selectively expanded and are finally subjected to differentiate into neuronal cells, of which the majority are DA cells with in vitro DA neuronal functions.

2. Materials

2.1. Media, Buffers, and Reagents

2.1.1. MS5 Medium

Modified Eagle's Medium alpha (α-MEM) (cat. no. 11900, Invitrogen/Gibco, Carlsbad, CA, USA) containing 20 mM sodium bicarbonate, 100 U penicillin/100 μg/ml streptomycin (cat. no. 15140, Invitrogen/Gibco), and 10% fetal bovine serum (FBS, cat. no. SH30397.03, Hyclone, Logan, UT, USA)

2.1.2. ITS Medium

Dulbecco's MEM (DMEM)/F12 (cat. no. 12500, Invitrogen/Gibco) containing 3 mM D(+) glucose (cat. no. G7021, Sigma, St. Louis, MO, USA), 2 mM L-glutamine (cat. no. G8540, Sigma), 5 mg/L insulin (cat. no. I1882, Sigma, in 0.1 N NaOH), 50 mg/L transferrin (cat. no. T2036, Sigma), 30 nM sodium selenite (cat. no. S5261, Sigma), 28.5 mM sodium bicarbonate (cat. no. S5761, Sigma), and 100 U penicillin/100 μg/ml streptomycin (cat. no. 15140, Invitrogen/Gibco).

2.1.3. Other Buffers and Reagents

Distilled water (cat. no. 15230-147, Invitrogen/Gibco).

100 mM ascorbic acid (AA; cat. no. A4544, Sigma, 500× stock).

10 µg/ml basic fibroblast growth factor (bFGF; cat. no. 233-FB, R&D, Minneapolis, MN, USA, 500× stock).

Ca^{+2}, Mg^{+2}-free HBSS (CMF-HBSS; cat. no. 14185-052, Invitrogen/Gibco, 10× stock).

Phosphate-buffered saline (PBS; cat. no. 70011, Invitrogen/Gibco, 10× stock).

200 µg/ml polybrene (hexadimethrine bromide, cat. no. H9268, Sigma, 100× stock).

1 mg/ml blasticidin (cat. no. R210-01, Invitrogen/Gibco, 100× stock).

15 mg/ml polyornithine (PO; cat. no. P3655, Sigma, 1000× stock).

1 mg/ml fibronectin (FN; cat. no. F1141, Sigma, 1000× stock).

10mg/ml (2500 U/ml) collagenase (collagenase Type IV, cat. no. 17104-019, Invitrogen/Gibco, 10× stock).

50 µg/ml sonic hedgehog (rmSHH-N; cat. no. 461-SH, R&D, 100 × –500× stock).

50 µg/ml fibroblast growth factor-8 (rmFGF-8b; cat.no. 423-F8, R&D, 500× stock).

0.1% gelatin (Sigma, cat.no. G1890, v/w in PBS).

2.2. Preparation of the MS5 Stromal Cell Feeder Layer

Prepare gelatin-coated dishes by adding gelatin solution (0.1%, v/w in PBS) to culture dishes and incubating for at least 5 min in a CO_2 incubator.

Irradiate MS5 cells at 6000 rad in a γ-irradiator (Gammacell 1000Elite, MDS Nordion, Canada).

Plate the γ-irradiated MS5 cells at 3×10^4 cells/cm^2 on gelatin-coated plates.

2.3. Generation of MS5 Cells Stably Overexpressing SHH

The human sonic hedgehog N-terminal region (ShhN) was amplified with primers 5′-CATATGCTGCTGCTGGCGAGAT-3′ and 5′-GTCGACTCAGCCT CCCGATTTGG-3′ using high-fidelity Taq polymerase (cat. no. 11304, Invitrogen/Gibco) with 30 cycles of 94 °C/0.5 min, 55 °C/min, and 68 °C/3 min. The ShhN polymerase chain reaction (PCR) product was cloned into the pGEM T-Easy TA cloning vector (cat. no. A 1360, Promega, Madison, WI, USA), digested with *Not*I/*Sal*I restriction enzymes, and cloned into the corresponding sites in the IRES-BsdEGFP-CL retroviral vector, which simultaneously expresses the antibiotic blasticidin resistance gene and the EGFP fusion protein (ShhN-BsdEGFP-CLBC3). The ShhN-BsdEGFP-CLBC3 plasmid was introduced into the 293gpg retrovirus packaging cell line *(9)* by transient transfection with Lipofectamine 2000 (cat. no. 11668, Invitrogen/Gibco). After 72 h, the supernatants were harvested and used to infect MS5 cells with polybrene ($2 \mu g/ml$) for 2 h. Two days later, transduced MS5 cells were selected by growing in the presence of $10 \mu g/ml$ blasticidin for 1 week.

2.4. Preparation of PO/FN-Coated Plates

Add 5–10 ml PO solution ($15 \mu g/ml$ in PBS) to culture plates, incubate for at least 4 h in a CO_2 incubator, and wash three times with PBS.

After washing, add FN solution ($1 \mu g/ml$ in PBS) and incubate for at least 1 h. Aspirate FN solution immediately before use and add cell solution to the dishes. The prepared PO/FN-coated dishes should be used within 1 week.

3. Methods

DA differentiation using the co-culture protocol with stromal cell lines MS5 and PA6 has been previously performed in the hES cell lines HSF6, SNU-hES-3, Miz-hES-1 *(8)*, H1, H9, HES-3 *(6)*, BG-01, and BG-03 *(7)*. The protocol described here is based on Park et al. *(8)*, with some modifications. Briefly, the protocol for hES differentiation into DA neuronal cells consists of four stages: propagation of undifferentiated hES cells (stage 0), neural induction of hES cells on the stromal feeder layer (stage I), further selection and expansion of neural precursor cells on PO/FN-coated plates (stage II), and terminal differentiation into DA neuronal cells (stage III) *(see* **Fig. 1**). Human ES-derived neuroepithelial cells or neural precursor cells after efficient neural induction or expansion (stage I or II) can be stored in liquid N_2 and recultured by simple freezing/thawing. Note that the efficiency of neural induction using the co-culture protocol is highly variable between hES cell lines and the different experimental conditions used *(see* **Note 1**).

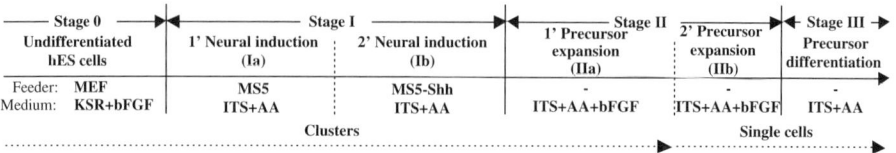

Fig. 1. General scheme of the co-culture protocol used for dopamine (DA) differentiation of human embryonic stem (hES) cells.

3.1. Culture for the Maintenance and Propagation of hES Cells (Stage 0)

In general, undifferentiated hES cells (*see* **Fig. 2**) should be maintained in specialized medium for hES maintenance on the layer of mouse fibroblast feeder cells or STO-1 cells. However, specific methods for the maintenance of each hES cell line should be followed according to the protocols provided by the cell line establishers.

3.2. Neural Induction on the Stromal Feeder Layer (Stage I)

1. Prepare the feeder layer of MS5 or PA6 stromal cells (*see* **Note 2**) on 10-cm culture dishes as described in **Subheading 2.2.**, 1 day before starting the neural induction of hES cells. Incubate the feeder cells in a CO_2 incubator until use.
2. Transfer hES colonies onto the MS5 (or PA6) feeder layer.

   a. Aspirate the medium from the cultures for undifferentiated hES cells and add 4 ml collagenase solution (1 mg/ml in DMEM/F12).

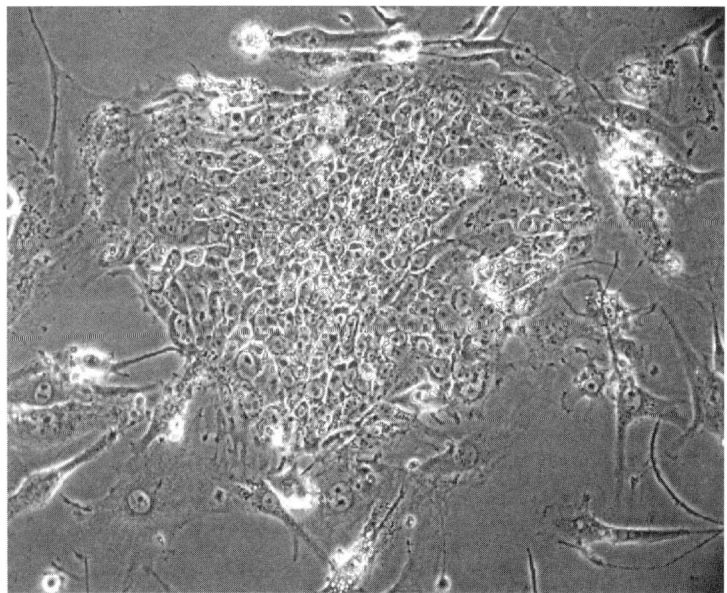

Fig. 2. Representative image of an undifferentiated human embryonic stem (hES) cell colony. Undifferentiated hES cells have morphologic characteristics including a high nucleus/cytoplasm ratio and prominent nucleoli.

b. After incubation for 10 min in a CO_2 incubator, gently aspirate the collagenase solution and wash the cells once with 5 ml DMEM/F12 medium (*see* **Note 3**).

c. Add 5 ml ITS medium to the collagenase-treated dishes and then harvest hES cell colonies by scraping the cells with a cell lifter (cat. no. 3008, Coring, Corning, NY, USA).

d. Place the cell suspension in a 15-ml centrifuge tube and wait for 5 min until the cell clusters sink down. Gently aspirate the transparent upper layer of the medium from the tube.

e. Split the cell clusters into smaller pieces (50–500 cells/cluster) by mechanical pipetting (*see* **Note 4**). Aspirate the upper medium again after waiting for 5 min.

f. During the collagenase treatment procedure, wash the MS5 feeder cells with DMEM/F12, and incubate in 5 ml ITS+AA medium until use.

g. Resuspend the chopped cell clusters in 1–2 ml ITS+AA and inoculate into one to five dishes of freshly washed MS5 feeder layers. Swirl the dishes gently to evenly distribute the cell clusters on the feeder layer. Return the dishes to the CO_2 incubator.

3. Repeat the procedure of subculturing onto freshly prepared MS5 (or PA6) feeders, as described in **step 2**, every 7 days, until > 80% of the cell colonies achieve a primitive neuroepithelial cell morphology (*see* **Fig. 3**), such as neural rosettes (arrow in **Fig. 3B**), on microscopic examination (*see* **Note 5**). Change the medium every other day. After the final round of subculture, cells can be harvested with collagenase treatment and frozen in ITS+AA containing 10% DMSO in liquid N_2.

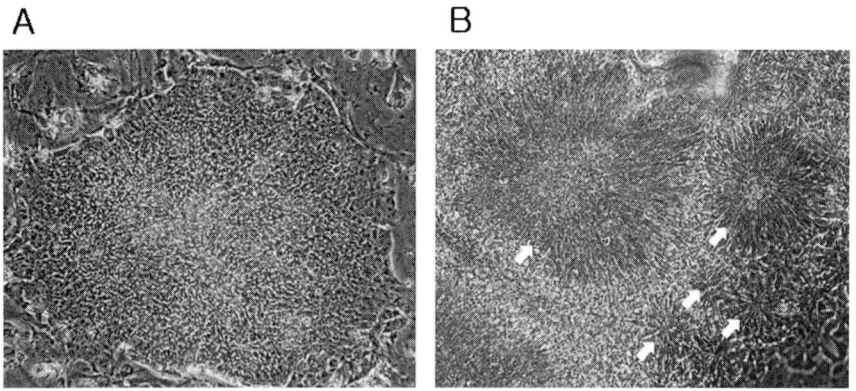

Fig. 3. A differentiated human embryonic stem (hES) cell cluster with a primitive neural structure. Compared to undifferentiated hES cells shown in **Fig. 2**, hES-derived neuroepithelial cells are much smaller, more compactly arranged, and have an abundant cytoplasm and barely visible nuclei. (**A**) subpopulation of the differentiated hES clusters contains neural tube-like primitive structures, neural rosettes (arrows in **B**).

3.3. Selection and Proliferation of hES-Derived Neural Precursor Cells (Stage II)

1. The differentiated hES colonies should be dislodged, chopped using collagenase treatment as described above, and then plated on PO/FN-coated plates (*see* **Subheading 3**.) in ITS+AA medium supplemented with 20 ng/ml bFGF (*see* **Note 6**). Alternatively, neuroepithelial cell colonies can be isolated by mechanical dissection under a microscope *(6)*, particularly in cases in which differentiated hES cell cultures comprise only a small percentage of the neuroepithelial cell clusters. Subculture onto freshly prepared PO/FN-coated plates every 7 days two to four times (*see* **Note 7**). Medium should be changed every other day, and bFGF should be added every day. During culture in bFGF-supplemented medium, the clusters of neuroepithelial or neural precursor cells will selectively survive and proliferate, whereas non-neural cells, including hES-derived cells with the other lineages, the stromal feeder cells, and differentiated neurons (or glia), will tend not to survive, particularly because of the mechanical procedures of detaching, chopping, and re-plating during the cell passages. As a result, a more uniform population of cells positive for nestin, a specific marker for neural precursor cells, will be obtained after subculturing.

2. Disrupt the cell clusters into single cells (*see* **Note 8**).

   a. Wash the cells one or two times for 5 min with PBS.
   b. Incubate the cells in 5 ml Ca^{+2}, Mg^{+2}-free-HBSS for 30–60 min in a CO_2 incubator. This will loosen cell–cell contacts because of the cell junction requirement for divalent metal ions.
   c. Using a 1-ml pipette, suck and squirt the HBSS solution onto the cells several times with the dish in a tilted position, so that the cell layer peels off. Scrape remnant cells with a cell lifter.
   d. Collect the cells in a 15-ml centrifuge tube and briefly spin down.
   e. Dissolve the cell pellet into single cells by pipetting in 1 ml ITS+AA+bFGF.

3. Inoculate the dissociated cells at $50,000$ cells/cm^2 on PO/FN-coated dishes (for maintenance and further propagation of the hES-neural precursor cells) or coverslips (12-mm diameter, Marienfeld 0111520; for immunocytochemical phenotype determinations and functional analyses). In this stage of the cell culture, hES-derived neural precursor cells can be maintained and propagated in ITS + AA supplemented with 20 ng/ml bFGF, a mitogen for neural precursor cells (*see* **Note 6**), for several months with serial passages (*see* **Note 9**). During the period of precursor cell expansion, cultures should be passaged every time 70–90% cell confluency is reached (usually every 4–7 days). The hES-derived neural precursor cells can be stored in bFGF-supplemented medium containing 10% DMSO in liquid N_2 and recultured by simple freezing/thawing. Midbrain precursor properties of hES-derived neural precursors in cell cultures at this stage have been determined in our previous study *(8)* by immunocytochemical and reverse transcriptase–polymerase

chain reaction (RT–PCR) analyses for various embryonic midbrain markers such as Pax2 and Engrailed-1.

3.4. Terminal Differentiation of hES-Derived Neural Precursor Cells into DA Neurons (Stage III)

The purpose of this step is to induce differentiation of hES-derived neural precursor cells in ITS+AA medium in the absence of bFGF. Such withdrawal of bFGF sufficiently induces the neural precursors to differentiate into neuronal and glial cells. Differentiation phenotypes can be assessed by immunologic and RT–PCR analyses for markers specific for neurons (tubulin β-III, TuJ1; microtubule-associated protein 2, MAP2), DA neurons (tyrosine hydroxylase, TH), astrocytes (glial fibrillary acidic protein; GFAP), and oligodendrocytes (CNPase), for example. In optimal culture conditions, up to 90% of cells are positive for the neuronal marker TuJ1, and 10–60% of the TuJ1$^+$ cells express the DA neuron marker TH after 10–15 days of differentiation (*see* **Fig. 4**). In in vitro functional analyses, avid depolarization-induced DA release and DA transporter-mediated DA uptake have been observed *(8)*.

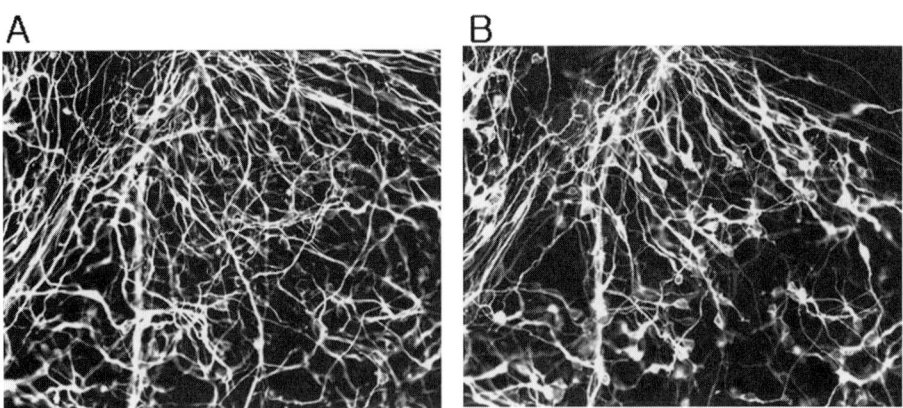

Fig. 4. Representative images of dopamine (DA) neurons derived from hES cells. In vitro-expanded hES-neural precursors (stage II) were induced to terminally differentiate by withdrawing basic fibroblast growth factor (bFGF) for 16 days (stage III). Phenotypic determination was performed using immunocytochemistry for TuJ1 (**A**, neuronal maker) and Tyrosine hydroxylase (TH) (**B**, DA neuronal marker).

4. Notes

1. Efficiency in the derivation of neural precursor cells and DA neurons is highly variable among hES cell lines. For instance, nestin-positive neural precursors and TH-positive DA neurons have been efficiently derived from hES cell lines HSF6 *(8)*, H1, and H9 *(6)* that were co-cultured with MS5 or PA6 stromal feeder cells. In contrast, only a minor proportion of Miz-hES-1 and SNU-hES-3 cells have been shown to differentiate into the midbrain neural precursor cells that give rise to DA neurons *(8)*. The degree of DA neuronal differentiation of hES cells also greatly varies depending on the number of hES cell passages, even within the same cell line, and depending on other experimental conditions, including cell density, the size of hES cell colonies, cell viability, and general conditions of the feeder layer.

2. Stromal feeder cell lines MS5 and PA6 may have similar efficacy in eliciting DA differentiation of hES cells. However, more synchronous differentiation of HSF6 hES cells, with homogenous populations of cells at the same developmental stage, has been observed with MS5 cells, whereas the period required for efficient differentiation of hES cells into a neural lineage is shorter using PA6 cells.

3. Collagenase treatment loosens cell–cell contacts, which facilitates cell dissociation during cell passaging. However, a 10-min collagenase treatment in most cases does not dislodge the cell colonies during washing, unless cells are mechanically scraped. Alternatively, cells can undergo a longer collagenase treatment to avoid the mechanical stress that occurs during cell scraping. However, this longer collagenase treatment is more lethal to the cells than the shorter collagenase treatment followed by scraping and most likely causes hES cells to have more chromosomal abnormalities *(10)*.

4. Neural differentiation of hES colonies is highly dependent on the size of the colony. The differentiation potential of hES cells assembled into smaller clusters tends to be greater, but cells are less viable if disrupted into clusters that are too small. Thus, the optimal conditions for cell disruption should be determined in each laboratory. We have found that the cell disruption procedure is more convenient and efficient when performed on an in vitro fertilization (IVF) dish (cat. no. 353653, BDF alcon, Franklin Lakes, NJ, USA).

5. Although 7 days is adequate for subculture, this period can be varied depending on the state of the cultures, including cell viability of the feeder or hES cells as well as the size and number of hES colonies. The percentage of differentiated hES colonies with a neuroepithelial cell morphology gradually increases during subculture. In general, four to eight subcultures (for 1–2 months) are required for >80% of the colonies to adopt a neuroepithelial cell morphology (*see* **Fig. 3**). The period required for such efficient differentiation is also highly variable depending on which hES cell lines and experimental conditions are used, as described in **Note 1**. DA neuronal yield is enhanced when cells are subcultured on an MS5-SHH feeder layer (see **Subheading 2**.) after two and three rounds of subculture *(8)*. However, the neural induction efficiency of hES cells greatly decreases when

cells are co-cultured on the MS5-SHH feeder layer from the beginning of the co-culture stage.

6. SHH (100–500 ng/ml) or SHH+FGF8 (100 ng/ml) can be added to bFGF-supplemented medium or substituted for bFGF to enhance DA neuronal yield *(6)*.

7. Both the size of the cell clusters and cell density are crucial factors for cell viability and uniformity of the cell population. In smaller clusters with low cell density, cell viability is low, but efficient selection for neural precursor cells can be achieved. The optimal conditions for cluster size and cell density, as well as the appropriate interval for each cell passage, should be determined by each experimenter.

8. If cultures are maintained in the cell cluster stage up to terminal differentiation, synchronous differentiation cannot be achieved: cells in the center of the cluster maintain their undifferentiated properties or undergo apoptosis because of high cell density, whereas cells in the periphery tend to have a higher differentiation potential. Thus, disruption of clusters into single cells and subsequent procedures for selective proliferation of the neural precursors by bFGF are required to obtain a uniform population of neural precursor cells. After several days of bFGF-induced expansion, > 95% of cells in cultures plated with single dissociated cells have been shown to be positive for the neural precursor cell marker nestin *(8)* (*see* **Fig. 5**). The majority (80–90%) of the precursors differentiate toward a neuronal fate under the differentiation conditions described in **Subheading 3.4** (see **Fig. 4**).

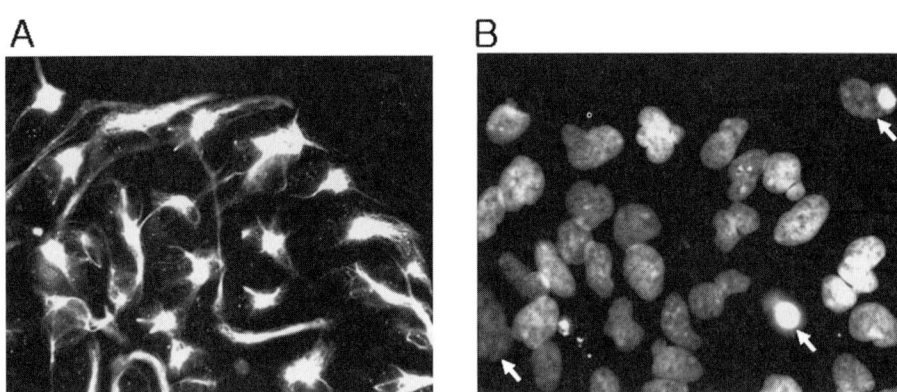

Fig. 5. Human embryonic stem (ES)-derived neural precursor cells. Differentiated hES cell clusters with neural structures were dissociated into single cells and cultured in medium supplemented with bFGF, a specific mitogen for neural precursors. The neural phenotype of these cells was confirmed by immunohistochemical analysis for expression of nestin, a marker specific to neural precursor cells (**A**) 3 days after bFGF expansion. Total cells are visualized by DAPI nuclear staining (**B**). Note that the majority of cells, with the exception of the cells marked with arrows, are nestin-positive.

In addition, the bFGF-induced increase in cell number tends to be much greater in cultures plated with single cells than in those plated with clusters.

9. hES-derived neural precursor cells that have undergone bFGF-expansion for several days predominantly differentiate into neurons. However, increased glial (astrocytic and oligodendrocytic) differentiation at the expense of neuronal differentiation has been observed in differentiated cultures in which precursors were allowed to expand for a longer period with several cell passages *(8)*. Although DA neural yield was also reduced in the hES-derived precursors with increased passage numbers, 10–20% of the total cells under optimal conditions were DA neurons in the differentiated cultures after cell expansion was allowed for > 2 months (data not shown). This contrasts greatly with the finding in primary rat midbrain precursor cultures that demonstrated a large decrease in DA differentiation after only one to two cell passages *(11–12)*.

Acknowledgments

This work was supported by SC-2150 (Stem Cell Research Center of the 21st Century Frontier Research Program) funded by the Ministry of Science and Technology, Republic of Korea.

References

1. Piccini P., Brooks D.J., Bjorklund A., Gunn R.N., Grasby P.M., Rimoldi O., Brundin P., Hagell P., Rehncrona S., Widner H. and Lindvall O. (1999) Dopamine release from nigral transplants visualized *in vivo* in a Parkinson's patient. *Nat. Neurosci.* **2**, 1137–1140.
2. Lee S.H., Lumelsky N., Studer L., Auerbach J. and McKay R.D. (2000) Efficient generation of midbrain and hindbrain neurons from mouse embryonic stem cells. *Nat. Biotechnol.* **18**, 675–679.
3. Kawasaki H., Mizuseki K., Nishikawa S., Kaneko S., Kuwana Y., Nakanishi S., Nishikawa S.I. and Sasai Y. (2000) Induction of midbrain dopaminergic neurons from ES cells by stromal cell–derived inducing activity. *Neuron* **28**, 31–40.
4. Reubinoff B.E., Itsykson P., Turetsky T., Pera M.F., Reinhartz E., Itzik A. and Ben-Hur T. (2001) Neural progenitors from human embryonic stem cells. *Nat. Biotechnol.* **19**, 1134–1140.
5. Zhang S.C., Wernig M., Duncan I.D., Brustle O. and Thomson J.A. (2001) In vitro differentiation of transplantable neural precursors from human embryonic stem cells. *Nat. Biotechnol.* **19**, 1129–1133.
6. Perrier A.L., Tabar V., Barberi T., Rubio M.E., Bruses J., Topf N., Harrison N. L. and Studer L. (2004) Derivation of midbrain dopamine neurons from human embryonic stem cells. *Proc. Natl. Acad. Sci. U.S.A.* **101**, 12543–12548.
7. Zeng X., Cai J., Chen J., Luo Y., You Z.-B., Fotter E., Wang Y., Harvey B., Miura T., Backman C., Chen G.-J., Rao M.S. and Freed W.J. (2004) Dopaminergic differentiation of human embryonic stem cells. *Stem Cells* **22**, 925–940.

8. Park C.H., Minn Y.K., Lee J.Y., Choi D.H., Chang M.Y., Shim J.W., Ko J.Y., Koh H.C., Kang M.J., Kang J.S., Rhie D.J., Lee Y.S., Son H., Moon S.Y., Kim K. S. and Lee S.H. (2005) In vitro and in vivo analyses of human embryonic stem cell-derived dopamine neurons. *J. Neurochem.* **92**,1265–1276.

9. Ory D.S., Neugeboren B.A. and Mulligan R.C. (1996) A stable human-derived packaging cell line for production of high titer retrovirus/vesicular stomatitis virus G pseudotypes. *Proc. Natl. Acad. Sci. U.S.A.* **93**, 11400–11406.

10. Draper J.S., Smith K., Gokhale P., Moore H.D., Maltby E., Johnson J., Meisner L., Zwaka T.P., Thomson J.A. and Andrews P.W. (2004) Recurrent gain of chromosomes 17q and 12 in cultured human embryonic stem cells. *Nat. Biotechnol.* **22**, 53–54.

11. Ko J.Y., Lee J.Y., Park C.H. and Lee S.H. (2005) Effect of cell-density on in-vitro dopaminergic differentiation of mesencephalic precursor cells. *Neuroreport* **16**, 499–503.

12. Yan J., Studer L. and McKay R.D. (2001) Ascorbic acid increases the yield of dopaminergic neurons derived from basic fibroblast growth factor expanded mesencephalic precursors. *J. Neurochem.* **76**, 307–311.

22

Isolation of Oligodendroglial Cells from Cultured Neural Stem/Progenitors

Frank Zeigler and Stephen G. Hall

Summary

The central nervous system (CNS) is composed of multiple cell types formed through a process of lineage commitment and phenotypic differentiation of stem-like progenitor cells into three key cell types: neurons, astrocytes, and oligodendrocytes. The ability to isolate and culture these CNS stem/progenitors has facilitated the characterization of the molecular mechanisms that regulate this process, in the hopes of providing therapeutically effective cells to treat disease and injury. Although astroglial, and to a lesser extent some neuronal, phenotypes are robustly generated when these cultured stem/progenitor cells are induced to differentiate, oligodendrocytes that form the myelin-rich sheath that allows nerves to conduct action potentials are only formed at a low frequency. This relatively low frequency has necessitated the development of methods for quantifying oligodendroglial phenotypes in vitro, with greater precision and accuracy than the standard technique of microscopic counting by hand. Here, we describe the isolation of neural stem cells and the application of intracellular flow cytometry to quantify oligodendroglial phenotypes in cultured CNS stem/progenitor cells using commercially available kits.

Key Words: CNS; Oligodendrocytes; Progenitor cell; Differentiation; Intracellular flow cytometry; Phenotype.

1. Introduction

Although the exact location of multipotent neural stem cells (NSCs) in the adult brain is somewhat controversial, it is accepted that one of the areas with the highest NSC concentration is in the subventricular zone (SVZ) of the lateral ventricular walls *(1–3)*. One can consistently distinguish multipotent NSCs

From: *Methods in Molecular Biology, vol. 407: Stem Cell Assays*
Edited by: M. C. Vemuri © Humana Press, Totowa, NJ

from other precursor cells as well as differentiated cells present in the central nervous system (CNS). The addition of epidermal growth factor (EGF) and basic fibroblast growth factor (bFGF) has been shown to expand neurospheres in undifferentiated state *(4)*. In contrast, retinoic acid (RA) can induce stem cells to differentiate into glial cells (astrocytes and oligodendrocytes) and neurons *(5)*.

The existence of a glial precursor in both the developing and the adult CNS was first demonstrated with the isolation and culture of a common glial progenitor to oligodendrocytes and type-2 astrocytes, termed O-2A progenitor or oligodendroglial progenitor cells (OPC) *(6)*. In culture, this precursor cell could directly give rise to either oligodendrocytes or type-2 astrocytes or divide to form another OPC. Subsequently, cells with similar attributes have been isolated from various regions of both rodent and human CNS *(7)*, and more recently, growth factor-responsive CNS stem/progenitor cells grown as non-adherent cultures of "neurospheres" have also been shown to contain OPC's *(8)*.

These studies have primarily relied on semi-quantitative determinations, such as reverse transcriptase – polymerase chain reaction (RT–PCR) for gene expression analysis and immunophenotyping with manual microscopic counting. Although RT–PCR is exquisitely sensitive, it cannot distinguish the precise number of cells of a given phenotype, whereas microscopic hand counting of immunostained cells is prone to both inaccuracy and variability, unless great effort is taken to count sufficient cell numbers to obtain a representative sample size. This type of quantitation is also subjective and often depends on individual operators whose relative experience and "trained eye" can vary widely. Additionally, manual immunophenotyping methods are not easily adapted to multi-parameter analyses, particularly when more than two markers are desired within a single population of cells. A method is therefore required to address the need for quantitative cellular phenotyping of cultured CNS stem/progenitors, OPCs, and terminally differentiated oligodendrocytes.

Flow cytometry provides an analytical technology that can supply the required phenotypic information, while also providing the necessary accuracy, sensitivity, and robustness. Its multi-parameter capability allows for the detection and enumeration of multiple phenotypes simultaneously on thousands of cells per second in just a few minutes. This provides a large amount of high-content information regarding a population of cells at the single-cell level and is ideally suited for studies of CNS stem/progenitor cells, which display a strong inherent potential for multi-lineage differentiation. However, many of the markers commonly used to phenotype OPCs and oligodendrocytes are in fact intracellular, for example, CNPase, glial fibrillary acidic protein (GFAP), nestin, and the transcription factors Olig 1/2.

Intracellular FACS requires that cells are first dissociated into a single-cell suspension. In our experience, trypsin is the best enzyme for efficient preparation of a viable single-cell suspension of cultured cells in many applications, including rodent and human cultured neural stem cells. However, care must be taken because trypsin can vary significantly in activity, and we have found that both considerable efforts can be required to optimize the use of trypsin, including optimization of reaction time and screening different manufacturers and production lots. CNS cells are fragile and have been reported to lose viability after trypsin dissociation, but as they will be fixed for intracellular FACS in any case, this is not problematic.

Once the method of cell dissociation has been optimized, cells must be fixed to preserve both cellular morphology and antigenicity and permeabilized to allow antibody penetration within the cell. Formaldehyde is the most widely used and effective fixative for preserving cell and tissue morphology and antigenicity, and although many different formulations and commercial kits exist for intracellular FACS, they generally rely on the covalent cross-linking of primary amines in cellular macromolecules, mostly proteins and nucleic acids, and generally contain between 1 and 4% formaldehyde. Some protocols use methanol as both fixative and permeabilizer; however, issues with cell losses and incompatibility can occur. Permeabilization can be accomplished using various detergents, such as saponin and Tween series non-ionic detergents, with effective concentrations of 0.01 to 0.1% in most cases.

With the basic techniques for cell dissociation and intracellular FACS preparation complete, attention must next be paid to the choice of both the phenotypic markers and the antibody probes used to identify them. Regarding the choice of phenotypic marker, each study may have its own unique requirements, and a full discussion of the relevance of the many possible markers is outside the scope of this work. However, a brief review of the literature shows that many phenotypic markers are used in the majority of studies involving cultured CNS stem/progenitors, OPCs, and oligodendrocytes derived from them. Nestin is commonly used to mark non-differentiated stem/progenitors, b-tubulin III identifies neuronal cells, GFAP for astroglial, and for committed oligodendroglial progenitors and terminally differentiated oligodendrocytes CNPase, MBP, and the O1/O4 glycolipid antigens. The biological relevance of each of these must of course be empirically determined through rigorous experimentation.

Many primary antibodies are available commercially; however, the degree to which they have been validated and in some cases even their specificity is variable or unknown. Each antibody–antigen combination needs to be evaluated for compatibility with the intracellular FACS reagents, as well as for specificity for the target antigen. Dual parameter flow cytometry can be helpful in this regard, for example, using both CNPase and MBP to qualify both antibodies by dual localization on primary spinal cord isolated oligodendrocytes. Cell lines that have been shown to express the antigen can also be useful in some cases, or more frequently, as a source of a non-expressing cell type for the particular antigen [i.e., if your myelin basic protein (MBP) antibody reacts with KG1a hematopoietic cells, then it cannot be trusted]. We have also found it important to visualize the pattern of reactivity with each antibody and cell type by immunofluorescence, to confirm the appropriate cellular localization (e.g., nestin and GFAP should look like intermediate filaments microscopically) when qualifying an antibody.

As FACS is an extremely sensitive detection methodology, and as one cannot visualize the phenotypic signal in intracellular FACS, the appropriate negative controls are absolutely required. Each cell type and preparation needs to have a negative control using either an isotype-matched immunoglobulin G (IgG) or an irrelevant IgG from the same species as the primary antibody and at same concentration, time, and so on. For multi-parameter studies using two different primary antibodies, tests should be run with the secondary fluorochrome-conjugated antibodies to confirm specificity of dual labeling.

Intracellular FACS for oligodendroglial phenotypes can be applied to the study of oligodendroglial lineage differentiation from cultured CNS stem/progenitor cells. Commercially available kits and reagents save time for assay development work, allowing researchers more time to focus on functional biology, and provide an excellent starting point for phenotyping oligodendroglial cell populations in vitro.

2. Materials

2.1. Isolation of Neural Stem Cells

1. Artificial cerebrospinal fluid (aCSF) (pH 7.4): 126 mM NaCl, 2.5 mM KCl, 2 mM $CaCl_2$, 1–2 mM $MgCl_2$, 1.25 mM NaH_2PO_4, 26 mM $NaHCO_3$, and 10–25 mM D-glucose
2. Add the appropriate amount of 50 × penicillin–streptomycin stock solution (cat.no. P4458, Sigma, St. Louis, MO, USA) to 100 ml of aCSF and sterile filter. Oxygenate the solution with 95% O_2 and 5% CO_2 for 20 min. A spinner flask may be adapted to facilitate oxygenation.
3. Enzymatic dissection solution: Mix 0.8 mg/ml hyaluronidase (cat.no. H6254, Sigma) and 0.5 mg/ml trypsin (cat.no. T1005, Sigma) and add to 50 ml of oxygenated aCSF at 30 °C.

4. Dulbecco's modified Eagle's medium (DMEM)/F-12 medium: Ham's F12 (50:50 v : v) (D/F) (cat.no. D8900, Sigma).
5. DMEM/F-12/EGF medium: DMEM/F-12 medium as described above plus 20 ng/ml EGF (cat.no. E4127, Sigma).

2.2. Preparation of Single Cell Suspension

1. Phosphate-buffered saline (PBS) (cat.no. P5368, Sigma) without calcium or magnesium.
2. 0.025% Trypsin/ethylenediaminetetraacetic acid (EDTA) (cat.no. T4174, Sigma).

2.3. Fixation and Permeabilization

1. Fixative: 2.0% (v/v) formaldehyde in PBS, prepared by dilution of a 10% EM grade stock (Ultrapure cat.no. 04018, Polysciences, Warrington, PA, USA).
2. Permeabilization/wash solution; PBS with 0.1% (v/v) added Tween-20 detergent (cat.no. P7949, Sigma).

2.4. FACS Staining

1. FACS storage buffer: 0.05% (v/v) formaldehyde in PBS, prepared by dilution of a 10% EM grade stock.
2. Secondary fluorochrome-conjugated antibodies; many good commercial suppliers, however, we prefer to use Jackson Immunochemicals products [e.g., R-PE AffiniPure F(ab′)2 Fragment donkey anti-mouse IgG (H + L), cat.no. 715-116-150 at 1:200 dilution/0.1 ml/1 × 10⁶ cells].

2.5. Commercial Kits

1. IntraCyte™ Intracellular FACS Reagent System (cat.no. 77017, Alpha Genix, Carlsbad, CA, USA) for the fixation, permeabilization, and detection of intracellular antigens by flow cytometry in single-cell suspensions. IntraCyte fixative solution is a formaldehyde-based fixative. IntraCyte wash solution permeabilizes and washes using a proprietary detergent blend that maximizes signal intensity and reduces background.
2. IntraCyte-rNSC™ FACS Phenotyping kit (cat.no. 77026, Alpha Genix, Carlsbad, CA, USA) for quantifying cultured rat neural stem and progenitor cell phenotypes through the fixation, permeabilization, and detection of intracellular antigens by flow cytometry in single-cell suspensions. This kit provides all the Ab reagents required, including primary Abs to nestin, GFAP, TUJ1, MBP, and CNPase, IgG controls, and fluorochrome-conjugated secondary antibodies. All antibody reagents come as ready-to-use pre-titered stocks sufficient for at least 20 cell samples stained for five different lineage markers. The kit provides everything required except for cells, PBS, staining tubes, centrifuge, and a flow cytometer.

3. Methods

3.1. Isolation of Neural Stem Cells

1. The SVZ of rat brain is dissected according to Paxino and Watson (*9*). Stereotactic coordinates representing the rat SVZ in reference to bregma are A/P 1.6 mm, M/L −1.3 mm, D/V, 5.3 mm.
2. Dissected brain tissue should be rinsed in oxygenated aCSF.
3. Using a dissecting microscope, dissect the striatum and cut the tissue into small pieces.
4. Transfer the tissue into a sterile tissue culture dish containing aCSF with enzymatic dissection solution added using a sterile wide-bore pipette.
5. Place on a rotating platform at low speed and allow to rotate for 1 h.
6. Transfer the tissue to a sterile 50-ml centrifuge tube containing 5 ml of DMEM/F-12 medium supplemented with ovomucoid (0.7 mg/ml) using a sterile, wide-bore Pasteur pipette. Triturate the tissue 50 times.
7. Centrifuge the suspension at 500 g for 5 min.
8. Aspirate the supernatant and resuspend single cells in DMEM/F-12/EGF complete medium.
9. Plate the cells onto T25 cell culture plates or flasks and culture at 37 °C in 5% CO_2
10. Passage the cells every 4 days with a half media change every second day. After about 7 days in the presence of EGF, the cells will form undifferentiated neurospheres. The neurospheres can be maintained for long periods of times by successive passages involving dissociation and proliferation or frozen in cryopreservation medium.
11. The cultures can be induced to differentiate in DMEM/F-12 medium containing 1.0 μM retinoic acid, 10 ng/ml T3 hormone, and 1% fetal bovine serum (FBS).

3.2. Preparation of Single-Cell Suspension

1. Aspirate cell culture medium and wash adherent cultured CNS cells with PBS once briefly.
2. Gently dispense approximately enough Trypsin solution to cover surface (e.g., 5 ml/10-cm dish or 1 ml per 6-well plate) and place in 37 °C incubator for 2–10 min. Monitor visually using phase-contrast microscopy every 2–3 min, and the earliest time that cells have begun to lift off the culture surface with a gentle tap of the vessel is usually after 5–10 min.
3. Terminate by the addition of an equal volume of protein-containing culture medium to the culture, followed by gentle pipetting up and down 3–5 times to dislodge cells.
4. Centrifuge at 300 g for 10 min at 4 °C.
5. Resuspend cell pellet in PBS and count using hemacytometer. Viability of trypan blue negative cells should be 90% or more.

3.3. Fixation and Permeabilization

1. Wash $1–5 \times 10^6$ viable cells in a single-cell suspension using 10–15 ml ice cold PBS by centrifugation at 300 g for 5 min.
2. Fix cells by re-suspension in fixative solution and incubate at room temperature for 30 min, or overnight at 4–8 °C. Fix $1–5 \times 10^6$ cells per 1.0 ml of fixative, scaling up volumes for larger cell numbers. Some cells and antigen/antibody combinations require longer fixation times at lower temperatures, whereas most antigens are efficiently preserved after 30 min at room temperature.
3. Wash fixed cells by addition of 10–15 ml of PBS and centrifugation at 300 g for 10 min. The formaldehyde and/or organic solvent contained in the fixative must be removed before permeabilization. All steps from this step forward can be performed at room temperature. After fixation, cells are less dense and therefore require slightly more g force to pellet effectively.
4. Resuspend in 1–2 ml using permeabilization/wash solution. Using a P1000 tip helps in resuspension, and a cell count can be performed as well to determine yield.

3.4. FACS Staining

1. It is generally convenient to stain $1–5 \times 10^5$ cells for FACS using 0.1–0.2 ml per test in a 1.5ml microfuge tube. 96-well format can also be used, and we recommend using 50 μl with $1–5 \times 10^5$ cells per well.
2. Carefully label tubes and dispense cells according to a prepared list. Remember to include the appropriate positive and negative controls.
3. Add appropriate concentration of primary Ab in 1–10 μl then mix gently. Dilute primary Ab as needed before addition using permeabilization/wash solution. Most primary Abs are effective at $1 μg/ml/0.1 \times 10^6$ cells; however, each antibody used should be carefully titered against test cells more specifically.
4. Incubate for 30 min at 4–8 °C, then wash cells by addition of 1.0–1.5 ml of IntraCyte-Wash and centrifugation at 300 g for 5 min. Use at least 1 ml per test for 1.5-ml tubes and 0.3 ml per test for 96-well plates. Only a single wash step is usually required, but sometimes washing twice can reduce background. When aspirating, remember to leave a small volume of wash fluid above the cell pellet.
5. Aspirate and resuspend in 0.1 ml/test wash/permeabilization solution.
6. Add appropriate concentration of pre-titered fluorochrome-conjugated secondary Ab at in 1–10 μl wash/permeabilization solution.
7. Incubate for 30 min, then wash cells as before in **step 3.4.4** by centrifugation.
8. Re-suspend in 0.1–0.2 ml per test of PBS or FACS storage buffer. Store in dark at approximately 4–8 °C for up to 1 week before FACS analysis.
9. Analyze on flow cytometer with appropriate filter set (*see* **Fig. 1**).

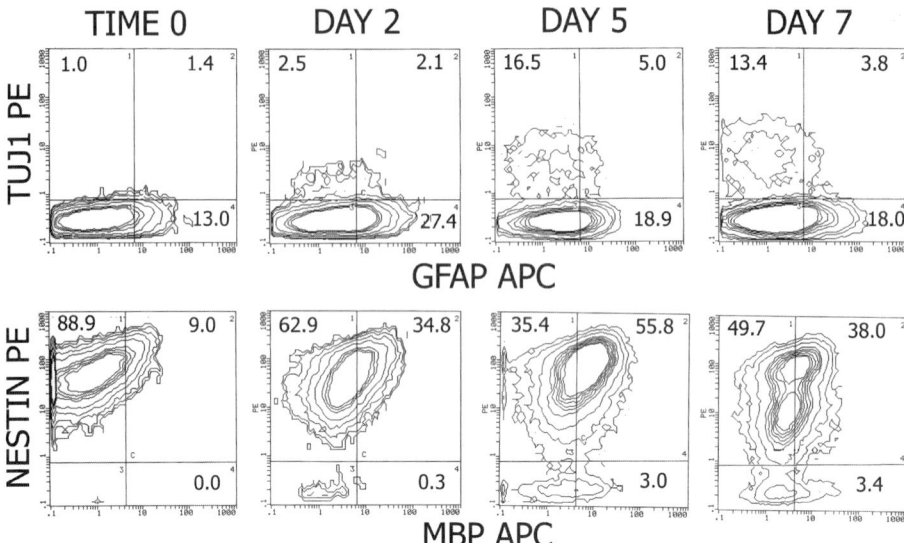

Fig. 1. Quantitative phenotyping of cultured rodent oligodendroglial cells using multi-parameter intracellular flow cytometry. NeuroCyte-Rat cultured rodent NSCs (cat.no. 77024, Alpha Genix, Carlsbad, CA, USA) were withdrawn from bFGF and exposed to 1.0 µM retinoic acid, 10 ng/ml T3 hormone, and 1% FBS to induce oligodendroglial differentiation and lineage commitment. Viable cells were identified retrospectively using oxidative metabolic activity as before. Cultures were analyzed for nestin, GFAP, TUJ1, and MBP at time 0, and days 2, 5, and 7 using IntraCyte-rNSC intracellular FACS kit. Controls included unstained samples and irrelevant IgGs from the appropriate species' primary Ab. FACS plots show the percent positive above the IgG controls.

4. Notes

1. Use sterile technique when performing the brain tissue dissection. Keep the temperature of the tissue and reagents between 30–34 °C.
2. Monitor the pH of the aCSF during the oxygenation process to insure that it stays relatively neutral. The presence of bicarbonate can cause shift in the pH of the aCSF as it converts to CO_2. If this is a problem, aCSF can be made without bicarbonate.
3. Be sure that the medium containing the cells does not have excessive debris in it. Centrifuge at 500 g for 5 min and resuspend in fresh DMEM/F-12/EGF medium if this is the case.
4. Protein must be removed before fixation. Bovine serum albumin (BSA) or serum for example will interfere with the fixation reaction. A 15-ml conical centrifuge tube is convenient, because subsequent washing volumes of 10–15 ml will be used. Hank's balanced salt solution can also be used.

5. One should determine the optimal fixation times and temperatures for your cells and antigen/antibody combinations.
6. The use of FACS for quantitative phenotyping can be performed whether or not the isolation of neural stem cell portion of the protocol is used.
7. Intracellular FACS generally gives higher levels of non-specific IgG binding; use an irrelevant IgG control as an "unstained" sample, reducing the voltage to bring this population into the first or second decade on a log scale, to correct for this effect.
8. Spiking some cells from your irrelevant IgG controls into your positive control sample can aid in getting the appropriate voltage and compensation parameters in the flow cytometer.
9. Both single-color controls for compensation and irrelevant IgG controls must be run for each experiment.
10. Titer all primary and secondary Abs, ranging from 0.1 to 10.0 ug/ml/1×10^6 cells.

References

1. Lois, C. and Alvarez-Buylla, A. (1993) Proliferating subventricular zone cells in the adult mammalian forebrain can differentiate into neurons and glia. *Proc Natl Acad Sci USA* **90**, 2074–2077.
2. Garcia-Verdugo, J. M., Doetsch, F., Wichterle, H., Lim, D. A., and Alvarez-Buylla, A. (1998) Architecture and cell types of the adult subventricular zone: in search of the stem cells. *J Neurobiol* **36**, 234–248.
3. Clarke, D. L, Johansson, C. B., Wilbertz, J., Veress, B., Nilsson, E., Karlstrom, H., Lendahl, U., and Frisen, J. (2000) Generalized potential of adult neural stem cells. *Science* **288**, 1660–1663.
4. Angénieux, B., Schorderet, D.F., and Arsenijevic, Y. (2006) Epidermal growth factor is a neuronal differentiation factor for retinal stem cells in vitro. *Stem Cells* **24**, 696–706.
5. Liu, S., Qu, Y., Stewart, T. J., Howard, M. J., Chakrabortty, S., Holekamp, T. F., and McDonald, J. W. (2000) Embryonic stem cells differentiate into oligodendrocytes and myelinate in culture and after spinal cord transplantation *Proc Natl Acad Sci USA* **97**, 6126–6131.
6. Raff, M. C., Miller, R. H., and Noble, M. (1983) A glial progenitor cell that develops in vitro into an astrocytes or an oligodendrocytes depending on the culture medium. *Nature* **303**, 390–396.
7. Scolding, N. J., Rayner, P. J., and Compston, D. A. (1999) Identification of A2B5-positive putative oligodendrocytes progenitor cells and A2B5-positive astrocytes in adult white matter. *J Neurosci* **89**, 1–4.
8. Zhang, S. C., Lundberg, C., Lipsitz, D., O'Connor, L. T., Duncan, I. D. (1998) Generation of oligodendroglial progenitors from neural stem cells. *J Neurocytol* **27**, 475–489.
9. Paxinos G. and Watson C. (ed.) (2005) *The Rat Brain in Stereotaxic Coordinates*, 5th ed. Elsevier Academic Press, San Diego, CA.

23

Differentiation of Human Embryonic Stem Cells Toward the Chondrogenic Lineage

Wei Seong Toh, Zheng Yang, Boon Chin Heng, and Tong Cao

Summary

Human embryonic stem cells (hESCs) have the ability to self-replicate and differentiate into cells from all three embryonic germ layers, thereby holding great promise for tissue regeneration applications. However, controlling the differentiation of hESCs and obtaining homogenous differentiated cell populations still remain a challenge. We present a highly efficient and reproducible experimental system that mimics the three-dimensional (3-D) environment of in vivo chondrogenesis and that supports the directed differentiation of human embryoid body (EB)-derived cells toward the chondrogenic lineage under serum-free chondrogenic culture conditions in the presence of bone morphogenetic protein-2 (BMP-2).

Key Words: BMP2; Chondrogenic; Differentiation; Embryonic stem cells; Human.

1. Introduction

Development of the vertebrate skeleton is a complex multi-step process that involves lineage commitment of mesenchymal cells, migration of these cells to the site of skeletogenesis, mesenchymal–epithelial interactions that result in cellular condensation, and differentiation of chondroblasts and/or osteoblasts *(1,2)*. Although many factors that have important function during this process are known, there is limited knowledge about the commitment of mesenchymal cells to differentiate into cartilage or bone *(3)*.

Embryonic stem cells (ESCs), derived from the inner cell mass of the blastocyst, represent a promising cell source for transplantation because of their

From: *Methods in Molecular Biology, vol. 407: Stem Cell Assays*
Edited by: M. C. Vemuri © Humana Press, Totowa, NJ

capacity for unlimited self-renewal and ability to differentiate into all somatic cell lineages *(4,5)*. Recent studies in mouse ESCs suggested that ESC differentiation through EBs parallels early embryonic development, and biochemical cues provided directly by growth factors can induce differentiation of ESC-derived embryoid bodies (EBs) toward the chondrogenic lineage *(6,7,8)*. Recent tissue engineering studies have also demonstrated potential chondrogenic differentiation of ESCs using three dimensional (3-D) scaffold systems incorporated with appropriate growth factors *(9,10)*. We have established a reproducible model system to study full-span chondrogenesis of human EB-derived cells and also demonstrated the temporal involvement of bone morphogenetic protein-2 (BMP-2) in early and late phase chondrogenesis *(11,12)*.

This chapter will focus on the effects of BMP-2 on chondrogenic differentiation of human EB-derived cells, culture systems for chondrogenic differentiation as well as characterization of human EB-derived chondrogenic cells.

2. Materials

2.1. Cell Culture

2.1.1. Embryoid Body Culture (see **Note** 1)

1. Dulbecco's Modified Eagle's Medium (DMEM)/F-12 media (cat. no. #11330-032, Gibco/BRL, Gaithersburg, MD).
2. Knockout™ Serum Replacement (KSR) (cat. no. #10828-028, Gibco/BRL)
3. MEM Non-Essential Amino Acids (NEAA) Solution, 10 mM; 100× (cat. no. #11140-050, Gibco/BRL)
4. Collagenase Type IV, Lyophilized (cat. no. #17104-019, Gibco/BRL). Prepare the collagenase IV splitting medium (1 mg/mL) with DMEM/F-12 media. Filter-sterilize before use.
5. EB differentiation medium: DMEM/F-12 supplemented with 20% KSR, 1% (v/v) NEAA, 1 mM L-glutamine (cat. no. #21051-016, Gibco/BRL), and 0.1 mM 2-mercaptoethanol (cat. no. #M7522, Sigma, St Louis, MO). EB media should be protected from light and stored in 4 °C.
6. 6-Well Ultra Low Attachment Microplates (cat. no. #3471, Costar® Corning, Nagog Park Acton, MA).

2.1.2. Culture Media for Chondrogenic Differentiation

1. DMEM high-glucose (4.5g/L; cat. no. #D1152, Sigma)
2. Fetal Bovine Serum (FBS) (cat. no. #CH30160.03, Hyclone, Logan, UT, USA).
3. Serum substitutes: KSR (cat. no. #10828-028, Gibco/BRL).
4. ITS^{+1} (6.25 µg/mL insulin, 6.25 µg/mL transferrin, 6.25 ng/mL selenium, 1.25 mg/mL bovine serum albumin (BSA), 5.35 µg/mL linoleic acid) (cat. no. #354352, BD Bioscience, Franklin Lakes, NJ).

5. MEM NEAA Solution, 10 mM; 100× (cat. no. #11140-050, Gibco/BRL).

6. Recombinant human BMP-2 (cat. no. #355-BM, R&D Systems, Minneapolis, MN) *(13)*. Prepare stock solution of 100 μg/mL by reconstituting in filter-sterilized 4 mM HCl containing 0.1% (w/v) BSA, follow by freezing in aliquots of working volume.

7. BSA (cat. no. #A9418, Sigma).

8. Gelatin, Type A, from Porcine Skin, approximately 300 Bloom (cat. no. #G1890, Sigma). 0.1% (w/v) gelatin coated 12-well tissue culture plates. Weigh out 0.1 g gelatin in an autoclavable bottle and add 100 mL of distilled water (dH₂O). Autoclave with cap tightened loosely, allow cooling to room temperature (RT), and pipet into the culture plates (1 mL/well of 12-well plate). Coat the plates overnight by incubating at 37 °C until use.

9. Dulbecco's Ca^{2+} and Mg^{2+} free phosphate-buffered saline (PBS).

10. 0.25% Trypsin/1 mM ethylenediaminetetraacetic acid (EDTA) solution (cat. no. #25200-072, Gibco/BRL).

11. Basic medium : DMEM high-glucose supplemented with 10% FBS, 10% KSR, and 100U/100 μg Penicillin/Streptomycin (P/S).

12. Basic serum-free chondrogenic medium: DMEM high-glucose supplemented with 1% ITS^{+1} (6.25 μg/mL insulin, 6.25 μg/mL transferrin, 6.25 ng/mL selenium, 1.25 mg/mL BSA, 5.35 μg/mL linoleic acid), 1% KSR (*see* **Note 2**), 2 mM L-glutamine (cat. no. #21051-016, Gibco/BRL), 40 μg/mL L-proline (cat. no. #P5607, Sigma), 50 μg/mL ascorbic acid 2-phosphate (AA2P) (cat. no. #A8960, Sigma), 1% sodium pyruvate (cat. no. #11360-070, Gibco/BRL), 1% NEAA, 10^{-7} M dexamethasone (cat. no. #D2915, Sigma), and 100U/100 μg P/S (cat. no. #15140-122, Gibco/BRL). Preparation of the basic serum-free chondrogenic differentiation media, the media components, and their storage conditions are summarized in **Table 1**.

13. 40-μm nylon cell strainer (Falcon cat. no. #352340, BD Biosciences).

14. 10-mL syringes (Becton Dickinson, Franklin Lakes, NJ, USA).

15. 22-G needles (Sterican B Braun).

2.2. Analysis of mRNA

1. Total RNA extraction performed using RNeasy® Mini Kit (cat. no. #74106, Qiagen, Chatsworth, CA, USA).

2. QIAshredder spin column (cat. no. #79654, Qiagen).

3. cDNA synthesis performed using iScript™ cDNA synthesis kit (cat. no. #170-8891, Bio-Rad, Hercules, CA).

4. Real-time reverse transcriptase–polymerase chain reaction (RT–PCR) using SYBR® Green PCR Master Mix System (cat. no. #204143, Qiagen).

5. Sterile RNase-free reagents, polypropylene tubes, tips, and other materials.

6. Nanodrop ND-1000 spectrophotometer (NanoDrop Technologies, Wilmingon, DE).

7. PCR thermal cycler (Bio-Rad).

Table 1
Preparation of Basic Serum-Free Chondrogenic Differentiation Media
(see Note 9)

Media components	Stock concentration/storage (see Note 10)	100 mL Preparation	Final concentration
DMEM-high glucose	NA; 4 °C	~ 92 mL	NA
L-Glutamine	200 mM (100×); −20 °C	1 mL	2 mM (1×)
L-Proline	5 mg/mL; −20 °C	0.8 mL	40 μg/mL
ITS	100×; 4 °C	1 mL	1%
KSR	100×; −20 °C	1 mL	1%
NEAA	10 mM (100×); 4 °C	1 mL	0.1 mM
P/S	$10,000 U/10,000 \mu g$ (100×); −20 °C	1 mL	$100 U/100 \mu g$ (1×)
Dexamethasone	$10^{-4} M$ (1000×); −20 °C	100 μL	$10^{-7} M$
AA2P	5 mg/mL (100×); −20 °C	1 mL	50 μg/mL
Sodium pyruvate	100 mM (100×); 4 °C	1 mL	1 mM (1×)

NA, not applicable.

AA2P, ascorbic acid 2-phosphate; DMEM, Dulbecco's Modified Eagle's Medium; KSR, Knockout™ Serum Replacement; NA, not applicable; NEAA, non-essential amino acids; P/S, Penicillin/Streptomycin.

8. RT–PCR thermocycler (MX3000P; Stratagene, La Jolla, CA).
9. Gel electrophoresis and standard apparatus (Bio-Rad).
10. Gel Doc EQ Imaging System (Bio-Rad).

2.3. Analysis of Matrix Protein Synthesis

2.3.1. Collagen Synthesis

2.3.1.1. IMMUNOFLUORESCENCE STAINING

1. Fixative: Methanol: Acetone (7:3).
2. Goat serum (cat. no. #S-1000, Vector Laboratories, Burlingame, CA, USA).
3. 2-Chamber tissue culture-treated glass slides (cat. no. #354102, Becton Dickinson).
4. Vetashield mounting medium with 4',6-diamidino-2-phenylindole (DAPI) for nuclear counterstaining (cat. no. #H-1200, Vector Laboratories).

2.3.1.2. IMMUNOHISTOCHEMISTRY

1. UltraVision HRP Detection System (cat. no. #TM-125-HL, Lab Vision, Fremont, CA, USA).

2. Pepsin (cat. no. #AP-9007-005, Lab Vision).
3. Hydrogen peroxide block (cat. no. #TA-125-HP, Lab Vision).

2.3.1.3. ANTIBODIES

1. Monoclonal antibody II-II6B3 (Chemicon, Temecuela, CA, USA).
2. Qdot 655 goat anti-mouse immunoglobulin G (IgG) antibody (Quantum Dot, Hayward, CA, USA).
3. Monoclonal antibody clone X-53 (Quartett Immunodiagnostika GmBH, Berlin, Germany).
4. Control mouse IgG isotype (Zymed Laboratories, San Francisco, CA, USA).

2.3.2. Proteoglycan Synthesis

2.3.2.1. SULFATED GLYCOSAMINOGLYCAN AND DNA QUANTITATION

1. Papain digestion buffer: Dissolve $125 \mu g/mL$ papain (cat. no. #P4762, Sigma) in sterile PBS with 5 mM cysteine hydrochloride (cat. no. #C6852, Sigma) and 5 mM Na_2EDTA (*see* **Note 3**).
2. Blyscan Sulfated Glycosaminoglycan (s-GAG) Assay kit (Biocolor, Newtownabbey, Ireland).
3. $10\times$ TNE buffer: Dissolve 12.11 g Tris base, 3.72 g Na_2EDTA, and 116.89 g NaCl in 900 mL of dH_2O. Adjust the pH to 7.4 with concentrated HCl and top up to 1000 mL. Stir well to dissolve and filter to remove particles. $1\times$ TNE: Dilute 10 mL $10\times$ TNE with 90 mL of dH_2O.
4. $1\times$ TE buffer: Dissolve 1.211 g Tris base and 0.372 g Na_2EDTA in 900 mL of dH_2O. Adjust the pH to 7.4 with concentrated HCl and top up to 1000 mL. Stir well to dissolve and filter to remove particles.
5. Hoechst 33258 dye (cat. no. #B2883, Sigma): Dilute 1 mL Hoechst 33258 (10 mg/mL solution) with 9 mL of dH_2O to make up the Hoechst 33258 stock dye solution (1 mg/mL). Protect from light and store at 4 °C for up to 6 months. Dilute $20 \mu L$ Hoechst 33258 stock solution (1 mg/mL) with $100 mL 1\times$ TNE to obtain $0.2 \mu g/mL$ ($2\times$) assay solution (*see* **Note 4**).
6. Calf thymus DNA (Cat No. #D1501, Sigma): Prepare 1 mg/mL stock solution of Calf Thymus DNA in $1\times$ TE.
7. Fluorescence plate reader (Safire; Tecan GmbH, Salzburg, Austria).
8. Multichannel pipetter (Eppendorf AG, Hamburg, Germany).

2.3.2.2. ALCIAN BLUE STAINING

1. 0.05% (w/v) Alcian Blue staining solution: Dissolve 4.5 g NaCl and 6.4 g $MgCl_2$ in 400 ml 3% acetic acid, add 0.25 g Alcian blue 8GX (cat. no. #A3157, Sigma), adjust to pH 1.5 and the volume to 500 mL. Stir for 2–3 h and clear by filtration.
2. 10% (v/v) neutral buffered formalin (NBF): To make 1 L of NBF, weigh out and dissolve 4 g NaH_2PO_4 and 6.5 g Na_2HPO_4 in 900 mL of dH_2O and then add 100 mL of 37% formaldehyde stock solution.

3. Methods

The methods pertaining to the (i) formation of EBs from hESCs, (ii) culture systems for chondrogenic differentiation of EBs, (iii) analysis of the mRNA expression, and (iv) synthesis of cartilage-specific matrix proteins are described below:

3.1. Formation of Embryoid Bodies

In vitro differentiation can be induced by culturing hESCs (*see* **Note 1**) in suspension to form EBs.

1. Aspirate medium from hESCs and add 1 mL/well of collagenase IV splitting medium into each well of the 6-well plate.
2. After incubation at 37 °C in incubator for 5 min, scrape the hESC colonies from the plate with a 5 mL pipet and transfer cells to a 15-mL sterile falcon tube.
3. Add 1 mL of EB differentiation medium (DMEM/F-12 supplemented with 20% KSR, 1% (v/v) NEAA, 1 mM L-glutamine, and 0.1 mM 2-mercaptoethanol) into each well. Scrape off the remaining cells in the well with a cell scraper and pool the cells together in the 15-mL falcon tube. Gently pipette the cells up and down a few times in the tube to further break up the colonies.
4. Pellet cells by centrifugation at $200\,g$ for 5 min. Aspirate the supernatant and then re-suspend the cell pellet with EB differentiation medium to give a final volume of 3 mL per well of 6-well ultra low attachment microplates. The splitting ratio is set at 1:1, where one 6-well plate of hESCs is split to one 6-well plate of EBs, maintaining at approximately 300 EBs per well.
5. After overnight culture in suspension, hESCs form floating aggregates known as EBs. Culture medium is changed every 2–3 days. Transfer EBs into 15-mL falcon tube and let the aggregates settle for 5 min. Aspirate the supernatant, replace with fresh EB differentiation medium (3 mL/well), and transfer back to the 6-well ultra low attachment microplates for further culture.
6. During the first few days, the EBs are small with irregular outline; they increase in size by day 4. At day 5, EBs are harvested for induction of chondrogenic differentiation.

3.2. Culture Systems for Chondrogenic Differentiation of EBs

3.2.1. EB-Outgrowth Culture

1. Transfer the 5"d" EBs into a sterile 50-mL falcon tube and allow the EBs to settle for 5 min. Aspirate the supernatant and wash once with PBS (*see* **Note 5**). Remove the PBS after EBs have settled and replace with pre-warmed basic medium (DMEM high-glucose supplemented with 10% FBS and 10% KSR and 100U/100 μg P/S).
2. For direct plating of EBs, split 1 well of EBs to 10 wells of the 12-well plate, maintaining at approximately 30 EBs per well (1:10 splitting ratio). The EB cultures were incubated at 37 °C for 24 h to enable cell attachment, before induction of differentiation.

3.2.2. EB-High Density Micromass Culture

1. 5"d" EBs are dissociated into single cells by trypsinization using 0.25% trypsin/EDTA for 5 min at 37 °C, followed by passing the cell suspension through a 22-G needle and then a 40-μm cell strainer to obtain single cells suspended in pre-warmed basic medium (DMEM high-glucose supplemented with 10% FBS and 10% KSR and 100 U/100 μg P/S) (*see* **Note 6**). The medium inactivates the trypsin.
2. Wash twice with the same medium by spinning down the cells at 250 g for 5 min and re-suspending in medium. At the second spin, resuspend the cells in a lower volume of medium for cell count.
3. Culture cells at a high density of 3×10^5 cells per 15 μL spot in a 12-well plate pre-coated with 0.1% gelatin (*see* **Note 7**). After incubation for at least 1 h, 1 mL of the same medium was carefully added to each well. High-density micromass (HDMM) cultures were incubated at 37 °C for 24 h to enable cell attachment, before induction of differentiation the next day.

3.2.3. Chondrogenic Induction

1. Chondrogenic differentiation of EB-derived cells plated directly as EB outgrowth or as HDMM is induced with basic serum-free chondrogenic medium. Preparation of the basic serum-free chondrogenic differentiation media, the media components, and their storage conditions are summarized in **Table 1**.
2. Following incubation for 24 h to allow cell attachment, replace the serum-containing medium with serum-free chondrogenic medium with or without 100 ng/mL BMP-2 (day 1 of differentiation) for a period of 21 days with medium change every alternate day. Cultures in the basic serum-free chondrogenic medium without growth factor supplementation will serve as the control.

3.3. Analysis of Cartilage-Specific Genes

3.3.1. Total RNA Extraction and cDNA Synthesis

1. Remove the differentiation medium from the cultures and wash once with PBS. Extract total RNA from the plated EBs or HDMM using the RNeasy Mini Kit, following the manufacturer's instructions. Scrape the cells with 350 μL of RNeasy Lysis Buffer per well of a 12-well plate.
2. Homogenize the cell suspension by passing through the QIAshredder column and spin down to keep the homogenate. The homogenate can be frozen at −80 °C to store the samples for RNA extraction.
3. Continue the RNA extraction protocol according to the manufacturer's instructions and elute the sample RNA with RNase-free water provided in the kit.
4. Determine the RNA quantity and quality by measuring the absorbance at 260 and 280 nm for each sample using the Nanodrop ND-1000 spectrophotometer. Calculate the RNA concentration of the sample: 1 unit at 260 nm corresponds to 40 μg of RNA per mL. The final preparation should give yields of range between 5–8 μg of RNA per sample by day 21 of differentiation with appropriate A260:A280 ratio

of approximately 2.0. At earlier time points of differentiation, the amount of RNA yield may be as low as 500 ng. Store the RNA samples at −80 °C if cDNA synthesis is not carried out on the same day.

5. RT reaction is performed using a PCR thermal cycler. cDNA synthesis is performed using 500 ng of total RNA per 20 μL reaction volume over a 30 min incubation time at 42 °C, with the addition of 5× iScript reaction mix and iScript reverse transcriptase, followed by enzyme inactivation at 85 °C for 5 min, following the manufacturer's instructions.

3.3.2. Semi-Quantitative RT–PCR

1. Perform PCR amplifications using 1 μL of the resulting cDNA samples by first denaturing at 95 °C for 5 min, followed by 35 cycles of 95 °C for 30 s, annealing at 58 °C for 45 s, and extending at 72 °C for 60 s, followed by a final extension at 72 °C for 5 min. PCR primers are summarized in **Table 2** *(14)*.

2. Analyze the PCR amplified products on 2% (w/v) agarose gel-electrophoresis. The results are evaluated by ultraviolet detection of the ethidium bromide-stained gel. Images are taken using the Biorad Light Imaging System.

Table 2
Primers Used for RT–PCR

Gene	Primer sequence	Product size (bp)
Sox 9	Sox9F:5′-GAACGCACATCAAGACGGAG-3′	631
	Sox9R:5′-TCTCGTTGATTTCGCTGCTC-3′	
Col 2a1	Col2F:5′-TTCAGCTATGGAGATGACAATC-3′	472
	Col2R:5′-AGAGTCCTAGAGTGACTGAG-3′	
Link protein	LPF:5′-CCTATGATGAAGCGGTGC-3′	618
	LPR:5′-TTGTGCTTGTGGAACCTG-3′	
β-Actin	β-ActinF: 5′-CCAAGGCCAACCGCGAGAAGATGAC-3′	587
	β-ActinR: 5′-AGGGTACATGGTGGTGCCGCCAGAC-3′	
Col 2[a]	Col2F: 5′-GGCAATAGCAGGTTCACGTACA-3′	79
	Col2R: 5′-CGATAACAGTCTTGCCCCACTT-3′	
GAPDH[a]	GAPDHF: 5′-ATGGGGAAGGTGAAGGTCG-3′	119
	GAPDHR: 5′-TAAAAGCAGCCCTGGTGACC-3′	

RT–PCR, reverse transcriptase-polymerase chain reaction.
PCR conditions are as follows: 95 °C/5 min, followed by 35 cycles of 30 s denaturation at 95 °C, 45 s annealing at 58 °C, and 1 min elongation at 72 °C, and final extension for 5 min at 72 °C.
   [a] Real-time RT–PCR conditions are as follows: 95 °C for 15 min followed by 40 cycles of 15 s denaturation at 94 °C, 30 s annealing at 55 °C, and 30 s elongation at 72 °C.

3. For analysis, mean pixel intensities of each band are measured using the NIH public domain imaging software, ImageJ, and normalized to mean pixel intensities of the β-actin band.
4. A profile of chondrogenic gene expression by human EB-derived cells cultured in basic chondrogenic medium with and without 100 ng/mL BMP-2 supplementation is shown in **Fig. 1**.

3.3.3. Quantitative Real-Time RT–PCR

1. *Col 2* gene expression is analyzed by real-time RT-PCR reactions using the SYBR® Green PCR Master Mix System on Real-Time PCR thermocycler (MX3000P; Stratagene).

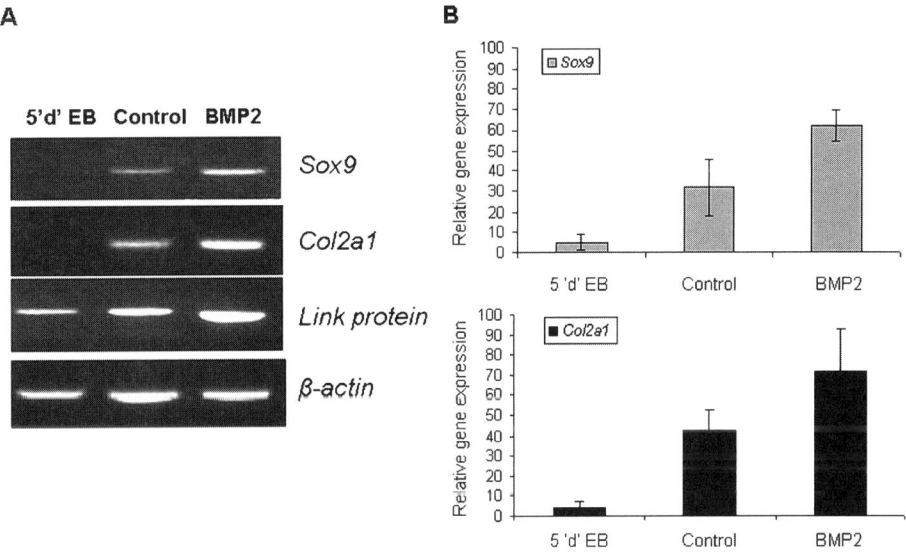

Fig. 1. Expression of chondrogenic markers in human embryoid body (EB)-derived chondrogenic cells. (**A**) Reverse transcriptase–polymerase chain reaction (RT–PCR) analysis of chondrogenic gene markers (*Sox9, Col2a1, Link protein*) and housekeeping gene, *β-actin* expressed by untreated 5"d" EBs and differentiated human EB-derived chondrogenic cells with and without bone morphogenetic protein-2 (BMP-2) treatment on day 21. (**B**) Expression of *Sox9* and *Col2a* was normalized to *β-actin* and expressed as the relative gene expression. There was induction of chondrogenic gene expression under basic chondrogenic condition, which was further enhanced by the presence of 100 ng/mL BMP-2. Values were calculated as mean values ± SD derived from duplicate analysis of at least two independent sets of experiments.

2. cDNA samples (1 μL for a total volume of 25 μL per reaction) were analyzed for gene of interest normalized to reference gene glyceraldehydes-3-phosphate dehydrogenase (GAPDH). The level of expression of each target gene is then calculated as $2^{-\Delta\Delta Ct}$, as previously described *(15)*.

3. Real-time RT–PCR is performed at 95 °C for 15 min followed by 40 cycles of 15 s denaturation at 94 °C, 30 s annealing at 55 °C, and 30 s elongation at 72 °C. PCR primers are summarized in **Table 2** *(16)*.

4. Human EB-derived cells cultured as EB outgrowth and HDMM in the presence of BMP-2 to induce chondrogenesis and processed for real-time RT–PCR for a period up to day 21, as shown in **Fig. 3A**.

3.4. Analysis of Matrix Protein Synthesis

3.4.1. Collagen II Immunofluorescence Staining

1. Rinse the EBs cultivated on chamber slides with PBS, fix with methanol : acetone (7:3) at −20 °C for 5 min, rinse three times in PBS, and incubate for 15 min in 10% goat serum for blocking at room temperature.

2. For detection of collagen II, incubate with the monoclonal antibody (II-II6B3), diluted 1:40 with PBS for 1 h at 37 °C in a humidified chamber.

3. Following incubation, rinse the specimens three times with PBS and incubate for 2 h at room temperature with Qdot 655 goat anti-mouse IgG antibody diluted 1:200 in PBS.

4. Slides are then washed three times in PBS and mounted with Vetashield mounting medium with DAPI for nuclear counterstaining. For negative control, primary antibody is omitted.

5. Single chondrogenic cells and nodules were observed in BMP-2-treated EB cultures, as shown in **Fig. 2**.

3.4.2. Collagen II Immunohistochemistry

1. Remove the culture medium and rinse the cultures twice with PBS. Fix with 10% NBF for 30 min before rinsing twice with PBS.

2. To facilitate antigen retrieval and antibody access, incubate cultures with pepsin at 37 °C for 20 min.

3. Rinse once with PBS, then block with hydrogen peroxide block for 15 min at room temperature to quench any endogenous peroxidase activity.

4. Wash four times with PBS, block with Ultra V Block before 1 h incubation of the cultures at room temperature with anti-collagen II monoclonal antibody, Mab II-II6B3 diluted 1:500 in PBS.

5. At the end of primary antibody incubation, rinse the cultures four times with PBS, before adding the pre-diluted biotin-conjugated goat derived anti-mouse secondary antibody and incubate at room temperature for 30 min.

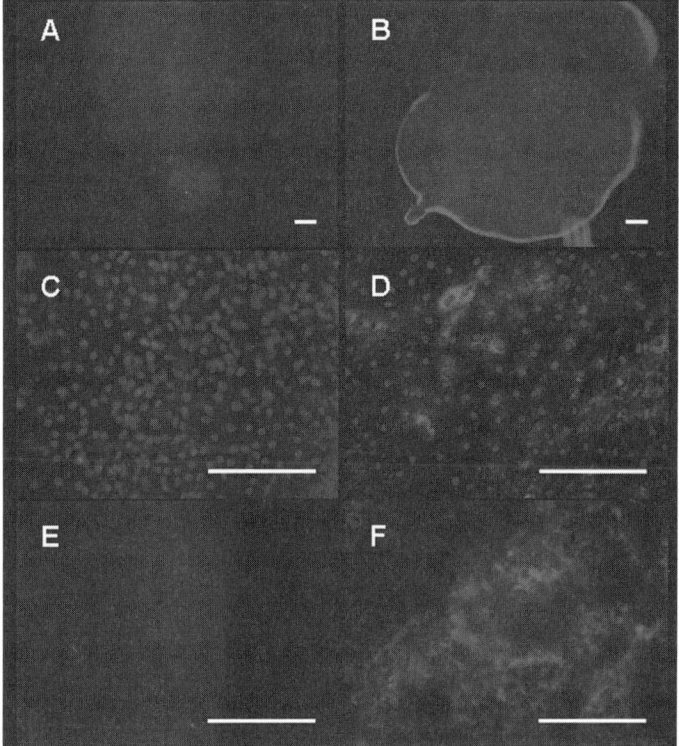

Scale bar = 100 μm

Fig. 2. Bone morphogenetic protein-2 BMP-2 induces chondrogenic differentiation of human EB-derived cells. Indirect immunofluorescence micrographs of control (**A, C,** and **E**) and BMP-2-treated (**B, D,** and **F**) EB cells 21 days after treatment. Expression of cartilage-specific collagen II protein is analyzed by immunofluorescence staining. Representative areas of the EB and its outgrowth are shown. BMP-2 induces chondrogenic differentiation of EB-derived cells with intense staining of collagen II-producing cells at the boundary of the EB (**B**) and outgrowth of chondrogenic cells that could exist as single cells in monolayer (**D**) or in three-dimensional nodular aggregates (**F**). Bar = 100 μm.

6. At the end of secondary antibody incubation, rinse the cultures four times with PBS, before incubating with streptavidin-conjugated horseradish peroxidase at room temperature for 45 min.
7. Wash four times with PBS before adding the diaminobenzidine (DAB) chromogen/substrate to visualize the antibody-antigen reaction.
8. Control mouse isotype is included as a negative control (*see* **Note 8**).

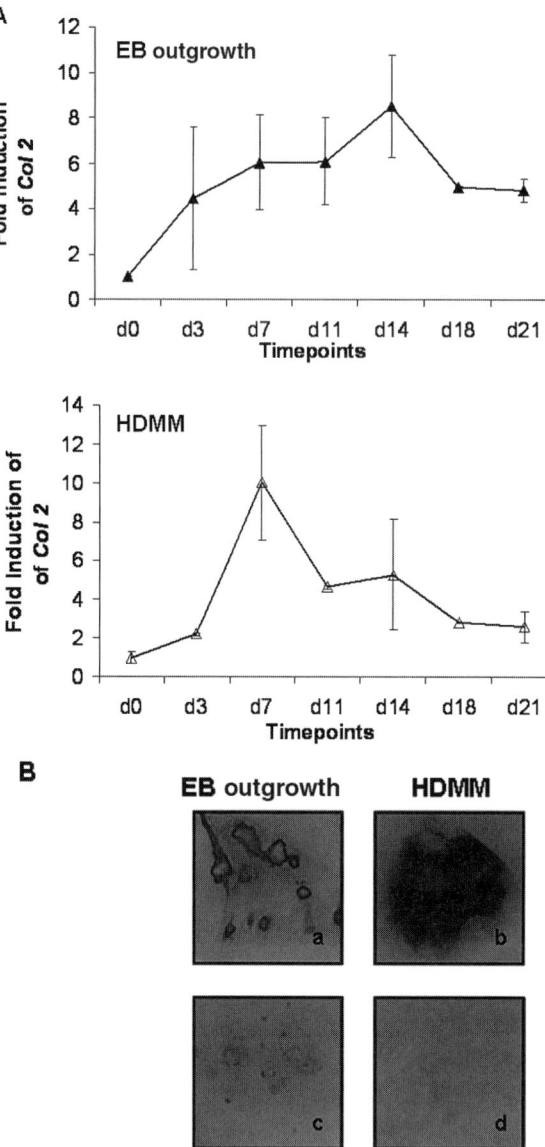

Fig. 3. Real-time polymerase chain reaction (PCR) analysis of *Col 2* expressed by human embryoid body (EB)-derived chondrogenic cells cultured as EB outgrowth and high density micromass (HDMM) along the course of differentiation. **(A)** Bone morphogenetic protein-2 BMP-2 stimulation of chondrogenic differentiation in EB outgrowth displays a slow and steady increase in *Col 2* mRNA levels up to day 14,

9. Human EB-derived cells cultured as EB outgrowth or HDMM in the presence of BMP-2 to induce chondrogenesis and processed for collagen II immunohistochemistry at the end of 21 days, as shown in **Fig. 3B**.

3.4.3. s-GAG and DNA Quantitation

1. Remove the culture medium from the cultures and wash once with PBS. Remove the PBS completely and scrape the cells with $500\,\mu L$ papain digestion buffer per well of a 12-well plate.
2. Collect the samples using sterile 1.5-mL eppendorf tubes and incubate the samples in water bath at $60\,^\circ C$ for 18 h.
3. The digests are collected and spun down at $16,000\,g$ for 5 min and sample supernatants can be assayed immediately or frozen down at $-20\,^\circ C$ for subsequent analyses.
4. s-GAG content is measured spectrophotometrically at 630 nm using the Blyscan Sulfated Glycosaminoglycan Assay kit *(17,18)*, and normalized to the DNA content, measured fluorometrically using the Hoechst 33258 method *(19)*.
5. Add $50\,\mu L$ of samples to $250\,\mu L$ of dimethyl-methylene blue (DMMB) dye reagent in 1.5 mL eppendorf assay tubes and incubate at room temperature for 30 min.
6. After incubation, spin the sample at $16,000\,g$ for 10 min. Discard the supernatant, but retain the dye-bound pellet, which is then dissociated with $200\,\mu L$ dissociation agent (supplied in the Blyscan Sulfated Glycosaminoglycan Assay kit) per sample to release the color.
7. Transfer the entire contents from the assay tubes to the wells of a 96-well microwell plate for spectrophotometric measurement at 630 nm. The dissociation agent is used as the blank. Standard curve of s-GAG is constructed using different concentrations of bovine trachea chondroitin sulfate (provided in the kit), according to the manufacturer's instructions.
8. Dilute $10\,\mu L$ of papain digests 10 times with PBS to a final sample volume of $100\,\mu L$ before adding to the wells containing $100\,\mu L$ of prepared Hoechst 33258 dye assay solution $(0.2\,\mu g/mL)$.

◀——

Fig. 3. before it decreases and plateaus, whereas in HDMM culture system, BMP-2 causes acute induction of *Col 2* by day 7 of chondrogenic differentiation in HDMM culture. Values are calculated as means $\pm$ SD from duplicate analysis of at least two independent experiments. **(B)** Collagen II immunohistochemistry indicates marked enhancement in collagen II synthesis in HDMM culture **(b)** at the end of day 21 differentiation. Appropriate isotype control is carried out in both EB outgrowth **(c)** and HDMM **(d)** cultures in parallel.

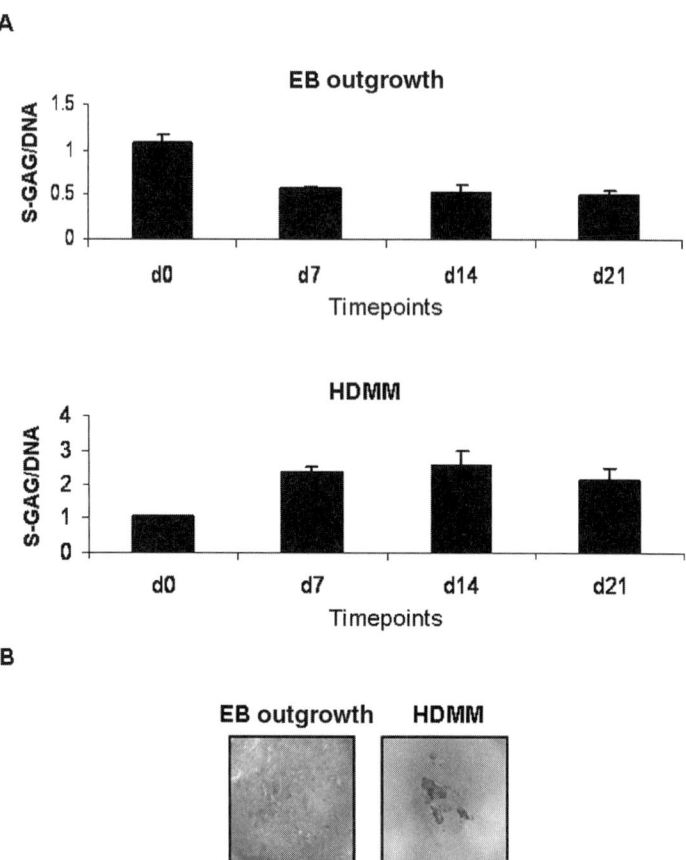

Fig. 4. (**A**) Analysis of sulfated glycosaminoglycan (s-GAG) synthesis by human embryoid body (EB)-derived chondrogenic cells cultured in EB outgrowth and high-density micromass (HDMM) along the course of differentiation. Bone morphogenetic protein-2 BMP-2 stimulation of chondrogenic differentiation in EB outgrowth culture system displays acute decrease in ratio of s-GAG/DNA from day 7 onward, whereas in HDMM culture system, BMP-2 induces time-dependent increase in ratio of s-GAG/DNA up to day 14, before it decreases, thereafter indicating hypertrophic maturation. Values are calculated as means $\pm$ SD from duplicate analysis of at least two independent experiments. (**B**) Alcian blue staining also indicates marked enhancement in s-GAG deposition in the HDMM culture, particularly in the regions of nodular growth.

9. The fluorescence measurement of Hoechst 33258 dye is performed using a fluorescence plate reader at 360 nm excitation and 460 nm emission. Calf thymus DNA is used for the construction of the standard curve for DNA.
10. The reagent blank (100 μL of dye +100 μL dH$_2$O) fluorescence value is subtracted from the sample data before analysis.
11. Human EB-derived cells are cultured as EB outgrowth or HDMM in the presence of BMP-2 to induce chondrogenesis and processed for s-GAG and DNA quantification along the course of differentiation, as shown in **Fig. 4A**.

3.4.4. Alcian Blue Staining

1. Remove the medium from the cultures and wash once with PBS. Fix the cell cultures with 10% (v/v) neutral buffered formalin for 30 min, followed by washing three times with PBS and rinsing in dH$_2$O.
2. Fixed cells are incubated with 0.05% (w/v) alcian blue solution overnight.
3. Remove excess stain by washing at least two times with PBS, followed by a further rinse in 5% acetic acid to remove non-specific staining and then a last rinse with PBS again. Store cultures in PBS for subsequent examination by light photomicrography.
4. Human EB-derived cells cultured as EB outgrowth or HDMM in the presence of BMP-2 to induce chondrogenesis and processed for alcian blue staining at the end of 21 days, as shown in **Fig. 4B**.

4. Notes

1. Culture of hESCs (H1 and H9; NIH stem cell registry: http://stemcells.nih.gov/research/registry, also see **refs 4** and **5**) follows exactly as recommended by the Wicell protocol. Refer to Web site http://www.wicell.org/ for protocols describing the expansion and propagation of hESCs on murine embryonic feeder cells.
2. 1% KSR instead of the strictly serum-free medium was used in this study, based on reports that mesodermal differentiation of ESCs can be inhibited under serum-free conditions **(20)**.
3. Always prepare the papain-digestion buffer fresh when needed. Allow the papain to completely dissolve and filter-sterilize before use.
4. Hoechst 33258 is a possible carcinogen and possible mutagen. Wear gloves and a mask and work under a fume hood. Keep 2× assay solution at room temperature. Prepare fresh solution whenever needed. Do not filter once dye has been added. Protect from light. 10× TNE stock buffer solution used in Hoechst dye DNA assay can be stored at 4°C for up to 3 months. Working solution is prepared fresh whenever needed.
5. Allow the EBs to settle down during the wash with PBS. Usually, it takes less than 3 min for all the EBs to settle. Avoid pipetting too many times or foaming because the cells should be collected in clumps.

6. Trypsinize the EBs into single cells for no more than 5 min and pass the cell suspension once through a 22-G needle and then 40-μm cell strainer. Prolong or repeated trypsinization is detrimental to cell viability.

7. The method for HDMM culture has been adapted from previously published methods *(21,22)*. After removal of 0.1% gelatin from the wells, ensure the plates are completely dry before use. This is to prevent dispersion of cells when dispensing the cells in a single drop onto the well in HDMM culture.

8. Isotype antibody control is necessary to ensure antibody specificity, to rule out any non-specific staining and to give better interpretation of the results.

9. Use chondrogenic differentiation media for not more than 2 weeks after preparation, as components such as L-glutamine and AA2P degrade over time. Store the chondrogenic differentiation media at 4°C and discard unused media after 2 weeks.

10. Prepare L-Proline, AA2P, dexamethasone in PBS/0.1% BSA, aliquot and freeze them at −20°C in working volumes till use. Also aliquot P/S, L-glutamine, and KSR in working volumes and freeze them at −20°C. Avoid repeated freeze-thaw of working solutions.

References

1. Hall BK, Miyake T. (2000) All for one and one for all: condensations and the initiation of skeletal development. *Bioessays* **22**, 138–147.

2. DeLise AM, Fischer L, Tuan RS. (2000) Cellular interactions and signaling in cartilage development. *Osteoarthr Cartilage* **8**, 309–334.

3. Heng BC, Cao T, Lee EH. (2004) Directing stem cell differentiation into the chondrogenic lineage in vitro. *Stem Cells* **22**, 1152–1167.

4. Thomson JA, Itskovitz-Eldor J, Shapiro SS, Waknitz MA, Swiergiel JJ, Marshall VS, Jones JM. (1998) Embryonic stem cell lines derived from human blastocysts. *Science* **282**, 1145–1147.

5. Reubinoff BE, Pera MF, Fong CY, Trounson A, Bongso A. (2000) Embryonic stem cell lines from human blastocysts: somatic differentiation in vitro. *Nat. Biotechnol.* **18**, 399–404.

6. Kramer J, Hegert C, Guan K, Wobus AM, Muller PK, Rohwedel J. (2000) Embryonic stem cell-derived chondrogenic differentiation in vitro: activation by BMP-2 and BMP-4. *Mech. Dev.* **92**, 193–205.

7. Hegert C, Kramer J, Hargus G, Muller J, Guan K, Wobus AM, Muller PK, Rohwedel J. (2002) Differentiation plasticity of chondrocytes derived from mouse embryonic stem cells. *J. Cell. Sci.* **115**, 4617–4628.

8. Kawaguchi J, Mee PJ, Smith AG. (2005) Osteogenic and chondrogenic differentiation of embryonic stem cells in response to specific growth factors. *Bone* **36**, 758–769.

9. Hwang NS, Kim MS, Sampattavanich S, Baek JH, Zhang Z, Elisseeff J. (2006) The effects of three dimensional culture and growth factors on the chondrogenic differentiation of murine embryonic stem cells. *Stem Cells* **24**, 284–291.

10. Tanaka H, Murphy CL, Murphy C, Kimura M, Kawai S, Polak JM. (2004) Chondrogenic differentiation of murine embryonic stem cells: effects of culture conditions and dexamethasone. *J. Cell. Biochem.* **93**, 454–462.

11. Toh WS, Yang Z, Heng BC, Cao T. (2006) New perspectives in chondrogenic differentiation of stem cells for cartilage repair. *ScientificWorldJournal* **6**, 361–364.

12. Toh WS, Yang Z, Liu H, Heng BC, Lee EH, Cao T. (2007) Effects of culture conditions and bone morphogenetic protein 2 on extent of chondrogenesis from human embryonic. *Stem Cells* **25(4)**: 950–960.

13. Wozney JM, Rosen V, Celeste AJ, Mitsock LM, Whitters MJ, Kriz RW, Hewick RM, Wang EA. (1988) Novel regulators of bone formation: molecular clones and activities. *Science* **242**, 1528–1534.

14. Sekiya I, Vuoristo JT, Larson BL, Prockop DJ. (2002) In vitro cartilage formation by human adult stem cells from bone marrow stroma defines the sequence of cellular and molecular events during chondrogenesis. *Proc. Natl. Acad. Sci. U.S.A.* **99**, 4397–4402.

15. Livak KJ, Schmittgen TD. (2001) Analysis of relative gene expression data using real-time quantitative PCR and the 2(-Delta Delta C(T)) method. *Methods* **25**, 402–408.

16. Martin I, Jakob M, Schafer D, Dick W, Spagnoli G, Heberer M. (2001) Quantitative analysis of gene expression in human articular cartilage from normal and osteoarthritic joints. *Osteoarthr Cartilage* **9**, 112–118.

17. Toh WS, Liu H, Heng BC, Rufaihah AJ, Ye CP, Cao T. (2005) Combined effects of TGFβ1 and BMP2 in serum-free chondrogenic differentiation of mesenchymal stem cells induced hyaline-like cartilage formation. *Growth Factors* **23**, 313–321.

18. Farndale RW, Buttle DJ, Barrett AJ. (1986) Improved quantitation and discrimination of sulphated glycosaminoglycans by use of dimethylmethylene blue. *Biochim. Biophys. Acta* **883**, 173–177.

19. Kim YJ, Sah RL, Doong JY, Grodzinsky AJ. (1988) Fluorometric assay of DNA in cartilage explants using Hoechst 33258. *Anal. Biochem.* **174**, 168–176.

20. Wiles MV, Johansson BM. (1999) Embryonic stem cell development in a chemically defined medium. *Exp. Cell. Res.* **247**, 241–248.

21. Ahrens PB, Solursh M, Reiter RS. (1977) Stage-related capacity for limb chondrogenesis in cell culture. *Dev. Biol.* **60**, 69–82.

22. Mello MA, Tuan RS. (1999) High density micromass cultures of embryonic limb bud mesenchymal cells: an in vitro model of endochondral skeletal development. *In Vitro Cell Dev. Biol. Anim.* **35**, 262–269.

24

Cartilage Tissue Engineering

Directed Differentiation of Embryonic Stem Cells in Three-Dimensional Hydrogel Culture

Nathaniel S. Hwang, Shyni Varghese, and Jennifer Elisseeff

Summary

The clinical goal of tissue engineering is to restore, repair, or replace damaged tissues in the body. Significant advances have been made in recent years using stem cells as a cell source for cartilage tissue engineering and reconstructive surgery applications. Embryonic stem cells have demonstrated the potential to self-renew and differentiate into a wide range of tissues including the chondrogenic lineage, depending on culture conditions. Three-dimensional scaffolds play an important role in tissue regeneration by providing attachment sites as well as bioactive signals for cells to grow and differentiate into specific lineages. The precise microenvironments required for optimal expansion or differentiation of stem cells are only beginning to emerge now, and the controlled differentiation of embryonic stem cells in tissue engineering remains a relatively unexplored field. Hydrogels are a class of polymer-based biomaterials that have been extensively utilized in tissue engineering as scaffolds. We have demonstrated that embryonic stem cells encapsulated within poly(ethylene glycol)-based (PEGDA) photopolymerizing hydrogels and cultured in an appropriate growth factor and medium conditions undergo chondrogenic differentiation with extracellular matrix deposition characteristic of neocartilage (Hwang et al., *Stem Cells* **24**, 284–291). Another hydrogel that has been widely used for encapsulating chondrocytes in cartilage tissue engineering is alginate. This hydrogel also has potential for tissue engineering applications using stem cells. Here, we describe the three-dimensional culture of embryonic stem cell-derived embryoid bodies in hydrogels and their differentiation toward chondrogenic lineage in chemically defined chondrogenic medium in the presence of TGF-β1 (chondrogenic inducing conditions). We also discuss various tools and assays used for characterizing the tissue-engineered cartilage.

From: *Methods in Molecular Biology, vol. 407: Stem Cell Assays*
Edited by: M. C. Vemuri © Humana Press, Totowa, NJ

Key Words: Tissue engineering; Cartilage; Chondrogenesis; Differentiation; Embryonic stem cells; Photopolymerizing hydrogels; Alginate hydrogels.

1. Introduction

Treatment of cartilage lesion remains an intractable problem because of the intrinsic biology of cartilage tissue, such as limited blood supply and lack of self-repair capacity *(1)*. Tissue engineering provides a potential solution for cartilage regeneration. However, one of the major obstacles in engineering tissues for clinical use is the large cell numbers that are often required for forming new tissues. The obvious advantage of using embryonic stem (ES) cells is that the cells are "immortal" and can potentially provide an unlimited supply of differentiated chondrocytes and chondro-progenitor cells for transplantation. Several protocols have recently been developed for the differentiation of ES cells along the osteo-chondral lineages, either through intact or dissociated embryoid bodies or through derivation of mesenchymal precursors *(2–6)*. Creation of embryoid bodies is the first step for differentiation of ES cells. Embryoid bodies, clusters of ES cells where differentiation occurs, can be formed by various methods such as the hanging drop method, mass suspension culture method, and differentiating in methylcellulose *(7)*.

Differentiation of stem cells is generally controlled by numerous cues in their microenvironment *(8,9)*. Porous 3D scaffolds provide many of those cues and a more physiological environment to stem cells that allows development of discreet tissues compared to 2D tissue culture *(2,10)*. Recent reports have indicated distinct cellular behavior in 3D culture that is not present in standard monolayer culture *(11,12)*. Designing a proper scaffold, which could provide necessary mechanical and biological environment to the encapsulated cells, is a pivotal issue in tissue engineering. A wide variety of biomaterials, including synthetic and naturally derived polymers, have been applied to cartilage tissue engineering. To this end, 3D hydrogels have been identified as excellent materials for encapsulating stem cells to yield functional cartilage tissues (*see* **Fig. 1**). When cells are seeded within hydrogels, the extracellular matrix (ECM) proteins produced by the cells are deposited within the scaffold, which then remodel to form tissue-like structures (*see* **Fig. 2**).

Hydrogels offer the unique ability to encapsulate stem cells, creating cell-laden scaffolds for developing new tissue. In addition to their ability to encapsulate cells, hydrogels have a high water content, which allows for efficient transport of nutrients and waste products, and mechanical properties that resemble the viscoelastic properties of native tissue, making them potentially useful as tissue engineering scaffolds. After encapsulation, mesenchymal stem

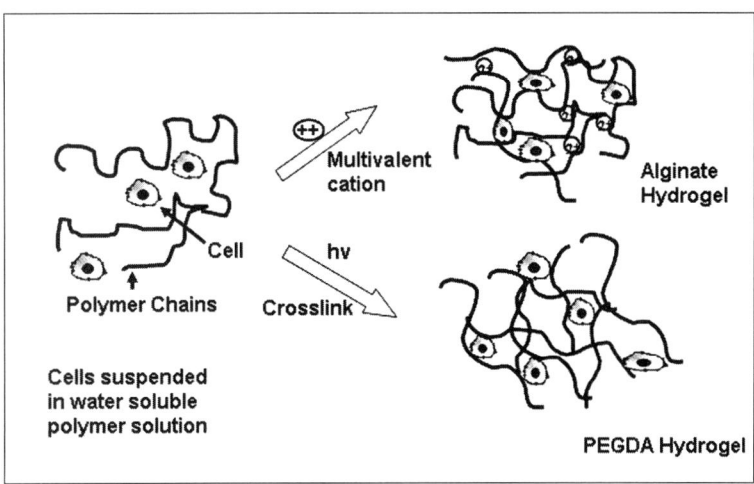

Fig. 1. Formation of calcium-induced alginate polysaccharide hydrogels and radical initiated poly(ethylene glycol)-diacrylate (PEGDA) hydrogels. In case of alginate hydrogel, crosslinking network is formed by the reaction of sodium alginate with calcium ions binding the L-guluronic acid subunits of alginate to form calcium alginate hydrogel. PEGDA hydrogel is formed by photopolymerization method where covalent crosslink of polymers is initiated by light and a photoinitiator.

cells remain spherical and differentiate into the chondrocyte phenotype in alginate hydrogels when incubated in appropriate conditions *(13)*. Embryonic stem cells and mesenchymal stem cells have also been differentiated in poly(ethylene glycol)-diacrylate hydrogels where they produced cartilage-like tissue *(2,14)*. Anchorage independent cells such as chondrocytes exhibit good cell viability within hydrophilic scaffolds such as hydrogels *(15)*. Moreover, many investigators have shown that the hydrophilicity of the scaffold facilitates the re-differentiation of de-differentiated human nasal chondrocytes *(16)*. Most hydrogels are poor substrates for cell and protein adhesion because of their hydrophilic nature. Therefore, adhesion-promoting oligopeptides such as RGD (arginine-glycine-aspartate) containing peptides have often been incorporated into the hydrogels to improve their adhesion properties, particularly for adhesion-dependent cells *(17–19)*. Hydrogels functionalized with specific proteins known to induce cell attachment have also been found to be useful in directing tissue-specific differentiation of stem cells *(20–22)*.

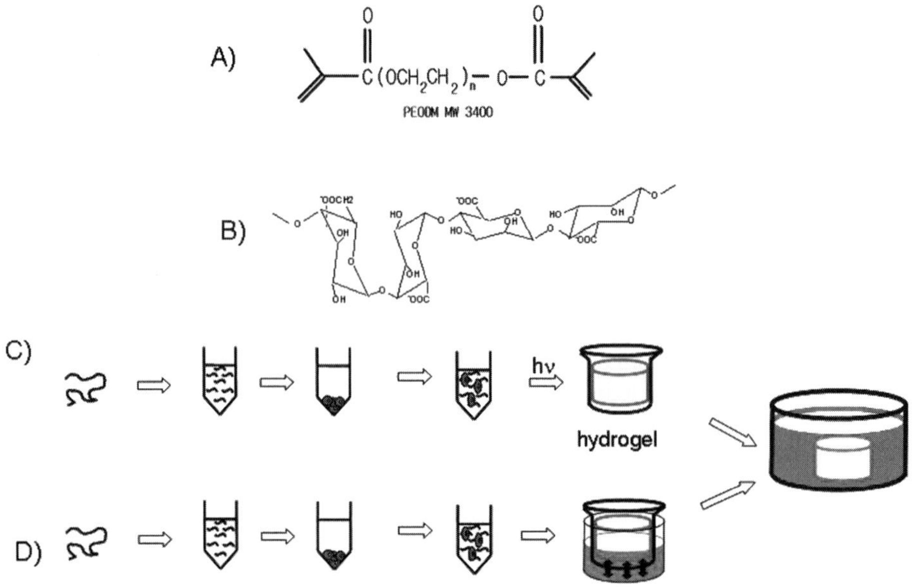

Fig. 2. Chemical structure of (**A**) poly(ethylene glycol)-diacrylate (PEGDA) and (**B**) sodium alginate monomer structure consisting of mannuronic and guluronic acid groups. Schematic representation of the method used to encapsulate stem cells in (**C**) PEGDA and (**D**) alginate hydrogels. In case of PEGDA hydrogels, cell-PEGDA polymer solution is polymerized using light in a cylindrical mold. Alginate hydrogel is created by pouring cell-alginate polymer solution in a cylindrical mold with semi-permeable membrane and subsequent placing of the mold in calcium chloride solution.

2. Materials

2.1. Tissue Culture Materials

All reagents and materials should be sterile.

1. 10-cm Tissue culture dishes (cat. no. 430167 Corning, Lowell, MA, USA).
2. Disposable pipettes with different sizes: 1ml, 5ml, 10ml, and 25 ml.
3. 0.1% gelatin solution: Autoclave 0.1% (w/v) gelatin (cat. no. G1890, Sigma-Aldrich, Saint Louis, MS, USA) suspended in de-ionized (DI) water.
4. 0.05% Trypsin/EDTA (cat. no. 25300054, GIBCO, Carlsbad, CA, USA).
5. Ca^{+2} and Mg^{+2}-free phosphate-buffered saline (PBS; cat. no. 10010-023, GIBCO).

2.2. ES Cell Culture Medium and Differentiation Medium

All reagents and materials should be sterile.

1. D3 murine ES cells (cat. no. SCRC-1003, American Type Culture Collections, Manassa, VA, USA) and mitomycin-C (cat. no. M0503, Sigma) treated mouse embryonic fibroblasts (MEFs).
2. ES cell medium: Dulbecco's modified Eagle's medium (DMEM; cat. no. 11965-118, GIBCO) supplemented with 15% fetal bovine serum (FBS; cat. no. SH30070, Hyclone, Logan, Utah), 2 mM L-glutamine (cat. no. 25030-081, GIBCO), 5×10^{-5} M β-mercaptoethanol (cat. no. M7522, Sigma-Aldrich), non-essential amino acids (cat. no. 11140-050, Gibco), and leukemia inhibitory factor (LIF; cat. no. L5158, Sigma-Aldrich).
3. Feeder cell medium: DMEM supplemented with 10% FBS, 2 mM L-glutamine.
4. Chondrogenic differentiating medium: 5% FBS (cat. no. SH30070, Hyclone) in addition to high-glucose DMEM supplemented with 100 nM dexamethasone (cat. no. D2915 Sigma-Aldrich), 50 μg/ml ascrobate-2-phosphate (cat. no. A8960, Sigma-Aldrich), 40 μg/ml proline (cat. no. P-5607, Sigma-Aldrich), 100 μg/ml sodium pyruvate (cat. no. 11360-070, Gibco), and 50 mg/ml ITS-Premix (cat. no. 354352, Collaborative Biomedical, Bedford, MA, USA; 6.25 ng/ml insulin, 6.25 mg transferin, 6.25 ng/ml selenious acid, 1.25 mg/ml bovine serum albumin, and 5.35 mg/ml linoleic acid).
5. Transforming growth factor-β1 (TGF-β1, cat. no. RDI-1021C, Research Diagnostics, Flandes, NJ, USA) is dissolved in PBS containing 2 mg/ml albumin so as to achieve a final concentration of 10 μg/ml and stored in single use aliquots at $-20\,°C$. It is recommended not to freeze thaw the aliquots frequently. Therefore, store the thawed aliquot at $4\,°C$ and consume within a month.

2.3. PEGDA Hydrogel

1. UV lamp with 365 nm light with an intensity of $4.5\,mW/cm^2$ (Glo-mark Systems, Broadband UV lamp, 365 nm, Upper Saddle River, NJ, USA).
2. Poly(ethylene glycol)-diacrylate (PEGDA, cat. no. 01010F12, Nektar, Huntsville, AL, USA). It should be UV-sterilized for at least 30 min before use. Keep the PEGDA on crushed ice, while UV sterilizing, to keep the temperature low.
3. Photoinitiator, Igracure 2959 (Product No. 1706673, Ciba Specialty Chemicals, Tarrytown, NY, USA).
4. Molds (refer to **Note 1** for further details)
5. UVX Digital Radiometer (UVP, Inc. Upland, Ca).

2.4. RGD-Modified PEG Macromer

1. Peptides, YRGDS (integrin-binding sequence) or YRDGS (control, non-binding sequence).
2. 50 mM Tris-HCl buffer (cat. no. BP152, Fisher Biotech, Pittsburgh, PA, USA), with pH 8.2.

3. Acryloyl polyethylene glycol-N-hydroxysuccinimide (ACRL-PEG-NHS, cat no. 014M0F02, Nektar).

4. RGD-modified PEG acrylate: Dissolve 70 mg of ACRL-PEG-NHS and 5 mg YRGDS or YRDGS in 2 ml of 50 mM Tris buffer at pH 8.2. Allow mixing at room temperature in dark for 2 h and 30 min. The dark environment can be achieved by covering the reaction vessel with aluminum foil and by switching off the lights in the hood.

5. Freeze the solution till it solidifies, and then lyophilize for at least 24 h. Store the lyophilized powder at $-20\,°C$.

2.5. Alginate Hydrogel

1. Alginate solution: Prepare 1.2% (w/v) alginate solution by dissolving 1.2 g of alginate (Kelton LV alginate, cat. no. 9005-38-3, Kelko Co, San Diego, CA, USA) in 100 ml of DI water containing 0.15 M NaCl, and 20 mM HEPES Buffer (cat. no. 15630-080, Gibco). Alginate dissolves slowly because of its high molecular weight. Therefore, dissolve the alginate overnight while constantly stirring it, followed by autoclave to sterilize.

2. Corning Costart Transwell culture inserts Model 3470.

3. Calcium Chloride solution: Dissolve 3 g of $CaCl_2$ powder (cat. no. C5670, Sigma-Aldrich) into 200 ml of sterilized dH_2O to make a 102 mM solution. Add 1 ml of 1M HEPES buffer (cat. no. 15630-080, Gibco) to the 200 ml of $CaCl_2$ solution to make a 2 mM HEPES/102 mM $CaCl_2$ solution. Sterilize the solution through a 0.22-micron filter and store at 4–8 °C.

4. Sterile forceps and spatula.

2.6. RGD-Modified Alginate polymers

1. MES buffer containing 0.3M NaCl and pH 6.5.

2. 1-ethyl-(dimethylaminopropyl) carbodiimide (EDC, cat. no. 39391, Sigma-Aldrich). Refer to **Note 2** about this reagent.

3. N-hydroxy sulfosuccinimide (sulfo-NHS; cat. no. 56485, Fluka). **Note 3** gives additional information about Sulfo-NHS.

4. RGD peptides (American Peptide Company, Sunnyvale, CA, USA).

5. Alginate solution: Make 1.2% (w/v) alginate solutions in 0.1M MES buffer containing 0.3M NaCl at pH 6.5. To get a homogeneous solution, stir using a magnetic bar on a stir plate.

6. RGD-modified alginate: To the 1.2% alginate homogenous solution, add 12.5 mg of sulfo-NHS while stirring. Add 25 mg of EDC to the above solution and stir for 5 min. Add 2.5 mg of RGD peptide to the above solution and perform the reaction for another 20 h under continuous stirring.

7. The resultant RGD modified alginate can be purified either by dialysis against distilled water or using a Sephadex™ column. **Note 4** gives additional information about the purification procedures.

8. Freeze the purified solution till it solidifies completely and then lyophilize for at least 24 h and store at −20 °C.

2.7. Live/Dead Cell Viability Assay

1. Sterile PBS (cat. no. 10010-023, Gibco).
2. DMEM cat. no. 11965-118, GIBCO).
3. Live/Dead Viability/Cytotoxicity Kit (cat. no. L3224, Molecular Probes, Carlsbad, CA, USA).
4. Sterile scalpel.

2.8. Fixing the Hydrogels for Histology

1. 4% paraformaldehyde solution in PBS (pH 7.4): Measure 45 ml DEPC-treated ddH$_2$O (cat. no. 351-068-131, Quality Biological, Gaithersburg, MD, USA) +25 μl 1N NaOH (made with DEPC-treated ddH$_2$O) into a measuring cylinder. Pour the above into a conical flask containing 2 g of paraformaldehyde (cat. no. P6148, Sigma-Aldrich). Cover the conical flask with parafilm, transfer into a fume hood and mix thoroughly. Take care not to splash paraformaldehyde around; it is a *rapid fixer* and *toxic*. Heat the solution to 65–70 °C with intermittent shaking until paraformaldehyde is completely dissolved. After cooling to room temperature, add 1 ml of 10× PBS and adjust pH to 7.3–7.4 with HCl.

2.9. Preparing for Biochemical Analysis

1. PBE buffer: Prepare 100 mM phosphate solution containing 10 mM EDTA by dissolving 7.1 g Na$_2$HPO$_4$ (cat. no. 3828, J.T.Baker) and 1.86 g NaEDTA (cat. no. BP120, Fisher biotech) into 500 mL of DI H$_2$O. Adjust pH to 6.5 with concentrated HCl, filter-sterilize and store this solution at 4 °C.
2. Make fresh papainase solution (Papain 125 μg/ml, cysteine 10 mM, phosphate 100 mM, EDTA 10 mM, pH 6.3). Dissolve 0.035 g cysteine in 20 ml PBE buffer, filter-sterilize. Then add 0.1 ml sterile papain stock (cat. no. P-3126, Worthington) to the cysteine (cat. no. C7352, Sigma Aldrich)/PBE solution. For better results make papainase solution fresh every time.

2.10. Fluorometric Assay of DNA Content Using Hoechst 33258

1. 10× TNE buffer stock solution.

   a. 12.11 g Tris base [Tris (hydroxymethyl) amino methane], (cat. no. BP152, Fisher Biotech).
   b. 3.72 g EDTA, disodium salt, dihydrate, (cat. no. BP120, Fisher Biotech).
   c. 116.89 g sodium chloride (cat. no. 4058, J.T. Baker).

   Dissolve the above components in 800 ml distilled water and adjust the pH to 7.4 by adding HCl. Add distilled water to make a 1000 ml solution. Filter the solution before use. This buffer can be stored at 4 °C for up to 3 months.

2. Hoechst 33258 low range dye: Low range assay solution typically detects DNA content around 10–500 ng/ml (refer to **Note 5**).

   a. 10 μl Hoechst 33258 Dye (cat. no. B2883, Sigma-Aldrich) of stock solution at 1 mg/mL concentration.
   b. 10 ml of 10× TNE buffer.
   c. 90 ml of distilled filtered water.

   Mix the above components to make assay solution at room temperature. Always use freshly prepared assay solutions for measurements. Do not filter the assay solution.

3. Calf Thymus DNA Standards: Prepare a 100 μg/mL stock solution of Calf thymus DNA (cat. no.15633-019, Invitrogen, Carlsbad, CA, USA) in 1× TNE buffer. Gently tap the tube to mix the solutions thoroughly.
4. Cuvette (square four-sided acrylin, cat. no. 67.7555, Starstedt, Newton, NC, USA).
5. Transfer pipette.
6. Hoefer DyNA Quant 200 Fluorometer.

2.11. Glycosaminoglycan Measurement Using Dimethylmethylene Blue Dye

1. PBE buffer: Refer to **Subheading 2.9., step 1**.
2. Chondroitin sulfate solution: Prepare chondroitin sulfate (CS) (cat. no. 27042, Sigma-Aldrich) stock solution by dissolving 50 mg of CS in 1 ml PBE/cysteine solution (0.105 g cysteine in 60 ml PBE). Store at −20 °C. Working CS standard solution is prepared by mixing 100 μl of 50 mg/mL CS stock solution to 49.9mL PBE/cysteine solution.
3. 1,9-Dimethylmethylene blue (DMMB) dye: Prepare an aqueous solution of 46 μm DMMB dye with pH 3.0 and A_{525} 0.31. To prepare 1 L dye solution, dissolve 16 mg DMMB (cat. no. 341088, Sigma-Aldrich) in 1 L water containing 3.04 g glycine (cat. no. 4059, J.T. Baker), 2.37 g NaCl, and 95 mL of 0.1M HCl.
4. 4.5 mL cuvette.
5. Spectrophotometer.

2.12. Collagen Content Measurement by Hydroxyproline Assay

1. Citrate-acetate buffer with pH 6.0: Add 17 g NaOH (cat. no. 3722, J.T. Baker), 25 g Citric Acid monohydrate (cat. no. C1909, Sigma-Aldrich), 60 g sodium acetate trihydrate (cat. no. S7670, Sigma-Aldrich), and 6 ml glacial acetic acid (cat. no. A6283, Sigma-Aldrich) to 250 ml of dH$_2$O while stirring. When dissolved, bring the volume to 500 ml. Transfer the solution to an appropriate glass storage container having a 750 ml total volume. Add 150 ml isopropanol, 100 ml dH$_2$O, bring pH to 6 with concentrated HCl while stirring. Add five drops of toluene as preservative and wrap with parafilm. This buffer is stable at room temperature for 3 months.

2. Chloramine T solution: Add 0.705 g Chloramine T (cat. no. 857319, Sigma-Aldrich), 40 ml of pH 6 buffer, and 5 ml isopropanol (cat. no. 278475, Sigma-Aldrich) into a 100-mL glass bottle. Shake well to mix, wrap with parafilm, and let it sit at room temperature for 24 h. Approximatley 0.5 ml of chloramine T solution is required per construct or sample. Make this solution fresh every time.

3. Hydroxyproline (HRP) standard: Prepare 100 mg/ml hydroxyproline standard stock solution by dissolving 10 mg of HRP (Cat. No.56250, Fluka, Milwaukee, WI, USA) in 100ml of DI H_2O. This can be stored at room temperature for 3 months.

4. Ehrlick's reagent: Dissolve 7.5 g p-dimethylaminobenzaldehyde (cat. no. 109762, Sigma-Aldrich) in 30 ml of isopropanol (cat. no. 278475, Sigma-Aldrich). Stir under fume hood and slowly add 13 ml of 60% percholoric acid (cat. no. 311413, Sigma-Aldrich) to it and keep stirring in the hood for an additional 10 min. Cap or wrap with parafilm and leave at room temperature for 24 h until use. Make this solution fresh every time.

5. 0.02% methyl red in methanol (cat. no. SI16, Fisher Scientific).

6. 2.5 M NaOH.

7. 0.5 M NaOH.

8. 0.5 M HCl.

9. Spectrophotometer.

10. 15 ml Pyrex tube (cat. no. 60827-533, VWR).

11. Disposable cuvettes.

2.13. RNA Extraction for RT–PCR

1. Commercial DEPC-treated water (cat. no. 351068, Quality Biological), aliquot and store at 4 °C.

2. 50 mM EDTA solution in 10 mM HEPES.

3. Distilled water. All the DI water used for RNA extraction procedures must be autoclaved and filtered under sterile condition. Store the water at 4 °C and keep it sterile.

4. RNase free pellet pestle tip (cat. no. 749521-1590, Fisher Scientific).

3. Methods

The method described over here uses the mass suspension cultures of embryonic stem cells to form EBs and subsequent chondrogenic differentiation of EBs either in its intact form or as dissociated cells (refer to **Note 6**). Embryonic stem cells can be encapsulated within the hydrogels prepared from polymer of both synthetic and natural origins. Because embryonic stem cells can differentiate into cells from all three germ layers, they provide a potentially powerful tool for tissue regeneration applications. Three dimensional microenvironments created by hydrogels, with appropriate growth factors and biological cues are useful in controlling ES cell differentiation (**Notes 7** and **8** on hydrogels). Differentiation patterns of ES cells in the hydrogel can be analyzed by reverse transcriptase–polymerase chain reaction (RT–PCR) analysis. Functional tissue formation by differentiated cells can be analyzed through histology and immunostaining. Biochemical analysis of ECM proteins provides quantitative understanding of biosynthetic activity of cells and their differentiation and tissue formation in 3D microenvironments.

3.1. Embryonic Stem Cell Culture and Formation of EBs

1. To prepare gelatin-coated tissue culture dishes, add 5 ml of 0.1% gelatin to each well incubate and aspirate out the solution after 20 min.
2. Thaw MEF cells and plate them onto the gelatin coated tissue dishes. Plate mitomycin-c-treated feeder cells (3×10^4 cells per cm^2) onto a gelatin-coated 10-cm Tissue culture dishes. MEF cells take around 12 h to attach to the culture dishes.
3. Thaw ES cells quickly at 37 °C in a water bath. Transfer the cells into 15-ml conical tubes and add pre-warmed ES medium in a drop-wise fashion. Centrifuge at 145g for 5 min. Aspirate the medium and resuspend the cell pellet in fresh ES cell media and plate the cells onto pre-plated feeders. When placing tissue culture dishes into the incubator, it is helpful to slide the dish gently back and forth to prevent cells from clumping in the middle of the dish.
4. Feed cells everyday with ES cell medium. ES cells should be passaged onto pre-plated MEFs when approaching confluence after trypisinizing using trypsin/EDTA.
4. To form embryoid bodies, remove medium from the culture wells. Add 0.05% Trypsin/EDTA and incubate for at least 5 min. Add culture medium twice the volume of Trypsin/EDTA, gently scrape cells, and transfer the cell suspension into a conical tube.
5. Centrifuge the cell suspension for 3 min at 116 g, resuspend cells in fresh media, and plate them onto 100-mm Petri dish with an approximate concentration of $1 \sim 3 \times 10^3$ cells/ml (10 ml per dish). Culture the cells for 5 days and supplement with additional 5 ml of medium at day 3.
6. After 5 days in culture, count number of EBs/dish and approximate number of cells/EB. (Refer to **Notes 9** and **10**).
7. Intact EBs as well as dissociated EBs can be encapsulated within the hydrogel. For the encapsulation of intact EBs, collect the floating EBs in 50-ml conical tube. Centrifuge the EBs for 5 min at 145 g and resuspend them in the desired polymer solution (e.g., PEGDA solution). Encapsulation of these EB's within hydrogels is outlined in **Subheadings 3.2.1** and **3.2.2**.
8. For dissociated EBs, collect the EBs in 50-ml conical tube. Centrifuge them for 5 min at 145 g and resuspend the cells in 1 mg/ml collagenase IV solution. Incubate for 30 min at 37 °C at 5% CO_2 with intermittent pipetting to break the EBs into single-cell suspensions. After 30 min of dissociation, add equal volume of ES cell

medium and centrifuge for 5 min at 145 g. Resuspend the dissociated cell pellets in the desired polymer solution.

3.2. Encapsulation of Embryonic Stem Cells within Hydrogels

3.2.1. Photo-Encapsulation of Embryonic Stem Cells in PEGDA or RGD-modified PEGDA Hydrogels

Photopolymerization and photo-initiator are described in **Note 11**.

1. Prepare PEGDA polymer solution or RGD-modified PEGDA polymer solution by mixing the macromer at 10% (w/v) in sterile PBS. Protect the polymer solution from light and that can be stored at −20 °C for 3 months. Refer to **Note 12** for detailed information on RGD-modified PEG.
2. Dissolve 100 mg of photo-initiator, Igracure 2959 (Product No. 1706673, Ciba Specialty Chemicals, Tarrytown, NY, USA), in 1 ml of 70% filter-sterilized ethanol.
3. Place both the PEGDA solution and the photo-initiator solution over crushed ice until their usage.
5. Add the photo-initiator to the PEGDA solution and mix thoroughly to make a final concentration of 0.05% (w/v). Make sure that the initiator is mixed very well with the macromer solution (5 μl of photo-initiator solution/ml of polymer solution).
6. Suspend the cells (20–30 million/ml) within the precursor (polymer with photo-initiator) solution by adding the PEGDA solution containing photo-initiator into a cell pellet and mix thoroughly using a pipette without creating bubbles.
7. Transfer 100 μl of EB-polymer solution to cylindrical mold and expose to long-wave, 365 nm light at 4.4 mW/cm² (Glowmark System, Upper Saddle River, NJ, USA), for 5 min to complete gel formation. (*see* **Fig. 3A** and **B** for gross and inverted microscope image of EBs-laden PEGDA hydrogel (EBs encapsulated within PEGDA hydrogel), respectively.

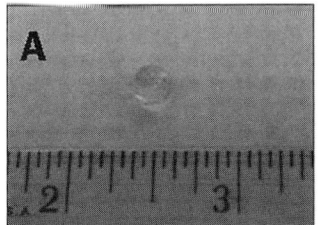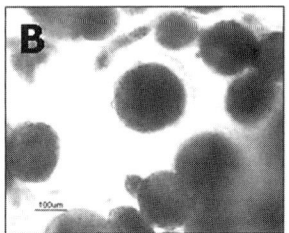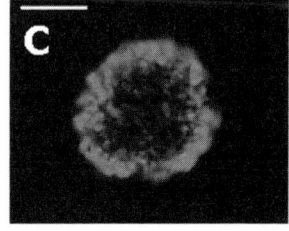

Fig. 3. Gross image of an acellular poly(ethylene glycol)-diacrylate (PEGDA) photopolymerzing hydrogel (**A**). Inverted light microscopy image of ES cell-derived EBs in PEGDA hydrogel immediately after photoencapsulation (**B**). Live-Dead assay immediately after encapsulation shows viability of encapsulated cells following photoencapsulation process (**C**).

8. Remove the "solidified" EBs-laden hydrogels from their mold and transfer them to 12-well plates with chondrogenic medium containing 10 ng/ml TGF-β1 and incubate at 37°C and 5% CO_2. Change medium every 2–3 days. Refer to **Note 13**.

3.2.2. Encapsulation of Stem Cells in Alginate Hydrogel

1. Collect the EBs or dissociated EBs in a 50-ml conical tube and centrifuge for 5 min at 145 g. Remove the supernatant and add the alginate polymer or RGD-modified alginate polymer solutions and gently suspend the cells using a (P-1000) Pipetteman. Avoid making bubbles in the liquid.
2. Take one of the Transwell tissue culture insert trays and fill the well with 1 ml of calcium chloride solution.
3. Using a Pipetteman, add 100 μl of the cell suspension to the tissue culture inserts. After all the inserts are filled, use sterile forceps to transfer the inserts into the wells containing calcium chloride solution. Incubate them at 37°C for 20 min in 5% CO_2.
4. Remove the constructs from the inserts using gentle prying motion with a thin, curved spatula. Place one construct in each well and incubate at 37°C at 5% CO_2 for the remainder of the experiment.

3.3. Live/Dead Cell Viability Assay

Cell death can be monitored using different techniques. One technique, which does not discriminate between necrosis and apoptosis, is the calcein AM and ethidium homodimer-1 (EthD-1)-based two-color fluorescence cell viability assay.

1. Live/Dead Cell viability/cytotoxicity kit (Molecular Probes, L-3224) can be used to analyze cell viability immediately after encapsulation. In this assay, the intracellular esterases of the live cells convert the non-fluorescent calcein acetoxymethyl (calcein AM) to a fluorescent calcien. EthD-1 enters the dead cells through the damaged membranes and becomes fluorescent when bound to nucleic acids. Calcien AM dye produces a bright green fluorescence in live cells, whereas EthD-1 produces a bright red fluorescence in dead cells.
2. It is recommended to check viability of the cells within 24 h of encapsulation to verify the effect of photo-polymerization on cell viability.
3. Prepare Live/Dead dye solution by mixing 0.5 μl of calcein-AM dye (for live cells) and 2 ml of EthD-1 dye (for dead cells) into 1 ml of serum-free DMEM.
4. Aspirate the culture medium from the constructs (cells encapsulated within hydrogels) of interest and wash with sterile PBS to remove the serum.
5. Using sterile scalpel, section the construct as thin as you can (approximately, < 1 mm thickness). The best images are achieved when the sections are the thinnest. Transfer the sectioned constructs to an eppendorf tube (1.8 ml) and wash once again in sterile PBS and aspirate the liquid.

6. Add Live/Dead dye enough to cover the sample (approximately 300–500 ml) and incubate them at 37°C for 30 min in 5% CO_2.
7. Aspirate out the Live/Dead dye and rinse the constructs with sterile PBS. Take the sliced constructs to fluorescent microscope for imaging. **Figure 3C** shows microscopic image of a live/dead assay carried out on EB-laden hydrogel (EBs encapsulated within 10% PEGDA hydrogel).
8. The percentage of live cells (cell viability) can be calculated as the number of live cells divided by the number of total cells.

3.4. Preparing the Constructs for Histology

1. Collect the constructs at desired time points and rinse with PBS.
2. Preserve the constructs in 4% paraformaldehyde for overnight.
3. Aspirate the paraformaldehyde and change to 70% EtOH on the following day.
4. The constructs are then processed for paraffin embedding according to standard histological methods. Refer to **Note 14** for more details. **Figure 4** provides images of cell-laden hydrogel stained with Safranin-O and Masson's trichrome.

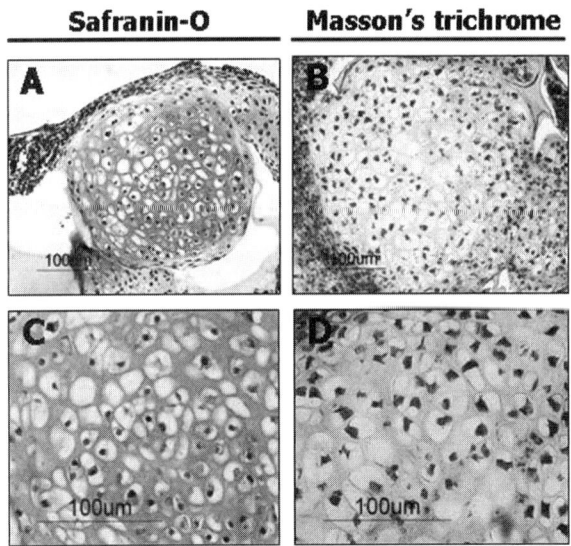

Fig. 4. Basophilic extracellular matrix (ECM) deposition characteristic of neocartilage from ES cell derived EBs in PEGDA hydrogel. EBs were cultured in chondrogenic medium in presence of TGF-β 1 for 17 days and stained with Safranin-O and Masson's trichrome for glycosaminoglycan and collagen detection, respectively.

3.5. Preparing Constructs for Biochemical Assay

Successful cartilage tissue engineering from embryonic stem cells depends on the cell density and accumulation of extracellular matrix molecules. Cartilage matrix proteoglycans, collagen, as well as cellular DNA can be quantified after ECM solubilization by papain digestion.

1. Collect the constructs at different time points and rinse with PBS.
2. Measure the wet weight of the constructs and place them in 1.8 ml eppendorf tubes.
3. Freeze-dry the hydrogel constructs using a lyophilizer for at least 48 h.
4. Measure the dry weight of the constructs.
5. For each lyophilized construct in an eppendorf tube, initially add 200 ml of papain solution and homogenize using a pestle. Then add an additional 800 µl of papain solution to the construct and incubate for 16 h at 60 °C using a water bath. Papain digested constructs can be stored at −20 °C for further analysis.
6. The supernatant of these papain-digested constructs can be used to conduct subsequent assays such as quantification of collagen, GAG, and DNA.

3.6. Fluorometric Assay of DNA Content Using Hoechst 33258

Hoechst 33258, weakly fluorescent, binds specifically into DNA, and therefore, its fluorometric reading can be used for a quantitative determination of DNA content.

1. Turn on the fluorometer and warm up the machine for 15 min before use. Set up parameters and calibrate fluorometer (refer to your Operating Manual for detailed instructions).
2. Generate standard curve, amount of DNA versus fluorescence intensity, using known concentration of DNA (e.g., Calf Thymus DNA) as shown in the following table. It is recommended to generate the standard curve every time you perform the assay.

Amount of DNA (100 µg/mL) (µL)	Amount of Hoechst dye assay solution (mL)	Amount of DNA (ng/mL)	Fluorometric reading
0	3	0	
1.5	3	50	
3	3	100	
4.5	3	150	
6	3	200	
9	3	300	
12	3	400	

3. At each measurement, zero the instrument with a "blank" cuvette containing 3 mL of Hoechst 33258 dye solution. To generate the standards curve, add appropriate amount of DNA as specified in the table. Mix the solution well using a transfer pipette and record the flourometric reading.

4. To measure the DNA content of the sample, add 30 μL of the supernatant of the papain digested solution into the blank cuvette containing 3 mL of Hoechst dye solution. Mix well and record the measurement. Refer to **Note 15** for more details.

5. On the basis of the DNA standard curve, one could obtain the approximate amount of DNA per sample by regression analysis.

3.7. Glycosaminoglycan Detection Using DMMB Assay

This is a modified version of colorimetric reactions to detect GAGs based on Farndale et al. *(23)*.

1. Set the spectrophotometer to OD_{525}.

2. To create a standard curve, amount of CS versus absorbance, vary the CS amount by diluting the CS working solution (*see* **Subheading 2.11., step 2**) in the range 0–100 μL using distilled water. This would result in linearly varying amount of CS within the range 0–10 μg. Add 2.5 mL of DMMB dye to these CS solutions with varying concentrations, cover with parafilm, mix well, and read the absorbance.

3. In 4.5 mL cuvette, combine 100 μL of papain-digested sample and 2.5 mL DMMB dye reagent solution. Shake well by covering the cuvette with parafilm and flipping it several times. Read the absorbance value.

4. From the standard curve of CS working solution one can estimate total amount of GAG by regression analysis.

3.8. Collagen Content Measurement by Hydroxyproline Assay

This is a modified version of measurement of hydroxyproline content of insoluble collagen based on Stegemann and Stalder *(24)*.

1. Combine 100 μL of papain-digested solution with 900 μL of 6N HCl solution and hydrolyze for 18 h at 110 °C in pyrex tubes with silicone-sealed gaskets.

2. Add 2–4 drops of methyl red (0.02% in methanol) to the acid hydrolyzates. The addition of methyl red imparts pink color to the solution, indicating an acidic solution. Refer to **Note 16** for more details on methyl red.

3. Neutralize with 2.5N NaOH, 0.5 M HCl, and 0.5 M NaOH. The amount of reagents required can be determined by observing the color change.

(a) Add enough 2.5N NaOH until pink color just disappears.

(b) Add 0.5 M HCl (drop wise) until pink color just reappears.

(c) Add few drops of 0.5 M NaOH to bring back the solution to the straw color (indicating a basic solution).

4. Add dH_2O to bring the volume to 15 ml so as to bring the salt concentration down to an acceptable range.

5. Transfer 1 ml of the above sample into a new 15 ml Pyrex tube for assay.

6. Prepare solutions for the standard curve by diluting hydroxyproline stock solution with H_2O to a total volume of 1 ml.

7. Transfer all experimental as well as standard samples to fume hood and add 0.5 ml Chloramine-T solution to each tube, vortex and let sit at room temperature for 20 min.

8. Add 0.5 ml Ehrlich's reagent and vortex vigorously until the white milky color disappears. Recap the tubes tightly and incubate them for 30 min at 60 °C using a water bath. The addition of Ehrlich's reagent followed by incubating at 60 °C enables the formation of chromophore, which could be measured at 550 nm.

9. Cool the samples on ice.

10. Read the absorbance at 550 nm in disposable plastic cuvettes. Measure the samples as quickly as possible because the color is stable only for 1 h.

11. On the basis of standard hydroxyproline curve, evaluate the amount of collagen per sample.

3.9. Preparing Constructs for RNA Extraction for RT–PCR

1. Approximately three constructs (cell-laden hydrogels) are required for sufficient RNA extraction.

2. Collect the constructs in different time points and place them in RNase-free 1.8-ml Eppendorf tubes. A spatula or RNase-free tweezers may be used to handle the constructs.

3. Add 1 ml of EDTA/HEPES solution (50 mM EDTA solution in 10 mM HEPES) and let it sit for 10 min.

4. In microcentrifuge, pellet the constructs 145 g for 10 min.

5. Aspirate medium and add 1 ml of PBS to rinse.

6. Pellet down the constructs again 145 g for 10 min.

7. Aspirate PBS and add 1 ml of Trizol (cat. no. 15596-018, Invitrogen, Carlsbad, CA, USA). Use the homogenizer tip manually to break the gel construct.

8. Leave the homogenized constructs at room temperature for 5 min to permit the complete dissociation of nuclear proteins. You can store the homogenate at this point in a −80 °C freezer or proceed with extraction and RT by following the standard manufacturer instructions.

3.10. Markers for Chondrogenic Differentiation

Chondrogenic differentiation of ES cells can be monitored by detecting specific chondrogenic lineage markers. Type II collagen, Sox-9, and aggrecan are cartilage-specific markers that can be detected by RT–PCR analysis. However, to confirm the lineage restricted differentiation of ES cells in the hydrogel (i.e., homogenous differentiation), analysis of ecto and endodermal markers is also required. RT–PCR analysis of β-tubulin mRNA is used as a control housekeeping gene to normalize samples. **Table 1** summarizes the sequence and annealing temperature for PCR primers and the size of expected cDNA product. An example of RT–PCR of chondrogenic-specific markers results is shown in **Fig. 5**.

Table 1

Sequences of Primers and Conditions for Chondrocyte-Specific Markers and Endodermal and Ectodermal Markers used in Reverse Transcription–Polymerase Chain Reaction (RT–PCR)

	PCR primers	
Gene	Sequence (forward and reverse)	Annealing temperature
Type II collagen	5′-AGGGGTACCAGGTTCTC CATC-3′ 5′-CTGCTCATCGCCG CGGTCCGA-3′	60°C; 432 bp (splice variant A) and 225 bp (splice variant B)
Sox9	5′-TGGCAGACCAGTACCCGCATCT-3′ 5′-TCTTTCTT GTGCTGCACGCGC-3′	57°C
Link protein	5′-TTCTGGGCTATGACCGCTG-3′ 5′-AGCGCCTTCTTGGTCGAGA-3′	60°C
Cytokeratin K18	5′-TTGTCACCACCAAGTCTGCC-3′ 5′-TTTGTGCCAGCTCTGACTCC-3′	60°C
AFP	5′-CCTTGGCTGCTCAGTACGACAAGG-3′ 5′-CCTGCAGACACTCCAGCGAGTTTC-3′	60°C
Nestin	5′-CGGCCCACGCATCCCCCATCC-3′ 5′-AGCGGCCTTCCAATCTCTGTTCC-3′	60°C
β-Tubulin	5′-GGAACATAGCCGTAAACTGC-3′ 5′-TCACTGTGCCTGAACTTACC-3′	54°C

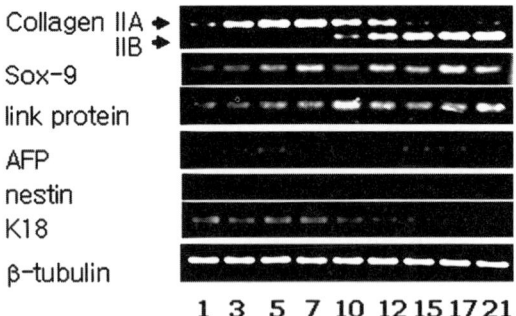

Fig. 5. Chondrogenic expression of EBs encapsulated in PEGDA hydrogels. EBs-hydrogel constructs cultured in chondrogenic medium demonstrated that splicing of type IIB collagen was initiated at day 7 of culture, and its expression was upregulated in a time dependent manner. Chondrogenic markers, *Sox9* and *link protein*, were upregulated, whereas endodermal and ectodermal markers such as *AFP, nestin*, and *cytokeratin K-18* were down-regulated or minimally expressed.

4. Notes

1. For cylindrical molds, you may use the cap of 500- or 1000-µL eppendorf tube.
2. The carboxylic acid functional groups present in the alginate polymer chain can be used for covalent modification with oligopeptides such as RGD motifs using aqueous carbodiimide chemistry. EDC enables the formation of an amide linkage between the carboxylic acid of the alginate and the amine group of the peptide, thus bridging the two reactants *(25)*.
3. Sulfo-NHS stabilizes the reactive intermediate product against the competitive hydrolysis reaction and thereby increases the efficiency of amide bond formation.
4. Dialysis is widely used to purify high molecular weight material from small molecular weight impurities. The main principle behind dialysis is the diffusion of solute along a concentration gradient across a semi-permeable membrane. The small molecular weight impurities diffuse across the membrane into the dialysis fluid, whereas the high molecule weight materials are left behind. For best results, the dialysis fluid needs to be removed consistently and renewed with fresh fluid. The other commonly used method for purification of high molecular weight materials is size exclusion chromatography (e.g., sephadex column). The mixture to be separated is first poured into the sephadex column. Sephadex beads, because of their porous nature, allow materials with small molecular weight to diffuse into them, whereas high molecular weight materials are excluded. The column is then eluted with solvent and the various fractions in the mixture come out of the column at different times depending on their molecular weight, with the highest molecular weight materials coming out first.

5. Hoechst 33258 Dye is a possible carcinogen and mutagen. Therefore, always wear gloves and other safety precautions while handling this reagent.

6. This chapter provides comprehensive techniques for encapsulating stem cells for two kinds of hydrogel scaffolds frequently employed in tissue engineering applications. We have specifically addressed the conditions for chondrogenesis. However, these encapsulation techniques can be utilized for studying other lineage differentiation in 3D microenvironment.

7. Hydrogels can be prepared from both synthetic and natural polymers. Common macromers for creating hydrogels are hydrophilic oligomers with crosslinking functional groups such as acrylates, methacrylates, acrylamides, and methacrylamides. Gelation of these vinyl macromers is initiated by radical initiators. Commonly used radical initiators comprise of thermal initiators, redox initiators, and photoinitiators. Macromers with single functional vinyl groups require crosslinking agents to create polymer networks *(26)*. The mechanical properties and water content of hydrogels are often controlled by varying the crosslink density, which is a function of initial oligomer or monomer concentration, molecular weight of the precursor, and amount of crosslinking agent *(26)*. Ionic polymers (generally known as polyelectrolytes) can be crosslinked using multivalent counterions. An example of an ionically crosslinked hydrogel is alginate, which is usually gelled using Ca^{2+} ions *(27,28)*. Hydrogels can also be created by reacting different functional groups present in the precursors *(29)*.

8. A significant advantage of using hydrogels for stem cell differentiation lies in the fact that ECM protein components can be easily added to the hydrogel. To prepare hydrogels with ECM components, make 20% PEGDA solution, mix 50/50 (v/v) with ECM protein solutions [such as hyaluronic acid (HA), collagen, human ECM, Matrigel, laminin, cartrigel, etc.] and photopolymerize. For alginate hydrogels, prepare 2.4% alginate solution and mix 50/50 (v/v) with ECM protein solutions and follow the subsequent ionic polymerization in Ca^{++} solution.

9. Embryoid body formation through suspension methods results in heterogeneous cell numbers per EB. Therefore, average number of cells per EB must be calculated before encapsulation.

10. Counting of EBs/dish can be performed with microscope under low power ($\times 4$ objective). Number of cells/EBs can be estimated by collecting EBs and counting the cells after trypsinization. Spin down the EBs at approximately 145 g for 5 min, aspirate medium, and resuspend the cells with 0.25% trypsin-EDTA. After 5–7 min of incubation, count the cells.

11. Photo-polymerization involves light-induced free-radical polymerization reaction that converts a liquid monomer or macromer solution into a solid network in the presence of a suitable photo-initiator *(30)*. Here, the photo-initiator absorbs light energy and forms free radicals, which initiate the polymerization. The commonly used light sources for photo-polymerization are visible light and ultraviolet. One of the main concerns with photo-polymerization is that the cytotoxicity

of photo-initiators increases with their concentration and exposure time *(30–32)*. Therefore, it is recommended to polymerize immediately after the cells are suspended in polymer solution. There are several varieties of photo-initiators available commercially (*see* **Table 2**).

12. YRGDS has MW of 632.6 and ACRL-PEG-NHS has MW of 3400. Reaction of 5 mg (7.9 μmol) of YRGDS with 70 mg (20.6 μmol, excess amount) of ACRL-PEG-NHS would result in 7.9 μmol of ACRL-PEG-YRGDS. To make 2.5 mM concentration of RGD in 10% total weight polymers, dissolve 23.75 mg of reacted products (of which 9.489 mg is ACRL-PEG-NHS) and 76.25 g PEGDA polymers in 1 ml of PBS.

13. Growth factor-dependent chondrogenic differentiation of murine embryonic stem cell-derived embryoid bodies has been evaluated previously. This chapter gives potential application of ES-derived cells for cartilage tissue engineering. The family of TGF-β growth factors induces chondrogenic response of ES cells.

14. Chondrogenic cells with basophilic ECM can be visualized through histological staining. Staining with safrainin-O and Masson's trichome, a stain for negatively charged proteoglycans and collagen, respectively, provides a good indication of chondrogenic differentiation.

15. Take extra care when mixing samples to avoid air bubbles. Air bubbles can cause scattering of light leading to inaccurate results. The sensitivity of this assay depends on limiting the background fluorescence by keeping the scatter to a minimum. Therefore, it is advisable to use disposable acrylic cuvettes. These cuvettes are generally dust- and scratch-free and therefore contribute little to light scatter.

16. Methyl red is an acid-base indicator, which changes its color in solution because of changes in pH. Methyl red exhibits red color in acidic pH and yellow color in basic pH.

Table 2
UV Sensitive Photo-Initiators That Have Been Studied for Cytotoxicity

Photoinitiators	Description
Irgacure 2959 (2-hydroxy-1-[4-(hydroxyethoxy)phenyl]-2-methyl-1-propanone	This photoinitiator is well tolerated by various cell types over a wide range of concentrations ≤ 0.05% (w/w)
Irgacure 907 (2-methyl-1-[4-(methylthio)phenyl]-2-(4-morpholinyl)-1-propoanone)	tolerable concentration ≤ 0.01% (w/w)
Irgacure 184 (1-hydroxycyclohexyl phenyl ketone)	tolerable concentration of ≤ 0.01% (w/w)

References

1. Buckwalter, J. A., and Mankin, H. J. (1998) Articular cartilage: degeneration and osteoarthritis, repair, regeneration, and transplantation. *Instr Course Lect* **47**, 487–504.
2. Hwang, N. S., Kim, M. S., Sampattavanich, S., Baek, J. H., Zhang, Z., and Elisseeff, J. (2005) Effects of three-dimensional culture and growth factors on the chondrogenic differentiation of murine embryonic stem cells. *Stem Cells* **24**, 284–291.
3. Kramer, J., Hegert, C., Guan, K., Wobus, A. M., Muller, P. K., and Rohwedel, J. (2000) Embryonic stem cell-derived chondrogenic differentiation in vitro: activation by BMP-2 and BMP-4. *Mech Dev* **92**, 193–205.
4. Kramer, J., Hegert, C., and Rohwedel, J. (2003) In vitro differentiation of mouse ES cells: bone and cartilage. *Methods Enzymol* **365**, 251–268.
5. Sottile, V., Thomson, A., and McWhir, J. (2003) In vitro osteogenic differentiation of human ES cells. *Cloning Stem Cells.* **5**, 149–155.
6. Nakayama, N., Duryea, D., Manoukian, R., Chow, G., and Han, C. Y. (2003) Macroscopic cartilage formation with embryonic stem-cell-derived mesodermal progenitor cells. *J Cell Sci* **116**, 2015–2028.
7. Dang, S. M., Kyba, M., Perlingeiro, R., Daley, G. Q., and Zandstra, P. W. (2002) Efficiency of embryoid body formation and hematopoietic development from embryonic stem cells in different culture systems. *Biotechnol Bioeng* **78**, 442–453.
8. Spradling, A., Drummond-Barbosa, D., and Kai, T. (2001) Stem cells find their niche. *Nature.* **414**, 98–104.
9. Streuli, C. (1999) Extracellular matrix remodelling and cellular differentiation. *Curr Opin Cell Biol* **11**, 634–640.
10. Liu, H., and Roy, K. (2005) Biomimetic three-dimensional cultures significantly increase hematopoietic differentiation efficacy of embryonic stem cells. *Tissue Eng* **11**, 319–330.
11. Cukierman, E., Pankov, R., Stevens, D. R., and Yamada, K. M. (2001) Taking cell-matrix adhesions to the third dimension. *Science* **294**, 1708–1712.
12. Schmeichel, K. L., and Bissell, M. J. (2003) Modeling tissue-specific signaling and organ function in three dimensions. *J Cell Sci* **116**, 2377–2388.
13. Ma, H. L., Chen, T. H., Low-Tone Ho, L., and Hung, S. C. (2005) Neocartilage from human mesenchymal stem cells in alginate: implied timing of transplantation. *J Biomed Mater Res A* **74**, 439–446.
14. Williams, C. G., Kim, T. K., Taboas, A., Malik, A., Manson, P., and Elisseeff, J. (2003) In vitro chondrogenesis of bone marrow-derived mesenchymal stem cells in a photopolymerizing hydrogel. *Tissue Eng* **9**, 679–688.
15. Bryant, S. J., Anseth, K. S., Lee, D. A., and Bader, D. L. (2004) Crosslinking density influences the morphology of chondrocytes photoencapsulated in PEG hydrogels during the application of compressive strain. *J Orthop Res* **22**, 1143–1149.

16. Miot, S., Woodfield, T., Daniels, A. U., Suetterlin, R., Peterschmitt, I., Heberer, M., van Blitterswijk, C. A., Riesle, J., and Martin, I. (2005) Effects of scaffold composition and architecture on human nasal chondrocyte redifferentiation and cartilaginous matrix deposition. *Biomaterials* **26**, 2479–2489.

17. Alsberg, E., Anderson, K. W., Albeiruti, A., Rowley, J. A., and Mooney, D. J. (2002) Engineering growing tissues. *Proc Natl Acad Sci USA* **99**, 12025–12030.

18. Rowley, J. A., Madlambayan, G., and Mooney, D. J. (1999) Alginate hydrogels as synthetic extracellular matrix materials. *Biomaterials* **20**, 45–53.

19. Yang, F., Williams, C. G., Wang, D. A., Lee, H., Manson, P. N., and Elisseeff, J. (2005) The effect of incorporating RGD adhesive peptide in polyethylene glycol diacrylate hydrogel on osteogenesis of bone marrow stromal cells. *Biomaterials* **26**, 5991–5998.

20. Lum, L. Y., Cher, N. L., Williams, C. G., and Elisseeff, J. H. (2003) An extracellular matrix extract for tissue-engineered cartilage. *IEEE Eng Med Biol Mag* **22**, 71–76.

21. Lutolf, M. P., and Hubbell, J. A. (2005) Synthetic biomaterials as instructive extracellular microenvironments for morphogenesis in tissue engineering. *Nat Biotechnol* **23**, 47–55.

22. Philp, D., Chen, S. S., Fitzgerald, W., Orenstein, J., Margolis, L., and Kleinman, H. K. (2005) Complex extracellular matrices promote tissue-specific stem cell differentiation. *Stem Cells* **23**, 288–296.

23. Farndale, R. W., Buttle, D. J., and Barrett, A. J. (1986) Improved quantitation and discrimination of sulphated glycosaminoglycans by use of dimethylmethylene blue. *Biochim Biophys Acta* **883**, 173–177.

24. Stegemann, H., and Stalder, K. (1967) Determination of hydroxyproline. *Clin Chim Acta* **18**, 267–273.

25. Hermanson, G. (1996) Bioconjugate Techniques. *San Diego, CA: Academy Press.*

26. Varghese, S., Lele, A. K., Srinivas, D., Sastry, M., and Mashelkar, R.A. (2001) Novel macroscopic self-organization in polymer gels. *Adv Mater* **13**, 1544–1548.

27. Elisseeff, J. H., Lee, A., Kleinman, H. K., and Yamada, Y. (2002) Biological response of chondrocytes to hydrogels. *Ann N Y Acad Sci* **961**, 118–122.

28. Drury, J. L., and Mooney, D. J. (2003) Hydrogels for tissue engineering: scaffold design variables and applications. *Biomaterials* **24**, 4337–4351.

29. Balakrishnan, B., and Jayakrishnan, A. (2005) Evaluation of an in situ forming hydrogel wound dressing based on oxidized alginate and gelatin. *Biomaterials* **26**, 3941–3951.

30. Bryant, S. J., Nuttelman, C. R., and Anseth, K. S. (2000) Cytocompatibility of UV and visible light photoinitiating systems on cultured NIH/3T3 fibroblasts in vitro. *J Biomater Sci Polym Ed* **11**, 439–457.

31. Williams, C. G., Malik, A. N., Kim, T. K., Manson, P. N., and Elisseeff, J. H. (2005) Variable cytocompatibility of six cell lines with photoinitiators used for polymerizing hydrogels and cell encapsulation. *Biomaterials* **26**, 1211–1218.
32. Temenoff, J. S., Shin, H., Conway, D. E., Engel, P. S., and Mikos, A. G. (2003) In vitro cytotoxicity of redox radical initiators for cross-linking of oligo(poly(ethylene glycol) fumarate) macromers. *Biomacromolecules* **4**, 1605–1613.

Index

375

Printed in the United States of America

Springer